MICROFOSSILS

Wonder is the first of all passions

René Descartes, 1645

MICROFOSSILS

SECOND EDITION

Howard A. Armstrong
Senior Lecturer in Micropalaeontology, Department of Earth Sciences, University of Durham, UK

Martin D. Brasier
Professor of Palaeobiology, Department of Earth Sciences, University of Oxford, UK

© 2005 Howard A. Armstrong and Martin D. Brasier

BLACKWELL PUBLISHING
350 Main Street, Malden, MA 02148-5020, USA
9600 Garsington Road, Oxford OX4 2DQ, UK
550 Swanston Street, Carlton, Victoria 3053, Australia

The right of Howard A. Armstrong and Martin D. Brasier to be identified as the Authors of this Work has been asserted in accordance with the UK Copyright, Designs, and Patents Act 1988.

All rights reserved. No part of this publication may be reproduced, stored in a retrieval system, or transmitted, in any form or by any means, electronic, mechanical, photocopying, recording or otherwise, except as permitted by the UK Copyright, Designs, and Patents Act 1988, without the prior permission of the publisher.

First edition published 1980 by George Allen & Unwin, © M.D. Brasier 1980
Second edition published 2005 by Blackwell Publishing Ltd

4   2009

Library of Congress Cataloging-in-Publication Data

Armstrong, Howard, 1957–
  Microfossils. – 2nd ed./Howard A. Armstrong and Martin D. Brasier.
    p. cm.
  Rev. ed. of: Microfossils / M.D. Brasier. 1980.
  Includes bibliographical references and index.
  ISBN 978-0-632-05279-0 (pbk. : alk. paper)
  1. Micropaleontology. I. Brasier, M.D. Microfossils. II. Title.

QE719.A76 2004
560–dc22

2004003936

A catalogue record for this title is available from the British Library.

Set in 91/2/12pt Minion
by Graphicraft Ltd, Hong Kong

The publisher's policy is to use permanent paper from mills that operate a sustainable forestry policy, and which has been manufactured from pulp processed using acid-free and elementary chlorine-free practices. Furthermore, the publisher ensures that the text paper and cover board used have met acceptable environmental accreditation standards.

For further information on
Blackwell Publishing, visit our website:
www.blackwellpublishing.com

Contents

Preface ... vii

Part 1 Applied micropalaeontology ... 1

Chapter 1   Introduction ... 3
Chapter 2   Micropalaeontology, evolution and biodiversity ... 8
Chapter 3   Microfossils in stratigraphy ... 16
Chapter 4   Microfossils, stable isotopes and ocean-atmosphere history ... 25
Chapter 5   Microfossils as thermal metamorphic indicators ... 35

Part 2 The rise of the biosphere ... 37

Chapter 6   The origin of life and the early biosphere ... 39
Chapter 7   Emergence of eukaryotes to the Cambrian explosion ... 48
Chapter 8   Bacterial ecosystems and microbial sediments ... 59

Part 3 Organic-walled microfossils ... 69

Chapter 9   Acritarchs and prasinophytes ... 71
Chapter 10   Dinoflagellates and ebridians ... 80
Chapter 11   Chitinozoa ... 96
Chapter 12   Scolecodonts ... 101
Chapter 13   Spores and pollen ... 104

Part 4 Inorganic-walled microfossils ... 127

Chapter 14   Calcareous nannoplankton: coccolithophores and discoasters ... 129
Chapter 15   Foraminifera ... 142
Chapter 16   Radiozoa (Acantharia, Phaeodaria and Radiolaria) and Heliozoa ... 188
Chapter 17   Diatoms ... 200
Chapter 18   Silicoflagellates and chrysophytes ... 210

Chapter 19	Ciliophora: tintinnids and calpionellids	215
Chapter 20	Ostracods	219
Chapter 21	Conodonts	249

Appendix – Extraction methods	273
Systematic Index	280
General Index	287

Preface

In the 25 years since the first, highly successful, edition of *Microfossils* was published there have been significant advances in all the areas of understanding of microscopic life and their fossil counterparts. Our new knowledge has led to major changes in the classification, applications and in some cases the biological affinities, of the major groups covered in this book. Greater understanding of species concepts, stratigraphical ranges and the completeness of the microfossil record means all of the Phanerozoic and parts of the Proterozoic can now be dated using microfossils. The high fidelity of the microfossil record provides the best test bed for numerous evolutionary studies. Microfossils remain an indispensable part of any sedimentary basin study, providing the biostratigraphical and palaeoecological framework and, increasingly, a measure of maturity of hydrocarbon-prone rocks. The rise of palaeoclimatology has given micropalaeontology a new impetus too, with calcareous-walled groups providing stable isotope and geochemical proxies for oceanographic, palaeoenvironmental and palaeo-climatic change. Indeed it is now widely accepted that some microscopic groups are responsible for maintaining the Earth as a habitable planet and have been doing so since the early Proterozoic and perhaps before. Micropalaeontology therefore now occupies a central role in the modern Earth and environmental sciences and increasingly a much wider group of Earth scientists are likely to come across the work of micropalaeontologists. We hope this second edition provides an inexpensive introductory textbook that will be of use to students, teachers and non-specialists alike.

We have not changed the main motivation of this book, which is to provide a manual for somebody with little micropalaeontological background working at the microscope. Morphology and classification lie at the core of the book, supported by more derivative information on geological history, palaeoecology and applications, with supporting references. An addition to this book are selected photomicrographs, which are not intended to give a comprehensive coverage of the taxa discussed but to supplement the line drawings.

Conscious of the adage that for every expert there is a different classification we have favoured the use of those schemes published in the *Fossil Record II* (Renton, M. (ed.), 1993, Chapman & Hall, London), a volume compiled by experts in the respective groups and a statement of the familial level classification at the time of publication. Students will therefore have access to a much more detailed treatment of family level stratigraphical ranges than can be provided by this text. Mindful also of the value of collecting and processing microfossil material, the section on preparatory methods has been retained. This focuses on techniques that are simple, safe and possible with a minimum of sophisticated equipment.

In order to compile this book we have relied on the work and advice freely given by our many colleagues past and present. We are particularly indebted to those who have commented on the various parts of the manuscript: Professor R.J. Aldridge, Professor D.J. Batten, Dr D.J. Horne, Professor A.R. Lord, Dr G. Miller, Dr S.J. Molyneau, Dr H.E. Presig, Dr J.B. Riding and Dr J. Remane. Mrs K.L. Atkinson prepared the diagrams and new line drawings. In addition, a special thankyou is offered to all these authors and publishers who have kindly allowed the use of their illustrations and photomicrographs; formal acknowledgement is provided throughout the text. Without all these people this project would never have been completed and we are most grateful for their help.

Blackwell Publishing and the Natural History Museum London are the publishers of *PaleoBase: Microfossils*, a powerful illustrated database of microfossils designed for student use. Please see **www.paleobase.com** for ordering details, or email ian.francis@oxon.blackwellpublishing.com

PART 1
Applied micropalaeontology

PART 1

Applied microplate reading

CHAPTER 1

Introduction

Microfossils – what are they?

A thin blanket of soft white to buff-coloured ooze covers one-sixth of the Earth's surface. Seen under the microscope this sediment can be a truly impressive sight. It contains countless numbers of tiny shells variously resembling miniature flügelhorns, shuttlecocks, water wheels, hip flasks, footballs, garden sieves, space ships and chinese lanterns. Some of these gleam with a hard glassy lustre, others are sugary white or strawberry coloured. This aesthetically pleasing world of microscopic fossils or microfossils is a very ancient one and, at the biological level, a very important one.

Any dead organism that is vulnerable to the natural processes of sedimentation and erosion may be called a **fossil**, irrespective of the way it is preserved or of how recently it died. It is common to divide this fossil world into larger **macrofossils** and smaller **microfossils**, each kind with its own methods of collection, preparation and study. This distinction is, in practice, rather arbitrary and we shall largely confine the term 'microfossil' to those discrete remains whose study requires the use of a microscope throughout. Hence bivalve shells or dinosaur bones seen down a microscope do not constitute microfossils. The study of microfossils usually requires bulk collecting and processing to concentrate remains prior to study.

The study of microfossils is properly called **micropalaeontology**. There has, however, been a tendency to restrict this term to studies of mineral-walled microfossils (such as foraminifera and ostracods), as distinct from **palynology** the study of organic-walled microfossils (such as pollen grains, dinoflagellates and acritarchs). This division, which arises largely from differences in bulk processing techniques, is again rather arbitrary. It must be emphasized that macropalaeontology, micropalaeontology and palynology share identical aims: to unravel the history of life and the external surface of the planet. These are achieved more speedily and with greater reward when they proceed together.

Why study microfossils?

Most sediments contain microfossils, the kind depending largely on the original age, environment of deposition and burial history of the sediment. At their most abundant, as for example in back-reef sands, 10 cm^3 of sediment can yield over 10,000 individual specimens and over 300 species. By implication, the number of ecological niches and biological generations represented can extend into the hundreds and the sample may represent thousands if not hundreds of thousands of years of accumulation of specimens. By contrast, macrofossils from such a small sample are unlikely to exceed a few tens of specimens or generations. Because microfossils are so small and abundant (mostly less then 1 mm) they can be recovered from small samples. Hence when a geologist wishes to know the age of a rock or the salinity and depth of water under which it was laid down, it is to microfossils that they will turn for a quick and reliable answer. Geological surveys, deep sea drilling programmes, oil and mining companies working with the small samples available from borehole cores and drill cuttings have all therefore employed micropalaeontologists to learn more about the rocks they are handling. This commercial side to micropalaeontology has undoubtedly been a major stimulus to its growth. There are some philosophical and sociological sides to the subject, however. Our understanding of the development and stability of the present global ecosystem has much to learn from the microfossil record, especially since many microfossil groups have occupied a place at or near to the base of the food web. Studies into the nature of evolution cannot afford to overlook the microfossil record either, for it contains a wealth of examples. The importance of understanding microfossils is further augmented by discoveries in Precambrian rocks; microfossils now provide the main evidence for organic evolution through more than three-quarters of the history of life on Earth. It is also to microfossils that science will turn in the search for life on other planets such as Mars.

The cell

A great many microfossils are the product of single-celled (**unicellular**) organisms. A little knowledge of these cells can therefore help us to understand their way of life and, from this, their potential value to Earth scientists. Unicells are usually provided with a relatively elastic outer **cell membrane** (Fig. 1.1) that binds and protects the softer cell material within, called the **cytoplasm** (or **protoplasm**). Small 'bubbles' within the cytoplasm, called **vacuoles**, are filled with food, excretory products or water and serve to nourish the cell or to regulate the salt and water balance. A darker, membrane-bound body, termed the **nucleus**, helps to control both vegetative and sexual division of the cell and the manufacture of proteins. Other small bodies concerned with vital functions within the cell are known as **organelles**. The whip-like thread that protrudes from some cells, called a **flagellum**, is a locomotory organelle. Some unicells bear many short flagella, collectively called **cilia**, whilst others get about by means of foot-like extensions of the cell wall and cytoplasm, known as **pseudopodia**. Other organelles that can occur in abundance are the **chromoplasts** (or chloroplasts). These small structures contain chlorophyll or similar pigments for the process of photosynthesis.

Nutrition

There are two basic ways by which an organism can build up its body: by heterotrophy or by autotrophy. In **heterotrophy**, the creature captures and consumes living or dead organic matter, as we do ourselves. In **autotrophy**, the organism synthesizes organic matter from inorganic CO_2, for example, by utilizing the effect of sunlight in the presence of chlorophyll-like pigments, a process known as **photosynthesis**. Quite a number of microfossil groups employ these two strategies together and are therefore known as **mixotrophic**.

Reproduction

Asexual (or vegetative) and **sexual** reproduction are the two basic modes of cellular increase. The simple division of the cell found in asexual reproduction results in the production of two or more daughter cells with nuclear contents similar in proportion to those of the parent. In sexual reproduction, the aim is to halve these normal nuclear proportions so that sexual fusion with another 'halved' cell can eventually take place. Information contained in each cell can then be passed around to the advantage of the species. This halving

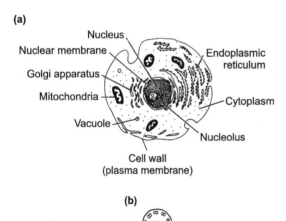

Fig. 1.1 The living cell. (a) Eukaryotic cell structure showing organelles. (b) Cross-section through a flagellum showing paired 9+2 structure of the microfibrils. (Reproduced with permission from Clarkson 2000.)

process is achieved by a fourfold division of the cell, called **meiosis**, which results in four daughter cells rather than two.

The empires of life

Living individuals all belong to naturally isolated units called **species**. Ideally, these species are freely interbreeding populations that share a common ecological niche. Even those lowly organisms that disdain sexual reproduction (such as the silicoflagellates) or do not have the organization for it (such as the cyanobacteria), occur in discrete morphological and ecological species. Obviously it is impossible to prove that a population of microfossils was freely interbreeding but, if specimens are sufficiently plentiful, it is possible to recognize both morphological and ecological discontinuities. These can serve as the basis for distinguishing one fossil species from another.

Whereas the species is a functioning unit, the higher taxonomic categories in the hierarchical system of classification are mere abstractions, implying varying degrees of shared ancestry. All species are placed within a **genus** that contains one or more closely related species. These will differ from other species in neighbouring genera by a distinct morphological, ecological or biochemical gap. **Genera** (plural of genus) tend to be more widely distributed in time and space than do species, so they are not greatly valued for stratigraphical correlation. They are, however, of considerable value in palaeoecological and palaeogeographical studies. The successively higher categories of **family**, **order** and **class** (often with intervening sub- or super-categories) should each contain clusters of taxa with similar grades of body organization and a common ancestor. They are of relatively little value in biostratigraphy and palaeoenvironmental studies. In 'animals' the **phylum** taxon is defined on the basis of major structural differences, whereas in 'plants' the corresponding **division** has been defined largely on structure, life history and photosynthetic pigments.

An even higher category is the **kingdom**. In the nineteenth century it was usual to recognize only the two kingdoms: Plantae and Animalia. Plants were considered to be mainly non-motile, feeding by photosynthesis. Animals were considered to be motile, feeding by ingestion of pre-formed organic matter. Although these distinctions are evident amongst macroscopic organisms living on land, the largely aqueous world of microscopic life abounds with organisms that appear to straddle the plant–animal boundary. The classification shown in Box 1.1 overcomes these anomalies by recognizing seven kingdoms: the Eubacteria, Archaebacteria, Protozoa, Plantae, Animalia, Fungi and Chromista.

The highest category is the **empire**. The classification of the empire Bacteria will be considered further in Chapter 8. The Bacteria are single celled but they lack a nucleus, cell vacuoles and organelles. This primitive **prokaryotic** condition, in which proper sexual reproduction is unknown, is characteristic of such forms as cyanobacteria. The empire is currently divided into two kingdoms, the Archaebacteria and the Eubacteria. The other five kingdoms are **eukaryotic**. That is their cells have a nucleus, vacuoles and other organelles and are capable of properly coordinated cell division and sexual reproduction. Attempts to divide unicellular eukaryotic organisms, often called **protists**, into plants or animals based on feeding style were abandoned when it was recognized that dinoflagellates, euglenoids and heterokonts have members that are both photosynthetic and heterotrophic, feeding by engulfing. Since the 1970s both ultrastructural analysis under the scanning electron microscope and molecular sequences have been used to elucidate protistan phylogenies and develop a large-scale classification. The new classification of Cavalier-Smith (1981, 1987a, 1987b, 2002) has put forward two new categories: the predominantly photosynthetic kingdom **Chromista** (brown algae, diatoms and their various relatives) and the primitive superkingdom **Archezoa** (which lack mitochondria (**amitochondrial**)). He has also proposed an ultrastructurally based redefinition of the kingdom **Plantae** which requires the exclusion of many aerobic protists that feed by ingestion (**phagotropy**). The kingdom **Protozoa** is now considered to contain as many as 18 phlya (Cavalier-Smith 1993, 2002) and their classification and phylogenetic relationships, which is in a state of flux, is largely based upon cell ultrastructure and increasingly sophisticated analyses of new molecular sequences. The kingdom

Empire	Superkingdom	Kingdom	Subkingdom(s)
BACTERIA		EUBACTERIA	NEGIBACTERIA
			POSIBACTERIA
		ARCHAEBACTERIA	
EUKARYOTA	ARCHEZOA		
	METAKARYOTA	PROTOZOA	GYMNOMYXA
			CORTICATA
		PLANTAE	VIRIPLANTAE (green plants)
			BILIPHYTA (red algae and glaucophytes)
		ANIMALIA	RADIATA
			BILATERATIA
		FUNGI	
		CHROMISTA	CHLORARACHINA
			EUCHROMISTA (cryptomonads, *Goniomonas*, heterokonts, haptophytes)

Fig. 1.2 The empires of life. (Modified from Cavalier-Smith 1993.)

Protozoa includes two subkingdoms, the Gymnomyxa and Corticata. Members of the Gymnomyxa have a 'soft' cell wall often with pseudopodia or axopodia (e.g. foraminifera). The Corticata are ancestrally biciliate (e.g. dinoflagellates).

Members of the superkingdom Archezoa differ from most Protozoa in having ribosomes, the RNA-protein structures on which messenger RNA is 'read' during protein synthesis, found in all other eukaryotes, and they also lack certain other organelles (e.g. mitochondria, Golgi bodies). The Archezoa comprise three phyla: the Archamoebae, Metamonada and Microsporidia. There is reasonable rDNA phylogenetic evidence to suggest that the latter two represent surviving relics of a very early stage in eukaryote evolution. The evolution of the eukayotes can thus be divided into two major phases. The origin of the eukaryote cell (the first archezoan) is marked by the appearance of the membrane-bounded organelles, cytoskeleton, a three-dimensional network of fibrous proteins that give order and structure in the cytoplasm, nucleus and cilia with a 9+2 structure (Fig. 1.1). This was apparently followed by the symbiotic origin of mitochondria and peroxisomes (Margulis 1981; Cavalier-Smith 1987c) to produce the first aerobically respiring protozoan. The change in their ribosomes may have occurred somewhat later in their evolution.

The kingdom Chromista is a predominantly photosynthetic category in which the chromoplasts are located in the endoplasmic reticulum but separated by a unique smooth membrane, thought to be a relic of the cell membrane of the photosynthetic eukaryotic symbiont that was 'engulfed' by the protozoan host, leading to the emergence of the Chromista (Cavalier-Smith 1981, 1987c). The Chromista contains a number of important microfossil groups such as the silicoflagellates, diatoms and calcareous nannoplankton.

The kingdon Plantae is taken to comprise two subkingdoms. The subkingdom Viriplantae includes the green plants including the green algae (Chlorophyta), the Charophyta and the 'land plants' or the Embryophyta. The subkingdom Biliphyta includes the red algae (Rhodophyta) and the Glaucophyta. It is not yet clear whether these two subkingdoms are correctly placed together in a single kingdom or should be separate kingdoms. The Viriplantae have starch-containing chloroplasts and contain chlorophylls a and b. The Biliphyta have similar chloroplasts but there is a total absence of phagotrophy in this group.

The kingdom Fungi comprises heterotrophic eukaryotes that feed by the adsorption of pre-formed organic matter. They are rarely preserved in the fossil record and have received little study as fossils and are not considered further in this book.

The kingdom Animalia comprises multicellular invertebrate and vertebrate animals that feed by the ingestion of pre-formed organic matter, either alive or dead. Invertebrates that are microscopic when fully grown, for example the ostracods, are considered as microfossils, but we are obliged to leave aside the microscopic remains of larger animals (such as sponge spicules, echinoderm ossicles and juvenile individuals). For more information on the macro-invertebrate fossil record the reader is referred to our companion volume written by Clarkson (2000).

Microfossils that cannot easily be placed within the existing hierarchical classification, for example acritarchs, chitinozoa and scolecodonts, are accorded the informal and temporary status of a **group** in this book.

REFERENCES

Cavalier-Smith, T. 1981. Eukaryote kingdoms: seven or nine? *Biosystems* 14, 461–481.

Cavalier-Smith, T. 1987a. Eukaryotes without mitochondria. *Nature (London)* 326, 332–333.

Cavalier-Smith, T. 1987b. Glaucophyeae and the origin of plants. *Evolutionary Trends in Plants* 2, 75–78.

Cavalier-Smith, T. 1987c. The simultaneous symbiotic origin of mitochondria, chloroplasts and microbodies. *Annals of the New York Academy of Sciences* 503, 55–71.

Cavalier-Smith, T. 1993. Kingdom Protozoa and its 18 phyla. *Microbiological Reviews* 57, 953–994.

Cavalier-Smith, T. 2002. The phagotrophic origin of eukaryotes and phylogenetic classification of protozoa. *International Journal of Systematic and Evolutionary Microbiology* 52, 297–354.

Clarkson, E.N.K. 2000. *Invertebrate Palaeontology and Evolution*, 4th edition. Blackwell, Oxford.

Margulis, L. 1981. Symbiosis in cell evolution. *Life and its Environment on the Earth*. Freeman, San Francisco.

CHAPTER 2

Micropalaeontology, evolution and biodiversity

Micropalaeontology brings three unique perspectives to the study of evolution: the dimension of time, abundance of specimens (allowing statistical analysis of trends) and long complete fossil records, particularly in marine groups. Despite these features giving special insights into the nature of evolutionary processes, micropalaeontologists have until recently concentrated mainly on documenting the ascent of evolutionary lineages, such are described in the separate chapters in this book.

Micro- and macroevolution are the two main modes of evolution. **Microevolution** describes small-scale changes within species, particularly the origin of new species. Speciation occurs as the result of **anagenesis** (gradual shifts in morphology through time) or **cladogenesis**, rapid splitting of a pre-existing lineage. Which of these is the dominant mode has remained one of the most controversial questions in palaeobiology in the last 30 years.

Some of the best recorded examples of anagenesis have been documented in planktonic foraminifera (Malmgren & Kennett 1981; Lohmann & Malmgren 1983; Malmgren *et al.* 1983; Hunter *et al.* 1987; Malmgren & Berggren 1987; Norris *et al.* 1996; Kucera & Malmgren 1998), whilst examples of cladogenesis (e.g. Wei & Kennett 1988; Lazarus *et al.* 1995; Malmgren *et al.* 1996) are less widely cited. Similar studies have been conducted on Radiolaria (Lazarus 1983, 1986) and diatoms (Sorhannus 1990a, 1990b). Where morphological change has been mapped onto an ecological gradient (such as temperature/depth gradients measured by oxygen isotope analysis) it appears that gradual morphological trends do not strictly reflect the rate of speciation or its mode. For example, Kucera & Malmgren (1998) showed that gradual change in the Cretaceous planktonic foraminifera *Contusotruncana fornicata* probably resulted in a shift in the relative proportion of high conical to low conical forms through time. High conical forms evolved rapidly and gradually replaced the low conical morphs, though at any one time the abundances of different morphs were normally distributed. Similarly, Norris *et al.* (1996) documented a gradual shift in the average morphology of *Fohsella fohsi* over ~400 kyr, suggesting only one taxon was present at any given time (Fig. 2.1), yet isotopic data indicated a rapid separation of the population, into surface- and thermocline-dwelling populations and reproductive isolation midway through the anagenetic trend. During the same interval keeled individuals gradually replaced unkeeled forms, a clear example of both anagenesis and cladogenesis occurring in the same population. Another 'classic' example of anagenetic change, that of *Globorotalia plesiotumida* and the descendant *G. tumida* (Malmgren *et al.* 1983, 1984), has been challenged by Norris (2000). *G. plesiotumida* ranges well into the range of *G. tumida* (e.g. Chaisson & Leckie 1993; Chaisson & Pearson 1997) and therefore cannot have given rise to *G. tumida* by the complete replacement of the ancestor population. An alternative explanation to this and probably all examples of anagenetic trends is that cladogenesis is quickly followed by a rapid change in the relative proportions of the ancestor and descendant populations. Apparently gradual changes in 'mean form' may be caused by natural selection operating on a continuous range of variation in populations living in environments lacking barriers to gene flow.

Macroevolution is concerned with evolution above the species level, the origins and extinctions of major groups and adaptive radiations. Microevolution and

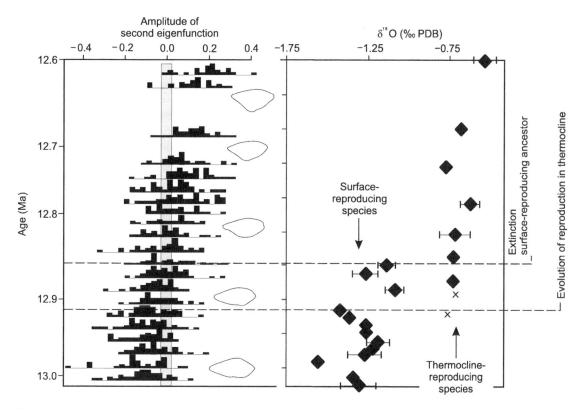

Fig. 2.1 Changes in morphology and habitat during the evolution of the planktonic foraminifera *Fohsella* from the mid-Miocene. On the left, frequency histograms show the gradual (anagenetic) change in the morphology of the shell outline. On the right, stable oxygen isotope data from the same specimens show an abrupt appearance of a new thermocline-reproducing species (cladogenesis). The ancestor became extinct ~70 kyr after the appearance of the descendant species. Morphological data suggest that no more than one population was present at any one time. (Redrawn from Norris *et al.* 1996 with permission.)

macroevolution processes are decoupled (Stanley 1979). This is because the individual is the basic unit of selection in microevolution whilst selection between species may occur at higher levels, although the notion of competition and natural selection occurring between higher taxonomic categories is not unanimously accepted (see Kemp 1999). New structures, body plans and biochemical systems, and the characters of high taxonomic categories, appear suddenly in the fossil record, for example the appearance of calcification in the calcareous nannoplankton in the Early Mesozoic. The evolutionary mechanisms behind these changes are the least well understood of evolutionary phenomena. Explanations invoke mutation in regulatory genes, which encode for hormones and other rate-effecting proteins and wholesale changes in chromosomal structure.

Mass extinctions are probably the most widely studied of the macroevolutionary patterns. These differ from 'background' extinction events in their speed (commonly <5 Myr) and intensity (where 20–50% of marine biodiversity may disappear in a single event). The Cretaceous–Tertiary boundary mass extinction provides the best-studied example of a mass extinction event. This been documented globally and has been attributed to a variety of terrestrial (including climate change) and extraterrestrial (meteorite impact) causes (see Hallam & Wignall 1997 for a review). A comprehensive review of the biological effects of the K–T mass extinction event is provided by MacLeod & Keller

(1996). Patterns of extinction in individual groups add little to the debate on the cause of the K-T mass extinction. For example, extinctions in planktonic foraminifera extend over an interval of 30 cm (<100,000 years) that spans the boundary and exhibit a preferential extinction of large ornate forms. Benthic foraminifera declined in diversity but were much less affected. Coccolithophorids were once thought to become almost extinct at this boundary, however Cretaceous species found in the lower Tertiary are now considered to have survived the event (Perch-Nielsen et al. 1982). Dinoflagellates were evidently less affected by events at the boundary. Dinoflagellate cysts are extremely abundant in the boundary clay, indicating that environmental conditions were ideal for stimulating dinoflagellate blooms. Diversity and species turnover rates are also high across the boundary. Plants on the other hand show major changes, Wolfe & Upchurch (1986) noted the decline in pollen and a sharp peak in fern spores, suggesting the influence of wildfires, though increasing humidity could also have caused an increase in fern abundance.

Mechanisms of cladogenesis

Models of cladogenesis rely upon the genetic isolation of a population. Random mutations in these small populations (**peripheral isolates**) are then quickly spread and eventually lead to the development of a new species, a process known as **allopatric** speciation (Fig. 2.2). In the marine realm genetic isolation would at first sight seem less probable. However a number of ecological barriers are present in the oceans. For example, ocean frontal systems, such as the Tasman Front, a boundary between tropical and subtropical water masses, have been proposed as effective barriers to dispersal and may have been important in promoting allopatric speciation in globoconelid planktonic foraminifera during the Pliocene (Wei & Kennett 1988). **Vicariant** models of speciation similarly subdivide an original population into smaller units through the development of physical barriers such as land barriers, sea-level fall and the strengthening of water mass boundaries. Knowlton & Weight (1998) have documented many examples of vicariant speciation in the marine realm following the separation of the Atlantic and the Pacific oceans through circulation changes during the Pleistocene. Low sea levels during the Pleistocene have also been implicated in the speciation of copepods on either side of the Indonesian Seaway (Fleminger 1986). However, many planktonic foraminifera species have the ability to cross such major barriers; *Pullenatina obliquiloculata* and other related species repeatedly reinvaded the tropical Atlantic from the Indo-Pacific during Pleistocene glacial cycles. Neither equatorial upwelling in the Atlantic nor the Isthmus of Panama were sufficient barriers to dispersal.

Many microfossil groups are planktonic and have high population sizes and high dispersal potential. These features would seem contrary to the conditions required for allopatric speciation. Species models that allow restricted genetic exchange may therefore be better explanations of speciation in these types of organisms.

Variation in morphology along geographical gradients (**clines**) can result in limited interaction between the ends of the cline and effective genetic isolation ('isolation-by-distance' or **parapatric speciation**). Clinal trends have been described in a wide range of marine planktonic organisms (van Soest 1975; Lohmann & Malmgren 1983; Lohmann 1992), though some believe these may represent geographical successions of distinct species (see below). Even the classical latitudinal morphological cline of *Globorotalia truncatulinoides*, originally described as continuous (Lohmann & Malmgren 1983) may contain distinct species (Healy-Williams et al. 1985; de Vargas et al. 2001).

Similarly 'isolation-by-ecology' appears common, and is particularly well documented for depth in foraminifera. Many forams reproduce by sinking (Norris et al. 1996), during which they cross the large number of physical and chemical barriers in the ocean. It seems plausible that speciation could occur by changes in the depth of reproduction, though confirmatory evidence is still rather sparse. Norris et al. (1993, 1996) used stable oxygen isotopes to show that the evolution of *Fohsella fohsi* in the mid-Miocene involved a rapid shift in reproductive depth habitat (Fig. 2.1). Using similar methodology Pearson et al. (1997) calculated 1–2°C differences in the temperature at which calcification occurred in closely related species, relating this to differences in either season or depth of growth. As

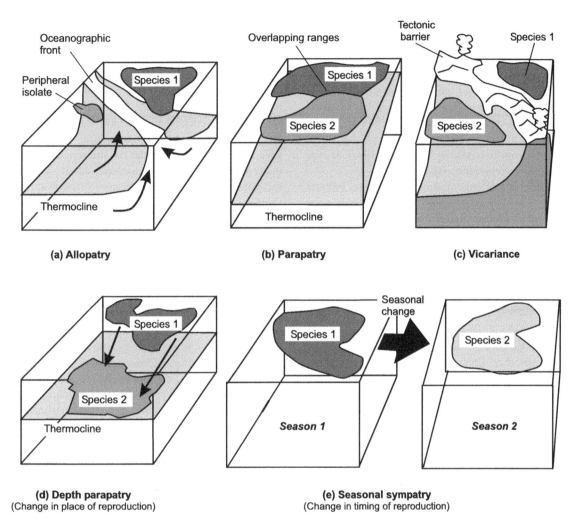

Fig. 2.2 Speciation models. (a) Allopatry, created by divergence on either side of a hydrographic boundary. (b,d) Parapatry in which species diverge along a gradual hydrographic gradient, for example a gradually changing thermocline depth (b) or depth (d). (c) Vicariance, occurs where a physical boundary creates isolation and the formation of a new species. (e) 'Seasonal sympatry' in which isolation is caused by a change in the timing of reproduction. In marine planktonic species complete genetic isolation as indicated in (a) and (c) is unlikely. (Redrawn after Norris 2000 with permission.)

the seasonal range in temperature of surface waters in the tropics and subtropics can be greater than this it is reasonable that divergence in these species could have occurred as the result of a shift in timing of reproduction and growth ('**seasonal sympatry**').

Theoretical and empirical studies (e.g. Howard & Berlocher 1998) have also indicated **sympatric speciation** may be more common in the marine realm than has been hitherto considered. Sympatry may have resulted from individuals evolving different strategies to avoid strong competition for a single food source (Dieckmann & Doebell 1999), or from **disruptive selection** which favours individuals with extreme characters, for example large and small predators at the behest of medium sized individuals (e.g. Kondrashov & Kondrashov 1999; Tregenza & Butlin 1999).

Biodiversity in the marine plankton

Briggs (1994) calculated there are approximately 12 million terrestrial multicellular species (approximately 10 million of which are insects!) but only 200,000 marine taxa. These are surprising numbers when models of ecosystem size, energy flow and environmental stability predict substantially higher numbers of marine to terrestrial taxa (Briggs 1994). Are the models or numbers incorrect?

Results of molecular phylogenetic analyses indicate there is a high cryptic biodiversity in the oceans. Numerous **sibling species** can be diagnosed using molecular sequence data but show few if any morphological differences (e.g. Bucklin 1986; Bucklin et al. 1996; Bucklin & Wiebe 1998), a feature that probably extends into the cyanobacteria (Moore et al. 1998) and bacterio-plankton (De Long et al. 1994). Cryptic speciation and high genetic diversity has also been documented for planktonic foraminifera (Huber et al. 1997; de Vargas & Pawlowski 1998; Darling et al. 1999; de Vargas et al. 1999) and, surprisingly, many morphologically similar taxa have ancient divergences. Distinguishing sibling species in the fossil record is extremely difficult and many previously defined ecological variants (**ecophenotypes**) may be distinct species; if this is the case then planktonic biodiversity has been grossly underestimated.

Reconstructing phylogeny

The higher classification (above species level) of a group of organisms should reflect their evolution. The taxonomic hierarchy is expressed as an upwardly inclusive **nested heirarchy**, similar species are grouped into genera, similar genera into families, families into orders, orders into classes and classes into phyla and where necessary subdivisions of these major categories, for example subfamily and superfamily, are also used. Higher taxonomic categories are distinguished by their suffix (i.e. -ae, -a, etc.) and many examples are included in subsequent chapters.

Defining higher taxonomic groupings is a largely subjective exercise. Until the 1970s classical taxonomists used a combination of **morphological** (or phenetic) similarity and **phylogenetic** (evolutionary) **resemblance**, based on ill-defined notions of ancestor–descendant relationships. Stratigraphical succession of species and their geographical distribution played an important role in establishing phylogenetic relationships. Since the 1970s an attempt has been made to reduce the subjectivity inherent in the classical method and two philosophical approaches have been followed. Phenetics (or **numerical taxonomy**) relies on scoring of characters. Cluster analysis and distance statistics can then be used on the resulting character matrix to quantify the similarities between taxa and **groupings** into higher taxonomic categories. **Cladistics** (or **phylogenetic systematics**), founded by W. Hennig (1966) has been much more widely applied to palaeontology though less so in micropalaeontology. The reader is referred to Smith (1994) for a comprehensive explanation of the methodology. At the heart of cladistics is the concept that organisms contain a combination of 'primitive' (**symplesiomorphic**) and evolutionary novelties (**synapomorphic**) or 'derived characters'. Closely related groups will share derived characters and these will distinguish them from other groups. For example, humans have a backbone, a primitive character of all vertebrates, and an opposable thumb, a derived character shared with our nearest relatives the great apes. A primitive character for all vertebrates, the backbone, is of course a derived character as compared to invertebrates. Synapomorphy and symplesiomorphy are therefore relative conditions of particular characters with reference to a particular phylogenetic reconstruction.

The results of a phylogenetic analysis are expressed in a **cladogram**, in which branching points are arranged in nested hierarchies. In the example in Fig. 2.3 C and D share a unique common ancestor, they are **sister groups** and share a synapomorphy not possessed by B. Thus B is the sister group of the combined grouping C + D and A is the sister group of B + C + D. If a large number of characters and taxa are being analysed the character matrix is routinely manipulated by computer programs such as PAUP (Phylogenetic Analysis Using Parsimony). This is a technique that makes the fewest assumptions (**parsimony**) to rank the set of observations and produce the cladogram. A cladogram

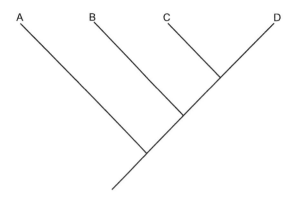

Fig. 2.3 A cladogram showing the phylogenetic relationship between A and D (see text for explanation).

is not an evolutionary tree but a hypothesis of relationships. Stratigraphical succession is explicitly ignored in the analysis. Once the cladogram has been produced stratigraphical succession can be used in the analysis of the cladogram (see Smith 1994) and to constrain the splitting of lineages in time. At this point the cladogram becomes a phylogenetic tree.

Distinguishing shared primitive (**sympleisiomorphic**) and shared derived (**synapomorphic**) characters is achieved by **outgroup analysis**. Here the **ingroup**, the group being studied, is compared to a closely related **outgroup**. In Fig. 2.3 B + C + D could be the ingroup and A the outgroup. Any character present in a variable state in the ingroup and found in the outgroup must be **plesiomorphic** (primitive). **Apomorphic** characters are those only found in the ingroup.

Three types of cladistic groups have been defined: **monophyletic** groups contain the common ancestor and all of its subsequent descendants; **paraphyletic** groups are descended from a common ancestor (usually extinct) but do not include all the descendants; **polyphyletic** groups are the result of convergent evolution. In the latter, their representatives are descended from different ancestors and though looking superficially similar, there is no close phylogenetic relationship.

Subjectivity cannot be entirely removed from phylogenetic reconstruction and higher taxonomic categories. In cladistics equally parsimonious cladograms can result from the analysis and choosing between these may become subjective. In numerical taxonomy the methods of measurement and the relative weighting given to characters are also subjective decisions. The possibility of morphological **convergence** during evolution is a problem for all taxonomic methods and ultimately molecular sequence data may be required to distinguish between polyphyletic and sibling species. Unfortunately such data are not available in extinct groups.

REFERENCES

Briggs, D.E.G. 1994. Species diversity: land and sea compared. *Systematic Biology* **43**, 130–135.

Bucklin, A. 1986. The genetic structure of zooplankton populations. In: Pierrot-Bults *et al.* (eds) *Pelagic Biogeography.* UNESCO, Paris, pp. 35–41.

Bucklin, A. & Wiebe, P.H. 1998. Low mitochondrial diversity and small effective population sizes in the copepods *Calanus finmarchicus* and *Nannocalanus minor*: possible impact of climatic variation during recent glaciation. *Journal of Heredity* **89**, 383–392.

Bucklin, A., Lajeunesse, T.C., Curry, E., Wallinga, J. & Garrison, K. 1996. Molecular diversity of the copepod: genetic evidence of species and population structure in the North Atlantic Ocean. *Journal of Marine Research* **54**, 285–310.

Chaisson, W.P. & Leckie, R.M. 1993. High-resolution Neogene planktonic foraminifer biostratigraphy of Site 806, Omtong Java Plateau (western Equatorial Pacific). *Proceedings of the Ocean Drilling Program, Scientific Results* **130**, 137–178.

Chaisson, W.P. & Pearson, P.N. 1997. Planktonic foraminifer biostratigraphy at Site 925: middle Miocene-Pleistocene. *Proceedings of the Ocean Drilling Program, Scientific Results* **154**, 3–31.

Clarkson, E.N.K. 2000. *Invertebrate Palaeontology and Evolution*, 4th edn. Blackwell, Oxford.

Darling, K.F., Wade, C.M., Kroon, D., Leigh Brown, A.J. & Bijma, J. 1999. The diversity and distribution of modern planktonic foraminiferal subunit ribosomal RNA genotypes and their potential as tracers of present and past ocean circulation. *Paleoceanography* **14**, 3–12.

Dieckmann, U. & Doebell, M. 1999. On the origin of species by sympatric speciation. *Nature* **400**, 354–358.

Fleminger, A. 1986. The Pleistocene equatorial barrier between the Indian and Pacific oceans and a likely cause for Wallace's Line. In: Pierrot-Bults *et al.* (eds) *Pelagic Biogeography*. UNESCO, Paris, pp. 84–97.

Hallam, A. & Wignall, P.B. 1997. *Mass Extinctions and their Aftermath*. Oxford University Press, Oxford.

Healy-Williams, N., Ehrlich, R. & Williams, D.F. 1985. Morphometric and stable isotopic evidence for subpopulations of *Globorotalia truncatuloides*. *Journal of Foraminiferal Research* 15, 242–253.

Hennig, W. 1966. *Phylogenetic Systematics*. University of Illinois Press, Urbana, IL. [Second edition published in 1979.]

Howard, D.J. & Berlocher, S.H. (eds) 1998. Endless forms: Species and Speciation. Oxford University Press, Oxford.

Huber, B.T., Bijma, J. & Darling, K. 1997. Cryptic speciation in the living planktonic foraminifer *Globigerinoides siphonifera* (d'Orbigny). *Paleobiology* 23, 33–62.

Hunter, R.S.T., Arnold, R.J. & Parker, W.C. 1987. Evolution and homoeomorphy in the development of the Paleocene *Planorotalites pseudomenardii* and the Miocene *Globorotalia* (*Globorotalia*) *margaritae* lineages. *Micropalaeontology* 31, 181–192.

Kemp, T.S. 1999. *Fossils and Evolution*. Oxford University Press. Oxford.

Knowlton, N. & Weight, L.A. 1998. New dates and new rates for divergence across the Isthmus of Panama. *Proceedings of the Royal Society of London Series B: Biological Sciences* 265, 2257–2263.

Kondrashov, A.S. & Kondrashov, F.A. 1999. Interactions among quantitative traits in the course of sympatric speciation. *Nature* 400, 351–354.

Kucera, M. & Malmgren, B.A. 1998. Differences between evolution of mean form and evolution of new morphotypes: an example from Late Cretaceous planktonic foraminifera. *Paleobiology* 24, 49–63.

Lazarus, D. 1983. Speciation in pelagic protista and its study in the planktonic microfossil record: a review. *Paleobiology* 9, 327–341.

Lazarus, D. 1986. Tempo and mode of morphologic evolution near the origin of the radiolarian lineage *Pterocanium prismatium*. *Paleobiology* 12, 175–189.

Lazarus, D., Hilbrecht, H., Pencer-Cervato, C. & Thierstein, H. 1995. Sympatric speciation and phylogenetic change in *Globorotalia truncatulinoides*. *Paleobiology* 21, 28–51.

Lohmann, G.P. 1992. Increasing seasonal upwelling in the subtropical South Atlantic over the past 700,000 yrs: evidence from deep-living planktonic foraminifera. *Marine Micropalaeontology* 19, 1–12.

Lohmann, G.P. & Malmgren, B.A. 1983. Equatorward migration of *Globorotalia truncatulinoides* ecophenotypes through the Late Pleistocene: gradual evolution or ocean change? *Paleobiology* 9, 414–421.

De Long, E.F., Wu, K.Y., Prizelin, D.D. & Jovine, R.V.M. 1994. High abundance of Archea in Antarctic marine picoplankton. *Nature* 371, 695–697.

MacLeod, N. & Keller, G. 1996. *Cretaceous–Tertiary Mass Extinctions: biotic and environmental changes*. Norton, New York.

Malmgren, B.A. & Berggren, W.A. 1987. Evolutionary changes in some Late Neogene planktonic foraminifera lineages and their relationships to palaeooceanograhic changes. *Paleoceanography* 2, 445–456.

Malmgren, B.A. & Kennett, J.P. 1981. Phyletic gradualism in a Late Cenozoic planktonic foraminiferal lineage: DSDP 284. Southwest Pacific. *Paleobiology* 7, 230–240.

Malmgren, B.A., Berggren, W.A. & Lohmann, G.P. 1983. Evidence for punctuated gradualism in the Late Neogene *Globorotalia tumida* lineage of planktonic Foraminifera. *Paleobiology* 9, 377–389.

Malmgren, B.A., Berggren, W.A. & Lohmann, G.P. 1984. Species formation through punctuated gradualism in planktonic foraminifera. *Science* 225, 317–319.

Malmgren, B.A., Kucera, M. & Ekman, G. 1996. Evolutionary changes in supplementary apertural characteristics of the Late Neogene *Spheroidinella dehiscens* lineage (planktonic foraminifera). *Palaios* 11, 96–110.

Moore, L.R., Rocap, G. & Chisholm, S.W. 1998. Physiology and molecular phylogeny of coexisting *Prochlorococcus* ecotypes. *Nature* 393, 464–467.

Norris, R.D. 2000. Pelagic species diversity, biogeography and evolution. *Paleobiology* 26(4), supplement, 236–259.

Norris, R.D., Corfield, R.M. & Cartlidge, J.E. 1993. Evolution of depth ecology in the planktic Foraminifera lineage *Globorotalia* (*Fohsella*). *Geology* 21, 975–978.

Norris, R.D., Corfield, R.M. & Cartlidge, J.E. 1996. What is gradualism? Cryptic speciation in globorotaliid planktic foraminifera. *Paleobiology* 22, 386–405.

Pearson, P.N., Shackleton, N.J. & Hall, M.A. 1997. Stable isotopic evidence for the sympatric divergence of *Globigerinoides trilobus* and *Orbulina universa* (planktonic foraminifera). *Journal of the Geological Society, London* 154, 295–302.

Perch-Nielsen, K., McKenzie, J. & He, Q. 1982. Biostratigraphy and isotope stratigraphy and the 'catastrophic' extinction of calcareous nannoplankton at the Cretaceous–Tertiary boundary. *Geological Society of America* 190, special paper, 291–296.

Smith, A.B. 1994. *Systematics and the Fossil Record. Documenting evolutionary patterns*. Blackwell, Oxford.

Sorhannus, U. 1990a. Punctuated morphological change in a Neogene diatom lineage: 'local' evolution or migration? *Historical Biology* 3, 241–247.

Sorhannus, U. 1990b. Tempo and mode of morphological evolution in two Neogene diatom lineages. In: Hecht, M.K., Wallace, B. & MacIntyre, R.J. (eds) *Evolutionary Biology*. Plenum, London, pp. 329–370.

Stanley, S.M. 1979. *Macroevolution: pattern and process*. Freeman, San Francisco.

Tregenza, T. & Butlin, R.K. 1999. Speciation without isolation. *Nature* 400, 311–312.

van Soest, R.M.W. 1975. Zoogeography and speciation in the Salpidae (Tunicata, Thaiacea). *Beaufortia* 23, 181–215.

de Vargas, C. & Pawlowski, J. 1998. Molecular versus taxonomic rates of evolution in planktonic foraminifera. *Molecular Phylogenetics and Evolution* 9, 463–469.

de Vargas, C., Norris, C.R., Zaninetti, L. Gibb, S.W. & Pawlowski, J. 1999. Molecular evidence of cryptic speciation in planktonic foraminifers and their relation to oceanic provinces. *Proceedings of the National Academy of Sciences of the USA* 96, 2864–2868.

de Vargas, C., Renaud, S., Hilbrecht, H. & Pawlowski, J. 2001. Pleistocene adaptive radiation in *Globorotalia truncatulinoides*: genetic, morphological and environmental evidence. *Paleobiology* 27, 104–125.

Wei, K-Y. & Kennett, J.P. 1988. Phyletic gradualism and punctuated equilibrium in the Late Neogene planktonic foraminiferal clade *Globoconella*. *Paleobiology* 14, 345–363.

Wolfe, J.A. & Upchurch, G.R. 1986. Vegetation, climatic and floral changes at the Cretaceous–Tertiary boundary. *Nature* 324, 148–154.

CHAPTER 3

Microfossils in stratigraphy

The stratigraphical column

The succession of rocks exposed at the surface of the Earth can be arranged into a **stratigraphical column**, with the oldest rocks at the base and the youngest ones at the top (Fig. 3.1). Although the absolute ages have been determined from studies of radioactive isotopes, it is customary to use the names of stratigraphical units, mostly distinguished on the basis of differences in their included fossils. These units are arranged into a number of hierarchies relating to rock-based stratigraphy (**lithostratigraphy**), fossil-based stratigraphy (**biostratigraphy**) and time-based stratigraphy (**chronostratigraphy**).

Lithostratigraphical units, such as beds, members and formations, are widely used in geological mapping but will not concern us further here. The **biozone** is the fundamental biostratigraphical unit and comprises those rocks that are characterized by the occurrence of one or more specified kinds of fossil known as zone fossils.

Formal chronostratigraphical time units are also important and include, in ascending order of importance, the age, epoch, period and era. For example we may cite the Messinian Age, of the Miocene Epoch, of the Neogene Period, of the Cenozoic Era. Rock units laid down during these times are properly referred to as stages, series, systems and erathems (i.e. the Messinian Stage, of the Miocene Series, etc.). Less formal divisions are also widely used so that we may talk of the lower Neogene rocks laid down during Early Neogene times. In the following text, these informal subdivisions are abbreviated as follows: lower (L.), middle (M.) and upper (U.) and their equivalents for chronostratigraphy early (E.), middle and late.

Microfossils and biostratigraphy

Biostratigraphy is the grouping of strata into units based on their fossil content with the aim of zonation and correlation. As such biostratigraphy is concerned primarily with the identification of taxa, tracing their lateral and vertical extent and dividing the geological column into units defined on their fossil content.

Microfossils are among the best fossils for biostratigraphical analysis because they can be extremely abundant in rocks (a particular consideration when dealing with drill cuttings) and they can be extracted by relatively simple bulk processing methods. Many groups are geographically widespread and relatively free from facies control (e.g. plankton, airborne spores and pollen). Many of the groups evolved rapidly, allowing a high level of subdivision of the rock record and a high level of stratigraphical resolution. It should also be emphasized that spores, pollen, diatoms and ostracods are indispensable for the biostratigraphy of terrestrial and lacustrine successions, where macrofossils can be scarce.

Detailed biostratigraphical zonations, using the groups mentioned in this book, have been developed for the entire Phanerozoic. Some areas of the column are better subdivided than others, for example the Cretaceous to Recent can be subdivided into approximately 70 biozones, based on calcareous nannoplankton and planktonic foraminifers, with an average duration of 2 million years per biozone. In comparison the Lower Palaeozoic has only been divided into 39 conodont biozones at an average duration of 3 million years. Detailed biostratigraphical zonations for the Mesozoic and Cenozoic are to be found in the two volumes of *Plankton Stratigraphy* (Bolli *et al.* 1985).

Chapter 3: Microfossils in stratigraphy 17

Age	540	650	850	1000	1200	1400	1600	1800	2050	2300	2500	2800	3200	3600
Systems period	Neoproterozoic III	Cryogenian	Tonian	Stenian	Ectasian	Calymmian	Statherian	Orosirian	Rhyacian	Siderian		No subdivision into periods		
Erathem era	Neoproterozoic			Mesoproterozoic			Palaeoproterozoic				Neoarchean	Mesoarchean	Palaeoarchean	Eoarchean
Eonothem eon	PROTEROZOIC PR										ARCHEAN AR			
	PRECAMBRIAN PC													

Abbrev.	Chx	Wuc	Cap	Wor	Rod	Kun	Art	Sak	Ass	Gzh	Kaz	Mos	Bash	Serp	Vis	Tou	Fam	Fra	Giv	Eif	Em	Pra	Loch	Lud	Gor	Hom	Shn	Tel	Aer	Rhud	Ash	Car	Lln	Arg	Trem	Dol	Mnt	Men	Sol
Ma	251.4	253.4	265				283			292			320	327	342	354	364	370	380	391	400	412	417 419	423			428		440				467.5		495	500	520	545	
Stage age	Changsingian	Wuchiapingian	Capitanian	Wordian	Roadian	Kungurian	Artinskian	Sakmarian	Asselian	Gzhelian	Kazimovian	Moscovian	Bashkirian	Serpukhovian	Visean	Tournaisian	Famennian	Frasnian	Givetian	Eifelian	Emsian	Pragian	Lochkovian	Ludfordian	Gorstian	Homerian	Sheinwoodian	Telychian	Aeronian	Rhuddanian	Ashgill	Caradoc	Llanvirn	Arenig	Tremadocian	Dolgellian	Maentwrogian	Menevian	Solvan
Series epoch	Lopingian		Guadalupian			Cisuralian				Pennsylvanian			Mississippian				Upper/Late		Middle		Lower/Early			Pridoli	Ludlow		Wenlock		Llandovery		Upper/Late		Middle	Lower/Early		Merioneth		St. Davids	
System period	PERMIAN									CARBONIFEROUS							DEVONIAN							SILURIAN							ORDOVICIAN					CAMB.			
Erathem era	PALAEOZOIC PZ																																						
Eonothem eon	PHANEROZOIC PH																																						

Abbrev.	Hol	Ple	Gel	Pia	Zan	Mes	Tor	Srv	Lan	Bur	Agt	Cht	Rup	Prb	Brt	Lut	Ypr	Tha	Sel	Dan	Maa	Cmp	San	Con	Tur	Cen	Alb	Apt	Brm	Hau	Vlg	Ber	Tth	Klm	Oxf	Clv	Bth	Baj	Aal	Toa	Plb	Sin	Het	Rht	Nor	Crn	Lad	Ans	Spa
Ma	0.01	1.18	2.58	3.60	5.32	7.12	11.2	14.8	16.4	20.5	23.8	28.5	33.7	37.0	41.3	49.0	55.0	57.9	61.0	65.5	71.3	83.5	85.8	89.0	93.5	98.9	112.2	121.0	127.0	132.0	136.5	142.0	150.7	154.1	159.4	164.4	169.2	176.5	180.1	189.6	195.3	201.9	205.7	209.6	220.7	227.4	234.3	241.7	244.8 250
Stage age		Calabrian	Gelasian	Piacenzian	Zanclean	Messinian	Tortonian	Serravallian	Langhian	Burdigalian	Aquitanian	Chattian	Rupelian	Priabonian	Bartonian	Lutetian	Ypresian	Thanetian	Selandian	Danian	Maastrichtian	Campanian	Santonian	Coniacian	Turonian	Cenomanian	Albian	Aptian	Barremian	Hauterivian	Valanginian	Berriasian	Tithonian	Kimmeridgian	Oxfordian	Callovian	Bathonian	Bajocian	Aalenian	Toarcian	Pliensbachian	Sinemurian	Hettangian	Rhaetian	Norian	Carnian	Ladinian	Anisian	Olenekian Induan
Series epoch	Holocene	Pleistocene		Pliocene			Miocene					Oligocene			Eocene					Palaeocene	Upper/Late						Lower/Early						Upper/Late				Middle			Lower/Early				Upper/Late			Middle	Lower/Early	
System period	Quaternary		NEOGENE									PALAEOGENE									CRETACEOUS												JURASSIC												TRIASSIC				
Erathem era	CENOZOIC CZ																				MESOZOIC MZ																												
Eonothem eon	PHANEROZOIC PH																																																

Fig. 3.1 The stratigraphical column (modified from the IUGS correlation chart). British stage/age names have been retained for the Ordovician and Cambrian systems/periods as these have to be internationally agreed. Whittaker *et al.* (1991) gives further information on stratigraphical terminology.

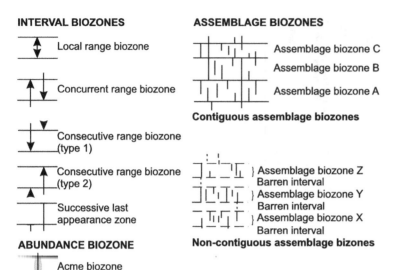

Fig. 3.2 Categories of biozones. (After Doggett in Briggs & Crowther 1987 with permission.)

The biostratigraphy of selected microfossil groups can be found in the 'Stratigraphic Index' series published by The Micropalaeontological Society and a host of specialist papers in scientific journals. The additional reading lists in this book provide an entry into this literature.

The basic unit of biostratigraphy is the biozone and fossils that characterize and give their names to a particular **biozone** are called **zone** or **index** fossils, for example the *Orbulina universa* Biozone of the Miocene. There are three basic types of biozone: the assemblage, abundance and interval biozones (Fig. 3.2). An **assemblage biozone** is based on the association between three or more species (though this concept is often more loosely applied) with little regard to the stratigraphical range of each. As species associations are strongly dependent upon local ecology, this type of biozone is most suitable for local or intra-basinal applications. The majority of defined biozones are **interval biozones** based upon the first appearance datum (**FAD**) and last appearance datum (**LAD**) of the named species. There are five types of interval biozone (Fig. 3.2), the most commonly used being the **local range zone** and the **concurrent range zone**. The latter comprises that interval which lies above the FAD of one species and below the LAD of a second species. The interval between two successive LADs is called a **successive last appearance zone** and is the most commonly used zone in commercial biostratigraphy where most of the samples are from borehole cores or cuttings and the FAD of a species cannot always be determined due to down-hole contamination ('caving').

Quantitative biostratigraphy

Because microfossils can occur in large numbers they are ideal for use in quantitative methods of biostratigraphy. Over the past 20 years a large number of techniques have become available for measuring biostratigraphical utility, defining and testing the error on a biozone and developing and testing correlations (Armstrong 1999). Typically quantitative methods are best applied to planktonic groups from continuous sections where FADs and LADs can be accurately determined. The most commonly used methods are semi-quantitative methods such as the **graphical correlation** technique developed by Shaw (1964). Details of this technique can be found in Armstrong (1999).

Graphical correlation uses a two-axis graph to compare the FADs and LADs of taxa found in common between two sections (Fig. 3.3). The heights of the first and last appearances of species are plotted as

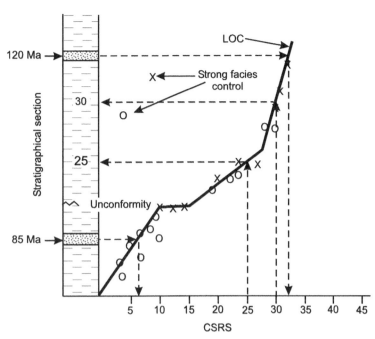

Fig. 3.3 Example of a graphical correlation. Shows the correlation of a new section with the composite standard reference section (CSRS). Sections have been correlated using the 25 and 30 standard time unit (stu) datum lines via a line of correlation (LOC) which exhibits changes in sedimentation rate and an unconformity plateau. The changing slope of the LOC Curve shows an increased rate of deposits above the unconformity, relative to the CSRS. Once the correlation has been made, other data, for example radiometric dates (85 Ma, 120 Ma) or isotope excursions, can be transferred into the CSRS via the LOC. Open circles, base of range; crosses, top of range.

coordinates in the field of the graph. A line of correlation (LOC) is then drawn through the scatter of points either by hand or using a variety of statistical techniques (e.g. least squares, linear regression or principal components analysis). The LOC is then used to transfer species range data from one section to the other. The latter becomes the composite standard reference section (CSRS). Additional sections are similarly correlated with the CSRS and included range data is also transferred to the composite, so that species ranges are progressively extended with the addition of new sections. When all the data from all available sections have been added, further rounds of correlation are undertaken to refine and stabilize the position of the LOCs. If only a small number of sections are to be correlated then the graphical correlation can be carried out by hand; computer packages are available for correlating large numbers of sections.

Species ranges within the CSRS should span the maximum within the included sections. Where sections are included that cover a wide range of geographical and palaeoecological settings, then these ranges should approach the full temporal span of that species. Lithological, geochemical and palaeomagnetic data can also be included in the CSRS and help strengthen the correlations.

The CSRS can be divided into units of equal length (**standard time units**-stu). The resultant chronometric timescale can then be transferred into the original sections using the LOCs. Standard time unit datum planes can be matched to provide a high resolution correlation of all the sections. This method of correlation is particularly useful for illustrating **diachronous** lithostratigraphical events: those that appear to be the same but occur at different times in different localities, between sections (e.g. progradation of sedimentary strata or facies or the diachronous nature of an unconformity).

The high resolution available using graphical correlation (limited only by the accuracy in placing the LOC) provides the only means by which the predictions of sequence stratigraphical correlation models can be independently tested (see below). The slope and geometry of the LOC is taken to reflect the relative rates of sedimentation between the two sections being compared. Strata that are missing, owing to faulting or a hiatus, or a highly condensed sequence, will appear as a plateau in the LOC (Fig. 3.3).

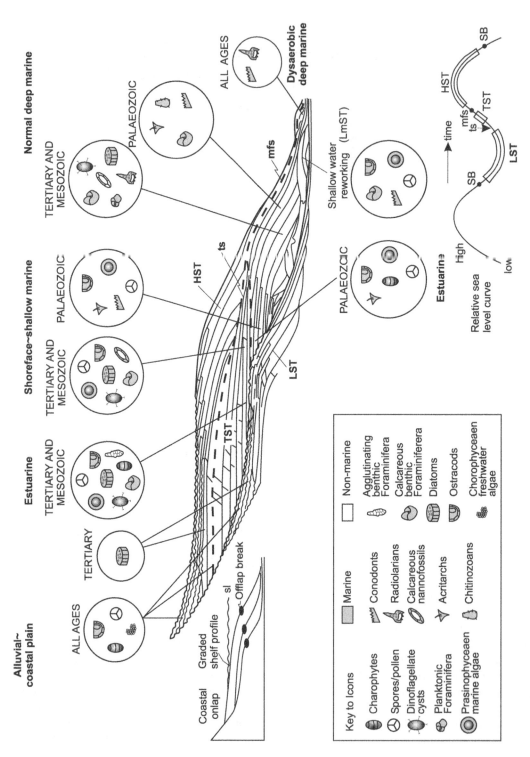

Fig. 3.4 Palaeoenvironmental distribution of some of the main microfossil groups through time. These are placed in a sequence stratigraphical framework. Insert shows the principle sequence statigraphical terms related to rising and falling sea level. HST, highstand systems tract; LST, lowstand systems tract; mfs, maximum flooding surface; sl, sea level; ts, transgressive surface; TST, transgressive systems tract; SB, sequence boundary. (After Hogg in Emery & Myers 1996 with permission.)

Microfossils in sequence stratigraphy

Sequence stratigraphy represents a powerful method for analysing familiar stratigraphical concepts such as **transgression**, **regression** and **eustatic cycles** and microfossils have a key role to play in sequence interpretation. The methods were largely developed as an extension of **seismic stratigraphy** and the need for correlation in the subsurface, but are equally applicable to outcrop geology where they have proved invaluable in understanding the influence of climate change on sedimentary successions. The reader is directed to Emery & Myers (1998) for a more detailed review of the principles of sequence stratigraphy. The basic philosophies of sequence stratigraphy are, firstly, that sediment accumulation occurs in discrete **sequences**, which are relatively conformable successions bounded by unconformities (or the correlative conformities in deep water). A sequence is considered to represent all the sediments deposited in an interval of time (0.5–5 Ma) and the **sequence boundaries** (intervals of no or very slow deposition) are considered effectively synchronous over large areas and can be used for matching sections. Secondly, the interaction of the rates of relative sea-level changes (eustasy), basin subsidence and sediment supply lead to variations in accommodation space, which is the space potentially available for sediment accumulation. The fundamental building blocks of sequences are **parasequences**, which generally represent shallowing or coarsening upwards cycles of short duration (10–100 kyr).

Every sequence comprises three **systems tracts** and potentially has a distinctive assemblage of microfossils (Fig. 3.4): a lower one representing periods of rapid but decelerating sea-level fall (LST, **lowstand systems tract**); a middle one relating to increasing acceleration in sea-level rise (TST, **transgressive systems tract**); and an upper one relating to a decreasing rate of sea-level rise and initial sea-level fall (HST, **highstand systems tract**). The base of each systems tract is defined as the **sequence boundary**, **transgressive surface** and **maximum flooding surface** respectively.

The interplay of environmental conditions, biological evolution, preservation potential of the microfossil group and cyclic changes in depositional style control the microfossil content of different sedimentary sequences. In a sequence stratigraphical analysis, it is the primary role of the micropalaeontologist to document changes in biofacies, and hence palaeoenvironment, and to provide a high-resolution biostratigraphical framework.

In the oil industry benthic foraminifera are commonly used to define marine benthic palaeoenvironments, although conodonts, ostracods and benthonic algae have also been used. **Palynofacies** analysis is most useful in defining fluvio-deltaic subenvironments (e.g. Brent Field, North Sea, Parry *et al.* 1981; see also Tyson 1995 for a review of palynofacies in sequence stratigraphy). Terrestrial microfossil assemblages can also provide a detailed record of climate changes around the margins of the sedimentary basin. With increasing knowledge of the ecological controls on microfossil groups, the relative abundances of different marine groups can be used to elucidate the changing palaeooceanography.

The transport or reworking of species into the marine environment by wind (e.g. bisaccate pollen) or rivers (e.g. miospores, charophytes, ostracods and woody material) or tides (e.g. foraminifera, dinoflagellates) can be problematic in biostratigraphy and palaeoenvironmental analysis. However the abundance gradients and size range of these derived fossils can be used to indicate the proximity of the source, location of palaeo-shorelines and exposure and uplift histories of the hinterland.

Few published studies have integrated the biostratigraphy, biofacies analysis and sequence stratigraphy. Exceptions include Armentrout (1987), Loutit *et al.* (1988), McNeil *et al.* (1990), Allen *et al.* (1991), Armentrout & Clement (1991), Armentrout *et al.* (1991), Jones *et al.* (1993) and Partington *et al.* (1993).

Sequence boundaries (SB) and correlative conformities

A sequence boundary is produced by a fall in relative sea level and may be associated with considerable erosion of the underlying sequence. It can be recognized by discrepancies in age and palaeoenvironment across the SB. The scale of these differences reflects the magnitude of the sea-level fall and location within the basin (McNeil *et al.* 1990). For example a SB can be characterized by a marked hiatus in nearshore sections or by subtle changes in biofacies across the correlative conformity within deep basinal settings. Our ability to resolve sequence boundaries biostratigraphically is limited by the biozonal resolution of the index fossils, commonly ~1 Ma or less if graphical correlation is used. Absence of preserved microfauna may mark the period of maximum regression. Reworking of specimens associated with erosion is commonplace above sequence boundaries.

Lowstand systems tract

This comprises two components, the **lowstand wedge** and **fan**. Both are produced by gravity sliding as sediment provided by rivers bypasses the shelf and upper slope through incised valleys and canyons which cut the continental shelf. Consequently both wedge and fan deposits will contain reworked terrestrially derived material and older, often polycyclic, marine microfossil assemblages when compared with adjacent shales with indigenous microfossil assemblages. Lowstand fan deposits in the Palaeogene of the North Sea, for example, only contained an impoverished microfauna comprising long-ranging agglutinated foraminifera.

The lowstand wedge is initiated as sea level begins to rise and can be **progradational** (sediment supply is greater than the rate of relative sea-level rise; facies belts migrate basinwards) or **aggradational** (sediment supply and relative sea-level rise are roughly balanced; facies belts thus stack vertically). In a complete vertical section through a prograding wedge microfossils will tend to indicate a shallowing upward signature from deep marine through to non-marine biofacies. Aggradational wedges will typically comprise a thick accumulation of the same biofacies. In nutrient-starved basins increased sediment supply to the basin during the lowstand will tend to bring additional nutrients which can result in increased plankton productivity and blooms. Thus, the distal parts of lowstand wedges may be represented by interbedded hemipelagic shales rich in marine palynomorphs that are similar to assemblages in the underlying highstand sediments.

Transgressive surfaces

The transgressive surface separates the lowstand and transgressive systems tracts. It is characterized by local winnowing and reworking of sediment. Glauconite- and/or phosphate-rich hardgrounds may also develop. The processes associated with the deposition and diagenesis at the transgressive surface may therefore result in poor preservation and selective removal of microfossils. The transgressive surface represents a **retrogradational** (i.e. sediment supply is less than the rate of sea-level rise and facies belts migrate landwards) diachronous boundary between terrestrial and marine biofacies. The presence of this surface can be inferred by the abrupt superposition of marine above terrestrial biofacies.

Transgressive systems tract (TST)

The TST contains retrogradational sequences which show an overall deepening upwards in the fossil biofacies. Transgression creates new shelf habitats that are rapidly colonized by opportunistic species. In addition they generate large areas of new wetland and saltmarsh habitat. The former may develop thick accumulations of peat and ultimately of coal. Diachronous shoreface deposits will contain shallowing marine biofacies.

As sediment supply becomes progressively reduced during the transgression, water turbidity decreases and clear water microfaunas (e.g. larger benthic foraminifera and seagrass species) will become more abundant (e.g. Van Gorsel 1988). Further sediment starvation will result in condensed sequences rich in well-preserved marine microfossils. These condensed sections will progressively onlap onto younger marine deposits up to the maximum flooding surface.

In deep basinal settings marine microfossil assemblages in pelagic condensed sections will contain abundant and diverse, typically cosmopolitan, planktonic species. The development of submarine fans formed by the regrading of the continental slope during transgression can be recognized by the presence of reworked shelf and upper slope microfossils within deep basinal condensed deposits (Galloway 1989). Shaffer (1987) used the presence and abundances of warm-water nannofossil assemblages to plot the progress of a transgression across an existing shelf.

Maximum flooding surface (mfs)

This surface separates the transgressive and highstand systems tracts and reflects the maximum landward development of marine conditions. Widespread condensed sections may occur across the shelf and basin due to sediment starvation. The mfs may also record a biostratigraphically distinctive event, usually with abundant planktonic fossils, and thus has the greatest potential for regional and global correlation. Partington *et al.* (1993) used palynomorph and microfossil assemblages at successive maximum flooding surfaces to produce a biochronostratigraphic framework for the Jurassic and Cretaceous of the North Sea.

At the basin margin the mfs is recognized by the sudden influx of low diversity marine plankton interbedded with shallower marine or terrestrial microfaunas. In the deep basin sediment starvation can produce highly fossiliferous deposits while the complete absence of clastic input means that calcareous or siliceous ooze, composed of the remains of diatoms, Radiolaria, planktonic forams and coccoliths, can accumulate.

Highstand systems tracts (HST)

An aggradational HST is characterized by thick, stacked shelf or terrestrial microfossil assemblages whereas a prograding system will exhibit a shallowing upward sequence of biofacies. Shelf assemblages will be strongly influenced by the buildup of deltas associated with rapid sedimentation. In nutrient-rich waters the microbenthos will be characterized by infaunal species with only rare calcareous planktonic species. Dinoflagellate cysts and acritarchs adapted to these conditions will be abundant. If progradation continues long enough to bring deltas to the shelf edge, then the result will be transport of terrestrial and shallow marine microfossils directly into deep basin environments.

The prograding highstand slope will be characterized by gravity flow deposits and considerable microfossil reworking. In vertical sections it may be possible to define highstand slope deposits on the gradual shallowing upward nature of benthonic organisms and the gradual decline in planktonic species (Van Gorsel 1988). In deep basinal settings starvation will result in deep-water species similar to those in the condensed parts of the TST. As the highstand slope migrates towards the basin the change from deeper to shallower marine environments can cause pseudo-extinction and diachronous correlation of these strata (Armentrout 1987).

REFERENCES

Allen, S., Coterill, K., Eisner, P., Perez-Cruz, G., Wornardt, W.W. & Vail, P.R. 1991. Micropalaeontology, well log and seismic sequence stratigraphy of the Plio-Pleistocene depositional sequences – offshore Texas. In: Armentrout, J.M. & Perkins, B.F. (eds) *Sequence Stratigraphy as an Exploration Tool: concepts and practices. 11th Annual Conference, Gulf Coast Section, SEPM*, pp. 11–13.

Armentrout, J.M. 1987. Integration of biostratigraphy and seismic stratigraphy: Pliocene–Pleistocene, Gulf of Mexico. In: *Innovative Biostratigraphic Approaches to Sequence Analysis: new exploration opportunities. 8th Annual Research Conference, Gulf Coast Section, SEPM*, pp. 6–14.

Armentrout, J.M. & Clement, J.F. 1991. Biostratigraphic calibration of depositional cycles: a case study in High Island–Galveston-East Breaks areas, offshore Texas. In: Armentrout, J.M. & Perkins, B.F. (eds) *Sequence Stratigraphy as an Exploration Tool: concepts and practices. 11th Annual Conference, Gulf Coast Section, SEPM*, pp. 21–51.

Armentrout, J.M., Echols, R.J. & Lee, T.D. 1991. Patterns of foraminiferal abundance and diversity: implications for sequence stratigraphic analysis. In: Armentrout, J.M. & Perkins, B.F. (eds) *Sequence Stratigraphy as an Exploration Tool: concepts and practices. 11th Annual Conference, Gulf Coast Section, SEPM*, pp. 53–58.

Armstrong, H.A. 1999. Quantitative biostratigraphy. In: Harper, D.A.T. (ed.) *Numerical Palaeobiology*. John Wiley, Chichester, pp. 181–227.

Bolli, H.M., Saunders, J.B. & Perch-Nielsen, K. 1985. *Plankton Stratigraphy*, vols 1, 2. Cambridge University Press, Cambridge.

Briggs, D.E.G. & Crowther, P.R. 1987. *Palaeobiology – a synthesis*. Blackwell Scientific Publications, Oxford.

Emery, D. & Myers, K. 1996. (eds) *Sequence Stratigraphy*. Blackwell Science, Oxford.

Galloway, W.E. 1989. Genetic stratigraphic sequences in basin analysis: architecture and genesis of flooding surface bounded depositional units. *Bulletin. American Association of Petroleum Geology* 73, 125–142.

Van Gorsel, J.T. 1988. Biostratigraphy in Indonesia: methods and pitfalls and new directions. In: *Proceedings. Indonesian Petroleum Association 17th Annual Convention*, pp. 275–300.

Jones, R.W., Ventris, P.A., Wonders, A.A.H., Lowe, S., Rutherford, H.M., Simmons, M.D., Varney, T.D., Athersuch, J., Sturrock, S.J., Boyd, R. & Brenner, W. 1993. Sequence stratigraphy of the Barrow Group (Berriasian-Valanginian) siliciclastics, Northwest Shelf, Australia, with emphasis on the sedimentological and palaeontological characterization of systems tracts. In: Jenkins, D.G. (ed.) *Applied Micropalaeontology*. Kluwer Academic, Dordecht, pp. 199–229.

Loutit, T.S., Hardenbol, J., Vail, P.R. & Baum, G.R. 1988. Condensed sections: the key to age determination and correlation of continental margin sequences. In: Wilgus, C.K., Hastings, C.G., Kendall, H.W., Posamentier, C.A., R. & Van Wagoner, J.C. (eds) *Sea Level Changes – an integrated approach*. SEPM, Tulsa 42, special publication, pp. 183–213.

McNeil, D.H., Dietrich, J.R. & Dixon, J. 1990. Foraminiferal biostratigraphy and seismic sequences: examples from the Cenozoic of the Beaufort-Mackenzie Basin, Arctic Canada. In: Hemelben, C., Kaminsji, M.A., Kuhny, W. & Scott, D.B. (eds) *Palaeoecology, Biostratigraphy, Palaeoceanography and Taxonomy of Agglutinated Foraminifera*. Kluwer Academic Publishers, Dordecht, pp. 859–882.

Parry, C.C., Whitley, P.K.J. & Simpson, R.D.H. 1981. Integration of palynological and sedimentological methods in facies analysis of the Brent formation. In: Illing, L. & Hobson, G.D. (eds) *Geology of the Continental Shelf of Northwest Europe*. Heydon, London, pp. 205–215.

Partington, M.A., Copestake, P., Mitchener, B.C. & Underhill, J.R. 1993. Biostratigraphic calibration of genetic stratigraphic sequences in the Jurassic-lowermost Cretaceous (Hettangian to Ryazanian) of the North Sea and adjacent areas. In: Parker, J.R. (ed.) *Petroleum Geology of Northwest Europe*. Geological Society of London, Bath, pp. 71–386.

Shaffer, B.L. 1987. The potential of calcareous nannofossils for recognizing Plio-Pleistocene climatic cycles and sequence boundaries on the shelf. In: *Innovative Biostratigraphic Approaches to Sequence Analysis: new exploration opportunities. 8th Annual Research Conference, Gulf Coast Section, SEPM*, pp. 142–145.

Shaw, A.B. 1964. *Time in Stratigraphy*. McGraw Hill, New York.

Tyson, R.V. 1995. *Sedimentary Organic Matter: facies and palynofacies*. Chapman & Hall, London.

Whittaker, A., Cope, J.W.C., Cowie, J.W., Gibbons, W., Hailwood, E.A., House, M.R., Jenkins, D.G., Rawson, P.F., Rushton, A.W.A., Smith, D.G., Thomas, A.T. & Wimbledon, W.A. 1991. A guide to stratigraphical procedure. *Journal of the Geological Society* 148, 813–824.

CHAPTER 4

Microfossils, stable isotopes and ocean-atmosphere history

Introduction

The skeletons of foraminifera and other $CaCO_3$ fossils take up chemical signals from sea water as they grow. The most important of these chemical signals are the stable isotopes of oxygen and carbon. These signals, when extracted from the $CaCO_3$ in a **mass spectrometer**, may be used to reconstruct past environmental changes such as temperature and ocean fertility, and to provide a high-resolution **chemostratigraphy**. The **oxygen isotope** technique was pioneered in the 1950s by Cesare Emiliani and the **oxygen isotope stages** that he initiated are now widely used as the basis for Quaternary and Tertiary stratigraphy (Figs 4.1, 4.2). The technique can also be used to estimate palaeotemperature, palaeosalinity and ice volume changes. The **carbon isotope** technique has been explored since the 1970s for carbon isotope stratigraphy and to provide information on the history of the carbon cycle and palaeoproductivity of the oceans.

Microfossils, especially foraminifera, are ideal for stable isotope research because they are easy to identify and readily checked for good preservation (using SEM), they have occupied a wide range of habitats and they can make up the bulk of oceanic sediments with a nearly continuous geological record.

Methodology

Both oxygen and carbon isotopes can be obtained during an analysis of a single sample of $CaCO_3$ (Box 4.1). The ratio between the **heavier** and **lighter** isotopes (i.e. $^{18}O/^{16}O$ and $^{13}C/^{12}C$) is expressed as the delta (δ) value in parts per thousand (‰). A standard sample is also

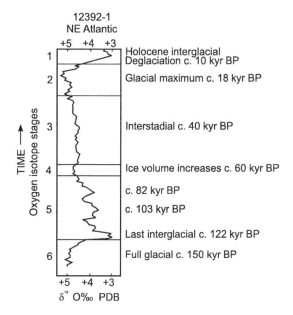

Fig. 4.1 Changes in oxygen isotope ratios of epibenthic foraminiferid calcite tests through the last 150,000 years, showing fluctuations related to changing ice volume. Core 12392–1 North East Atlantic. (Modified from data in Brasier 1995.)

run so that comparisons can be made between runs or between machines. In carbonates, this standard was originally the calcite guard of a belemnite from the Late Cretaceous Pee Dee Formation of South Carolina, USA. Samples are now compared with the Pee Dee Belemnite (or **PDB**) via a second, usually in-house, standard such as Carrara Marble. Oxygen isotopes from modern oceanic waters are more usually calibrated against **SMOW** (standard mean oceanic water). The terms, **heavy/light**, **positive/negative** and

26 Part 1: Applied micropalaeontology

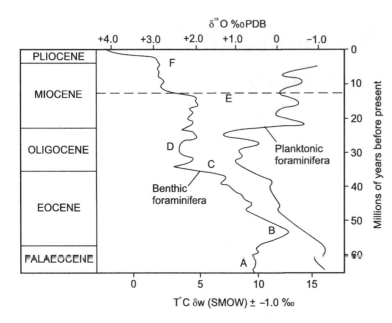

Fig. 4.2 Changes in oxygen isotope ratios of both benthic and planktonic foraminiferid calcite tests through the Tertiary, showing fluctuations related to changing water temperature and/or ice volume. Temperature estimates depend on assumed values for δw in each period. Letters A–F refer to features discussed in the text. SMOW, standard mean ocean water. (Modified from data in Hudson & Anderson 1989.)

Box 4.1 Stable isotope analysis

1. Sample with microfossils is disaggregated (e.g. using ultrasound) and dried.
2. Species of known habitat (e.g. surface water planktonic, infaunal benthic) are picked out for ecological studies. (Bulk samples of calcareous nannofossil carbonate from the <63 μm fraction can also give stratigraphically useful results.)
3. Specimens with evidence for secondary alteration (e.g. calcite overgrowths, pyrite) are rejected.
4. Specimens of the same size range are selected. Older machines may analyse from 1 to 40 planktonic foraminifera. Newer laser machines may analyse a single chamber.
5. The sample is dried at <50°C.
6. The $CaCO_3$ of the test is reacted with phosphoric acid:

$$H_3PO_4 + CaCO_3 \rightarrow CaHPO_4 + H_2O + CO_2 \quad (1)$$

7. Liquid nitrogen is used to freeze the water and the CO_2 gas. The frozen CO_2 is transferred to the mass spectrometer at −100°C to draw off water and other impurities.
8. CO_2 molecules are ionized and separated into ion beams of three different masses: $44 = {}^{12}C\,{}^{16}O\,{}^{16}O$, $45 = {}^{13}C\,{}^{16}O\,{}^{16}O$, $46 = {}^{12}C\,{}^{16}O\,{}^{18}O$.
9. The ratio between different ion beams is measured: 46/44 gives the ratio ${}^{18}O/{}^{16}O$, 45/44 gives the ratio ${}^{13}C/{}^{12}C$.
10. These ratios are then compared with those in a reference CO_2 gas.
11. The ratios are expressed as δ values, according to equations (2) and (3) below:

$$\delta^{18}O = \frac{({}^{18}O/{}^{16}O)\,\text{sample} - ({}^{18}O/{}^{16}O)\,\text{standard}}{({}^{18}O/{}^{16}O)\,\text{standard}} \times 1000 \quad (2)$$

$$\delta^{13}C = \frac{({}^{13}C/{}^{12}C)\,\text{sample} - ({}^{13}C/{}^{12}C)\,\text{standard}}{({}^{13}C/{}^{12}C)\,\text{standard}} \times 1000 \quad (3)$$

12. To allow for global comparison, $\delta^{18}O$ results are expressed relative to the universal PDB standard or to the SMOW standard. These can be converted as follows:

$$\delta^{18}O\,(\text{calcite SMOW}) = 1.03086\,\delta^{18}O\,(\text{calcite PDB}) + 30.86 \quad (4)$$

$$\delta^{18}O\,\text{PDB} = 0.97006\,\delta^{18}O\,\text{SMOW} - 29.94 \quad (5)$$

enriched/depleted indicate a relative increase/decrease in the heavy isotope (i.e. either ^{18}O or ^{13}C).

Oxygen isotopes

Five main factors affect the ratio between the stable isotopes ^{16}O and ^{18}O in $CaCO_3$ skeletons (Box 4.2). For the influence of one of these to be calculated, the other five will need to be estimated or known. The results obtained have been applied to a wide range of geological problems, as discussed below.

The Quaternary icehouse

Microfossils from deep sea sediments have played a crucial role in the reconstruction of palaeotemperature and ice volume changes over the last 100 million years. Emiliani (1955) used the δ^{18}O in planktonic foraminifera from deep sea cores to outline oxygen isotope stages for the Quaternary, believing these to reflect only surface temperature changes alone. It later became apparent that δ^{18}O can also be influenced by ice volume changes (Shackleton & Opdyke 1973). This is because expanding ice sheets lock up more of the lighter isotope ^{16}O that falls as precipitation, and prevent it from returning to the sea (Box 4.2). In theory, the ice volume signal can be obtained from the δ^{18}O record of deep-water smaller benthic foraminifera, such as *Uvigerina* and *Fontbotia* (formerly called *Cibicidoides*), if it is assumed that stable temperatures prevailed in deep waters through the glacial–interglacial cycles (e.g. Shackleton 1982). It has been argued that the deep ocean has also experienced drops in temperature (Prentice & Matthews 1991), which makes assumptions about the volume of land ice, and about the δw of ancient waters, more of a problem.

Figure 4.1 shows the δ^{18}O record obtained from deep sea benthic foraminifera in a DSDP core spanning the last 150,000 years. Isotope stages for glacial intervals take even numbers (e.g. 2, 4, 6) while those for warmer phases take odd numbers (e.g. 1, 3, 5 and

---

Box 4.2 Surface processes affecting oxygen isotopes

1 **Isotopic composition of the water** (δw, mean δ^{18}O). More ^{16}O than ^{18}O **evaporates** in H_2O from the ocean, and more ^{16}O than ^{18}O **precipitates** as rain from clouds. In the standard model clouds tend to form from evaporation at low latitudes and move towards the poles, so that there is a continuous **Rayleigh distillation**, leading to enrichment in ^{16}O of high latitude clouds and snow. A similar distillation takes place with altitude. $H_2^{16}O$ is therefore preferentially stored in polar icecaps. Carbonates precipitated in sea water at a time of higher ice volume therefore have a more positive δ^{18}O than found at times of lower ice volume. Salinity is similarly affected on a regional scale: fresh water has much more ^{16}O than does sea water, for the reasons given above. Carbonates precipitated in fresh water therefore tend to incorporate more ^{16}O and less ^{18}O (and hence have a more negative δ^{18}O) than those precipitated in normal sea water. Carbonates precipitated in hypersaline waters generally have a more positive δ^{18}O.
2 **Temperature**. Carbonates precipitated in warmer water incorporate more ^{16}O and less ^{18}O (and hence have a more negative δ^{18}O) than those precipitated in cooler water. This results in a **fractionation** of about 0.22‰ PDB per 1°C.
3 **Mineral phase**. Aragonitic foraminiferid tests are enriched by 0.6‰ relative to calcitic benthic foraminifera, owing to differences in the vibrational frequencies of the carbonate ions. Mg calcite is also enriched in ^{18}O relative to calcite by 0.06‰ per mole %$MgCO_3$, at 25°C.
4 **Vital effect**. Many species do not secrete their $CaCO_3$ in isotopic equilibrium with sea water owing to metabolic processes. This 'vital effect' varies between taxa from the same habitat. Smaller benthic and planktonic foraminifera and calcareous nannofossils are generally closer to equilibrium values than are larger benthic foraminifera, echinoderms and corals.
5 **Diagenesis**. δ^{18}O is easily reset by meteoric and burial diagenesis. Fluids tend to carry lighter isotopes and therefore make the ratios more negative. Specimens selected for study must therefore be free from diagenetic overgrowths (see Marshall 1992; Corfield 1995). Microfossils from ODP and DSDP cores are often but not invariably better preserved than those from exposed cratonic successions.

so on, back through time). Both glacial maxima and low sea level are inferred at c. 150,000 years BP (stage 6), and 20,000 years BP (stage 2). Rapid changes to full interglacial conditions with high sea levels took place at c. 122,000 years (stage 5e) and again at 10,000 years BP (stage 1). The increase in ice volume appears to have been prolonged, with episodic improvements to interstadial conditions at c. 103,000 years (stage 5c), 82,000 years (stage 5a) and 40,000 years (stage 3). A wide range of evidence, from reef terraces in New Guinea to ice cores in Antarctica, has supported this story.

Hays et al. (1976) showed that the regularity of Quaternary climatic oscillations was driven by changes in solar insolation brought about by the 'Milankovitch' orbital parameters of precession (~19 kyr), obliquity (~41 kyr) and orbital eccentricity (~100 kyr). Similar oscillations have been convincingly demonstrated back into the Mesozoic, and have been calibrated against the magnetostratigraphical scale for the Tertiary.

The Tertiary oxygen isotope record

The Tertiary oxygen isotope record of benthic and plankonic foraminifera (Fig. 4.2) reveals general ^{18}O enrichment through time. Low $\delta^{18}O$ values in benthic foraminifera from the Palaeocene (Fig. 4.2A) suggest that bottom waters were relatively warm, with a marked 'climatic optimum' in the Early Eocene (Fig. 4.2B). The fall in $\delta^{18}O$ of both surface and bottom waters through the Middle to Late Eocene, and the rapid fall in temperature at the Eocene–Oligocene boundary (Fig. 4.2C), has been attributed to falling temperatures. Part of the fall, however, may have been due to initial growth of the Antarctic ice cap (e.g. Zachos et al. 1992).

Both bottom-water and surface temperatures remained relatively cool though the Oligocene (Fig. 4.2D). The divergence between bottom and surface $\delta^{18}O$ values in the Middle Miocene (Fig. 4.2E) implies warmer surface waters and/or an expansion of ice sheets such as those in Antarctica (e.g. Prentice & Matthews 1988). The steep fall in bottom-water $\delta^{18}O$ in the Pliocene (Fig. 4.2F) has been taken to indicate the build-up of Northern Hemisphere land ice.

A major problem concerns assumptions about the δw of sea water prior to the Middle Miocene.

Shackleton & Kennett (1975) have inferred a lack of ice prior to this time, and that δw was about −1‰. This assumption, however, gives cool tropical sea surface temperatures contrary to evidence from fossil distributions. Prentice & Matthews (1988, 1991) have argued that there is little evidence on which to base ice-volume estimates and suggest that the benthic $\delta^{18}O$ values mainly record changes in the temperature of bottom waters. This problem is not yet resolved.

Oxygen isotope records have also been obtained from well-preserved microfossil materials in the Late Cretaceous (Jenkyns et al. 1994) when bottom waters appear to have been much warmer than at present. Diagenesis and burial metamorphism have generally reset the $\delta^{18}O$ values in older rocks exposed in cratonic successions. The emphasis therefore shifts away from microfossils towards robust and well-preserved macrofossils and cements in sedimentary rocks.

Palaeosalinity

In rivers and lakes, δw depends on the altitude and temperature of the precipitate plus the effects of humidity and evaporation (Box 4.2, Fig. 4.3). Ostracod carapaces from glacial lakes, for example, show strongly negative $\delta^{18}O$ that can be used to reconstruct climate change through the Late Quaternary (e.g. Hammerlund & Keen 1994).

In brackish water estuaries and deltas, the δw (mean $\delta^{18}O$) of sea water is diluted by isotopically light ^{16}O from rivers, so that $\delta^{18}O$ values of $CaCO_3$ skeletons generally become more negative than in coeval sea water. Glacial meltwater, for example, brought in negative $\delta^{18}O$ values to the Gulf of Mexico via the Mississippi delta during the Pleistocene (e.g. Williams et al. 1989).

In hypersaline lakes, lagoons and restricted seas such as the Mediterranean Sea, the increased ratio of evaporation to precipitation means that ^{16}O is removed, leaving both waters and $CaCO_3$ enriched in ^{18}O (e.g. Thunell et al. 1987). Seasonal evaporation of fresh water can produce a similar trend, as seen in larger benthic foraminifera across modern Florida Bay (Brasier & Green 1993). Primary productivity in marginal settings can be high, with considerable nutrient

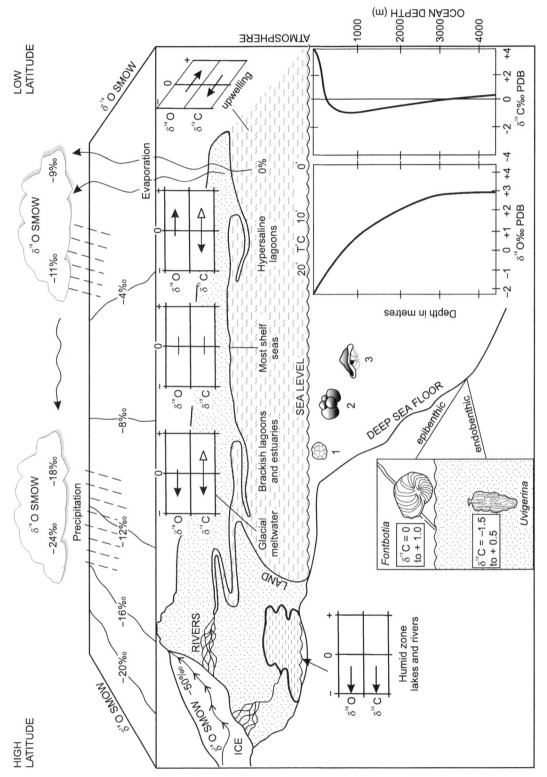

Fig. 4.3 Diagram illustrating how the stable isotopes of oxygen and carbon in microfossil skeletons will tend to vary with depth and salinity. Some typical genera are shown: 1, coccolithophorid in surface waters; 2, *Globigerinoides* in surface waters; 3, *Globorotalia* in intermediate waters; *Fontbotia* is epibenthic; *Uvigerina* is endobenthic. Black arrows indicate the most typical isotopic trends seen as environments become more extreme.

inputs from the land, so that bottom sediments tend to be organic rich and $\delta^{13}C$ values also tend to become more negative and highly variable, though exceptions to this rule are known.

Palaeotemperature

Calculation of the palaeotemperature from skeletal carbonate (**palaeothermometry**) can be determined from the following equations:

Calcite: $t(°C) = 16.9 - 4.4(\delta c - \delta w) + 0.10(\delta c - \delta w)^2$
(after Grossman & Ku 1986)

Aragonite: $t(°C) = 21.8 - 4.69(\delta Ar - \delta w)$

where δc and δAr are the mean $\delta^{18}O$ of CO_2 produced from calcite or aragonite respectively, by the reaction of phosphoric acid at 25°C, and δw is the $\delta^{18}O$ of CO_2 in equilibrium with water at 25°C, both versus PDB.

These equations assume that the value of δw is known (i.e. that the salinity and ice volume are known) and that the vital effect is zero. Since temperature may vary seasonally, and some organisms may vary their water depth with growth (e.g. planktonic foraminifera), it is clear that bulk samples provide crude estimates of palaeotemperature but microsamples can give great precision.

Carbon isotopes

Carbon is not only an essential building block for life, it also modulates the climate of the planet (through CO_2) and allows for oxygenation of the atmosphere (through photosynthesis and carbon burial). At the Earth's surface, carbon is mainly found in either the **oxidized reservoir** (as CO_2, HCO_3^- and carbonate minerals) or in the **reduced reservoir** (as organic matter, fossil fuels and native C). In the oxidized reservoir, the amount of dissolved CO_2 and HCO_3^- in the oceans is vastly greater (‰) than that of CO_2 in the atmosphere. The 'mixing time' for CO_2 to pass through the atmosphere and ocean is about 1000 years. Carbon isotopic studies are beginning to reveal that the interchange between these reservoirs has seldom achieved a stable balance (Box 4.3).

There are two stable isotopes of carbon: ^{12}C (98.9%) and ^{13}C (1.1%). The $^{13}C/^{12}C$ ratio in the atmospheric CO_2 gas (currently −7‰ PDB) is isotopically lighter than that of dissolved CO_2 and HCO_3^- in the oceans (currently −1‰ PDB) but an isotopic equilibrium is maintained between them because of the mixing effect of wind and waves. In an inert world, the ratio of $^{13}C/^{12}C$ in marine HCO_3^- would closely reflect that of primordial mantle carbon, which still escapes in the form of volcanic CO and CO_2 (−5‰ $\delta^{13}C$ PDB). In the living world, however, the $^{13}C/^{12}C$ ratio inclines towards a heavier value because autotrophs preferentially select the lighter isotope ^{12}C during photosynthesis. Living organic matter therefore has an average $\delta^{13}C$ value of −26‰ PDB (i.e. strongly negative), and the $\delta^{13}C$ of the ocean and atmosphere are correspondingly depleted in ^{12}C (i.e. positive).

Calcareous nannoplankton and some foraminifera living in surface waters secrete $CaCO_3$ tests in which the $\delta^{13}C$ value is more or less in equilibrium with surface water HCO_3^- (c. +2‰ PDB). A number of factors cause $\delta^{13}C$ values to vary, as shown in Box 4.3 and Figure 4.3. Beneath the photic zone, both the degradation of phytoplankton (especially by heterotrophic bacteria) and the release of respiratory CO_2 result in the return of ^{12}C to the water column. This can be seen in the more negative $\delta^{13}C$ of deeper water benthic foraminifera in the Atlantic (+1 to +0.5‰ PDB). In the modern Pacific, where the ocean width is large and the bottom waters are comparatively old, much oxygen has been removed during respiration so that the **apparent oxygen utilization** (AOU) index and the $\delta^{13}C$ values are correspondingly lower (−0.5 to +0.0‰ PDB). On the deep sea floor of modern oceans, the bottom waters, which have originated from shallow polar regions, help ventilation and also bring in more ^{13}C. Beneath the sediment–water interface, bacterial decay of organic matter releases ^{12}C-enriched CO_2 back into the pore waters.

Gradients are therefore found in $\delta^{13}C$ through both the water column and the sediment (Fig. 4.3). These gradients tend to show an inverse relationship with oxygen and phosphorus concentrations, which is because organic degradation removes oxygen from the

> **Box 4.3 Surface processes affecting carbon isotopes**
>
> 1. **Surface water productivity.** Where primary productivity is high, ^{12}C is preferentially removed from the ocean and atmosphere. Raised productivity produces an increase in the $\Delta\delta^{13}C$ gradient between benthos and plankton and can result in temporal shifts towards more positive $\delta^{13}C$.
> 2. **Biological oxidation.** The respiration of organic matter in mid-waters and on the sea floor results in a return of ^{12}C to the water column. Increased rates of biological oxidation will produce a decrease in the $\Delta\delta^{13}C$ gradient and can result in temporal shifts towards more negative $\delta^{13}C$.
> 3. **Upwelling and mixing.** Where ^{13}C-depleted waters are brought up to the surface, as by upwelling, then $\delta^{13}C$ values of surface waters are correspondingly lowered (e.g. off Peru). A similar reduction in $\delta^{13}C$ can be brought about seasonally by the influence of summer stagnation on the open shelf (e.g. east coast USA) and by the influence of humic-rich fluvial or swamp waters in coastal regions (e.g. north of Florida Bay, USA).
> 4. **Microhabitat effect.** A further gradient is found in sediments. Here, the $\delta^{13}C$ of pore waters becomes increasingly out of equilibrium with sea water values as depth below the sediment–water interface increases. This is because of the build up of respiratory CO_2 and HCO_3^- in pore waters.
> 5. **Carbon burial.** Factors which tend to raise the global proportion of organic matter buried in sediment are liable to raise the $\delta^{13}C$ of the whole ocean-atmosphere system. Such factors include raised primary productivity, increased mid- to bottom-water stagnation, and raised rates of sediment accumulation.
> 6. **Vital effect.** Taxa are known to differ in the proportion of HCO_3^- taken in from sea water (c. 0‰ PDB) and from cytoplasm (c. −28‰ PDB). Although foraminifera show much less vital fractionation than seen in echinoderms and corals, many species show a vital effect (e.g. due to photosymbiosis). Fractionation may even change during growth (e.g. larger rotaliids become more positive; larger miliolids become more negative; Murray 1991). Where possible, a single species and a single size fraction should be used for studies of trends through time.
> 7. **Diagenesis.** $\delta^{13}C$ is much less easily reset by meteoric and burial diagenesis than is $\delta^{18}O$. Even so, most diagenetic fluids tend to carry ^{12}C and can therefore make the ratios more negative. Specimens selected for study must therefore be free from diagenetic overgrowths and cements (see Marshall 1992).

water column but returns both ^{12}C and P. Such environmental gradients can be measured by calculating the difference in $\delta^{13}C$ ($\Delta\delta^{13}C$; called the 'delta del ^{13}C') between surface water microfossils (e.g. *Globigerinoides* spp. or calcareous nannoplankton) and a coexisting epibenthic species (e.g. *Fontbotia wuellerstorfi*), or between the latter and an infaunal taxon (e.g. *Uvigerina* sp.; Fig. 4.3).

The Quaternary carbon pump

The role of CO_2 in climate change has been suspected since the nineteenth century. W.S. Broecker first suggested that carbon isotopes could provide a proxy for changing CO_2 through the ice ages and Shackleton *et al.* (1983) were able to reveal the nature of this record. They found that the $\Delta\delta^{13}C$ of planktonic-benthic foraminifera has varied markedly over the last 130,000 years in a way that can be tied to changes in ice volume revealed by the $\delta^{18}O$ record. $\Delta\delta^{13}C$ proves to be greatest during glacial phases and least during interglacial phases (Fig. 4.4). This may be taken to infer that the partial pressure of atmospheric CO_2 was least during glacial phases and greatest during interglacials, which has since been confirmed by direct measurements from ice cores. The changes in $\Delta\delta^{13}C$ may also indicate major changes in primary productivity through the climatic cycle.

The Tertiary carbon isotopic record

Figure 4.5 shows the carbon isotopic record for much of the Tertiary. Note that this pattern is very different from the oxygen isotope record (Fig. 4.2) and shows no long-term trend. High $\delta^{13}C$ values of about 3‰ PDB are found in the Late Cretaceous. A rapid fall to 2‰ took place across the K-T boundary and into the basal Palaeocene. At the K-T boundary, the $\Delta\delta^{13}C$ fell

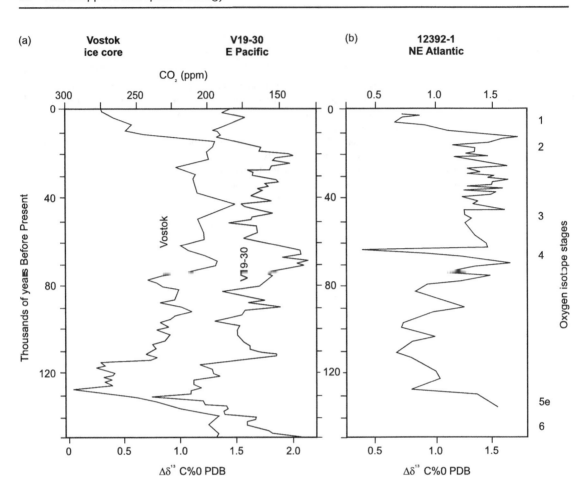

Fig. 4.4 Changes in the difference (Δ) between $\delta^{13}C$ values of planktonic and benthic foraminiferid tests ((a) core V19–30) and epibenthic and endobenthic foraminiferid tests ((b) core 12392–1) through the last 150,000 years, showing fluctuations are related to changes in atmospheric CO_2 ((a) Vostock ice core). (Modified from data in Brasier 1995.)

to about 1‰ (Fig. 4.5A), which has been taken to indicate the devastating effect of an extraterrestrial cometary impact on primary productivity (e.g. Hsu et al. 1982).

The Tertiary carbon isotope record shows evidence for two long-term cycles (Shackleton & Kennett 1975) with peaks in the Late Palaeocene and Middle Miocene. The Palaeocene provides a maximum $\delta^{13}C$ for the Tertiary of c. +3.5‰ (Fig. 4.5B) accompanied by a rise in the $\Delta\delta^{13}C$ between planktonic and benthic foraminifera. This may have been due to high rates of productivity and carbon burial under greenhouse conditions.

A sharp fall in $\delta^{13}C$ occurred across the Palaeocene–Eocene transition (Fig. 4.5C) which was even greater than that across the K-T boundary. A mass extinction of about 50% of deep sea benthos took place at this time. The mid–late Eocene boundary interval was accompanied by a marked divergence in planktonic and benthic values and an increase in diatom abundance and diatom and dinoflagellate diversity. Together, these may be taken to indicate the effects of an increased thermal gradient on surface water productivity and carbon burial. Values were moderate during the Oligocene (Fig. 4.5D).

Chapter 4: Microfossils, stable isotopes and ocean-atmosphere history

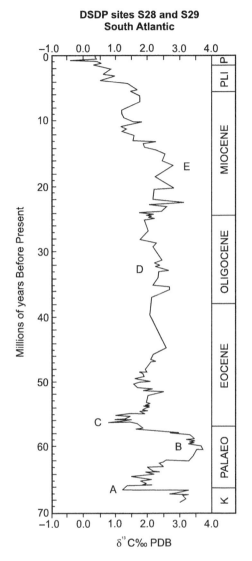

Fig. 4.5 The carbon isotopic record of the Late Cretaceous (K) to Tertiary, obtained mainly from planktonic microfossil carbonates in DSDP sites S28 and S29 of the South Atlantic. Letters A–E refer to features discussed in the text. P, Pleistocene; PLI, Pliocene. (Modified from Shackleton 1987.)

The $\delta^{13}C$ peak in the early–mid Miocene (Fig. 4.5E) coincided with widespread diatomites around the Pacific (the Monterey Event). The subsequent fall of c. 2.5‰ to Recent values may owe much to the greater oxidation of organic matter brought about by cooler glacial oceans.

Further background information on isotopes can be found in the books on isotope geology by Faure (1986) and Hoefs (1988). Tucker & Wright (1990) and Marshall (1992) provide overviews from a sedimentological perspective. Williams *et al.* (1989) give an expanded discussion of Cenozoic isotope stratigraphy. Hudson & Anderson (1989) and Corfield (1995) review some of the achievements of oxygen isotope studies, while Murray (1991) reviews oxygen and carbon isotope data from benthic foraminifera. Brasier (1995) brings together stable isotope and other data used to interpret palaeoclimates and nutrient levels, while Purton & Brasier (1999) show how stable isotopes can be used to estimate changes in seasonality, ocean stratification, growth rate and life span.

REFERENCES

Brasier, M.D. 1995. Fossil indicators of nutrient levels. 1: Eutrophication and climate change. *Geological Society Special Publication* 83, 113–132.

Brasier, M.D. & Green, O.R. 1993. Winners and losers: stable isotopes and microhabitats of living Archaiadae and Eocene *Nummulites* (larger foraminifera). *Marine Micropalaeontology* 20, 267–276.

Corfield, R.M. 1995. An introduction to the techniques, limitations and landmarks of carbonate oxygen isotope palaeothermometry. *Geological Society Special Publication* 83, 27–42.

Emiliani, C. 1955. Pleistocene temperatures. *Journal of Geology* 63, 538–575.

Faure, G. 1986. *Principles of Isotope Geology*. John Wiley, New York.

Grossman, E.L. & Ku, T.L. 1986. Oxygen and carbon isotope fractionation in biogenic aragonite: temperature effects. *Chemical Geology* 59, 59–74.

Hammerlund, D. & Keen, D.H. 1994. A Late Weichselian stable isotope and molluscan stratigraphy from southern Sweden. *GFF* 116, 235–248.

Hays, J.D., Imbrie, J. & Shackleton, N.J. 1976. Variations in the Earth's orbit: pacemaker of the ice ages. *Science* 194, 1121–1132.

Hoefs, J. 1988. *Stable Isotope Geochemistry*. Springer-Verlag, Berlin.

Hsu, K.J., McKenzie, J.A. & He, Q.X. 1982. Terminal Cretaceous environmental and evolutionary changes. *Geological Society of America* 190, special paper, 317–328.

Hudson, J.D. & Anderson, T.F. 1989. Ocean temperatures and isotopic composition through time. *Transactions of the Royal Society of Edinburgh: Earth Sciences* **80**, 183–192.

Jenkyns, H.C., Gales, A.S. & Corfield, R.M. 1994. Carbon and oxygen-isotope stratigraphy of the English chalk and Italian Scaglia and its palaeoclimatic significance. *Geological Magazine* **131**, 1–34.

Marshall, J.D. 1992. Climatic and oceanographic signals from the carbonate rock record and their preservation. *Geological Magazine* **129**, 143–160.

Murray, J.W. 1991. *Ecology and Palaeoecology of Benthic Foraminifera.* Longman, Harlow.

Prentice, M.L. & Matthews, R.K. 1988. Cenozoic ice volume history: development of a composite oxygen isotope record. *Geology* **17**, 963–966.

Prentice, M.L. & Matthews, R.K. 1991. Tertiary ice sheet dynamics: the snow gun hypothesis. *Journal of Geophysical Research* **96**(B4), 6811–6827.

Purton, L.M.A. & Brasier, M.D. 1999. Giant protist *Nummulites* and its Eocene environment: life span and habitat insights from $\delta^{18}O$ and $\delta^{13}C$ data from *Nummulites* and *Venericardia*, Hampshire Basin, UK. *Geology* **27**, 711–714.

Shackleton, N.J. 1982. The deep sea sediment record of climate variability. *Progress in Oceanography* **11**, 199–218.

Shackleton, N.J. 1987. The carbon isotope record of the Cenozoic history of organic carbon burial and oxygen in the ocean and atmosphere. In: Brooks J.R.V. & Fleet A.J. (eds) *Marine Petroleum Source Rocks.* Published for the Geological Society by Blackwell Scientific Publications, Oxford, pp. 423–435.

Shackleton, N.J. & Kennett, J.P. 1975. Paleotemperature history of the Cenozoic and the initiation of Antarctic glaciation: oxygen and carbon isotope analyses of DSDP sites 277, 279 and 281. *Initial Reports Deep Sea Drilling Project* **29**, 743–755.

Shackleton, N.J. & Opdyke, N.D. 1973. Oxygen isotope and paleomagnetic stratigraphy of Equatorial Pacific core V28–238. Oxygen isotope temperatures and ice volumes on a 10^5 and 10^6 year scale. *Quaternary Research* **3**, 39–55.

Shackleton, N.J., Hall, M.A., Line, J. & Shuxi, C. 1983. Carbon isotope data in core V19-30 confirm reduced carbon dioxide in the ice age atmosphere. *Nature* **306**, 319–322.

Thunell, R.C., Willims, D.F. & Howell, M. 1987. Atlantic–Mediterranean water exchange during the Late Neogene. *Paleoceanography* **2**, 661–678.

Tucker, M.E. & Wright, V.P. 1990. *Carbonate Sedimentology.* Blackwell Scientific Publications, Oxford.

Williams, D.G., Lerche, I. & Full, W.E. 1989. *Isotope Chronostratigraphy: theory and methods.* Academic Press Geology Series, San Diego.

Zachos, J.C., Breza, J. & Wise, S.W. 1992. Early Oligocene ice-sheet expansion on Antarctica: sedimentological and isotopic evidence from Kerguelen Plateau. *Geology* **20**, 569–573.

CHAPTER 5

Microfossils as thermal metamorphic indicators

Microfossils with a mineral skeleton are commonly composed of high or low magnesium calcite or calcium phosphate whereas palynomorphs are composed of organic materials such as **sporopollenin**, **chitin** and **pseudochitin**. These materials, though highly resistant, are susceptible to weathering, reworking by erosion, oxidation and to thermal metamorphism. Pristine palynomorphs have transparent to very pale green-yellow walls and often have to be stained to see the material under the microscope. Less well preserved fossil material can range in colour from yellow to black. It is this colour change which can be used as an indicator of the thermal metamorphic history of a rock, as outlined below. The primary factors affecting the colour of fossil palynomorphs are oxidation during weathering, heat related to depth of burial or contact metamorphism, and length of exposure to heat. Oxidation can initially remove the fine detail and ornamentation, ultimately removing the palynomorphs entirely. Oxidation is the prime cause of the absence of palynomorphs from reddened shales and sandstones.

When subjected to heating, owing to increased depth of burial or to contact metamorphism, organic matter undergoes a series of irreversible chemical and physical changes best seen in the transformation of peat to coal. Similarly, dispersed organic matter in sediments undergoes similar changes with the loss of H and O and the concomitant increase in C during **diagenesis** (up to 50°C), or **catagenesis** (50–150°C) and **metagenesis** (150–200°C) and finally **metamorphism** above 250°C. Experimental heating of organic matter in inert or reducing atmospheres requires higher temperatures than in air to affect the same colour change. Pressure alone does not cause carbonization. These physical and chemical processes also affect organic matter trapped within the mineralized skeletal components of all fossils. This is particularly true for the conodonts which contain sufficient organic material to colour the biogenic apatite yellow.

Experimentally derived temperature ranges have been assigned to the rather subjective scale of colour changes for various groups of palynomorphs and conodonts (Fig. 5.1). Palynomorphs associated with these chemical changes show a change in colour of the wall from transparent through yellows, to browns and finally to black. Small differences do occur between groups. For example, pristine acritarchs and dinoflagellates are almost transparent and take more heat to darken than do the already darker spores and pollen. The conodont colour alteration index (CAI: Epstein *et al.* 1977) scale for conodonts has been extended beyond black (300°C) through grey (CAI 6–7, 360–720°C) to colourless at temperatures greater than 600°C (Rejebian *et al.* 1987).

Figure 5.1 compares the various palynomorph thermal maturity indices against other commonly measured thermal parameters. The economically important oil window is indicated by mid-brown colours in most organic indices. Darker colours than this indicate petroleum source rocks will have generated gas. Although mainly used in hydrocarbon exploration microfossil coloration has been successfully applied in unravelling the geological histories of sedimentary basins (e.g. Robert 1988) and ancient orogens (e.g. Bergström 1981), in base metal and mineral exploration, and for the tracking of ancient hotspots and geothermal energy studies (e.g. Nowlan & Barnes 1987). A comprehensive review on sedimentary organic matter can be found in Tyson (1995).

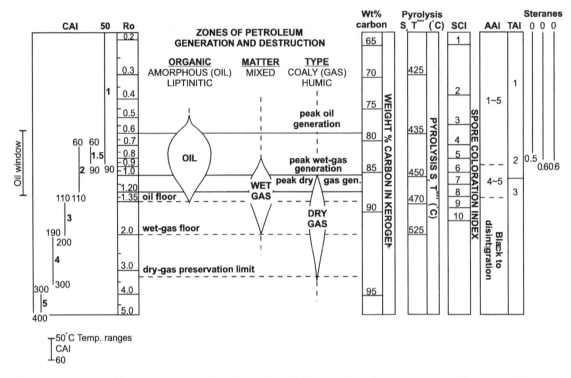

Fig. 5.1 Comparision of the main indicators of thermal maturity with the zones of petroleum generation and destruction. AAI, acritarch alteration index; CAI, conodont colour alteration index; SCI, spore coloration index; TAI, thermal alteration index.

REFERENCES

Bergström, S.M. 1981. Conodonts as paleotemperature tools in Ordovician rocks of the Caledonides and adjacent areas in Scandinavia and the British Isles. *Geol. Fören. Stockholm Förhandl* 102, 337–392.

Epstein, A.G., Epstein, J.B. & Harris, L.D. 1977. Conodont Color Alteration – an index to organic metamorphism. *US Geological Survey Professional Paper* 995, 27pp.

Nowlan, G.S. & Barnes, C.R. 1987. Thermal maturation of Paleozoic strata in eastern Canada from conodont colour alteration (CAI) data with implications for burial history, tectonic evolution, hotspot tracks and mineral and hydrocarbon exploration. *Bulletin. Geological Survey of Canada* 367, 47pp.

Rejebian, V.A., Harris, A.G. & Hueber, J.S. 1987. Condont color and textural alteration – an index to regional metamorphism and hydrothermal alteration. *Bulletin. Geological Society of America* 99, 471–479.

Robert, P. 1988. *Organic Metamorphism and Geothermal History*. Elf Aquitaine/D. Reidel, Dordrecht.

Tyson, R.V. 1995. *Sedimentary Organic Matter: organic facies and palynofacies*. Chapman & Hall, London.

PART 2
The rise of the biosphere

CHAPTER 6

The origin of life and the early biosphere

Planet Earth is believed to have formed from the coalescence of dust particles at some time close to 4.55 Ga. While this accretion and the ensuing phase of catastrophic impacts would have caused a molten surface, the crust appears to have been cool by about 3.85 Ga. If any life forms were synthesized before this date they must have been **hyperthermophile** heat-tolerant bacteria, similar to those found living around volcanic vents or deep in the Earth's crust today. The oldest rocks on the Earth are found in Western Australia and northern Canada dated at ~4 Ga and the Isua Group from western Greenland, dated at ~3.8 Ga. The Isua rocks are a mix of abiogenic limestones, sandstones and pillow lavas. These rocks formed under water and indicate a crust had stabilized and oceans were present (Fig. 6.1).

Origins of life

The **Oparin-Haldane hypothesis** for the origins of life (Fig. 6.2) envisaged that the primitive atmosphere was reducing and contained CO_2, CO, H_2, NH_3, CH_4 and H_2O but no O_2. It is now thought that NH_3 and CH_4 would have been unstable in the early atmosphere. A scarcity (but not a lack) of oxygen is a reasonable assumption given the existence of pyrite conglomerates before 2 Ga (Figs 6.1, 6.2) and the derivation of nearly all O_2 in the modern atmosphere from photosynthesis. Experiments by Miller & Urey (Miller 1953) showed that amino acids may be synthesized from a mixture of these gases and water, through which ultraviolet light or electric discharge (cf. lightning) has passed, especially if temperatures are kept below 25°C. In fact, temperatures close to freezing can conserve nucleic acids much better, and it has been suggested that nucleic acids and ultimately DNA could have been synthesized in as little as 10,000 years. It is difficult to reconcile glacial conditions, however, with other indications for a very warm greenhouse world at this time.

The **panspermia hypothesis** (Fig. 6.2) suggests that **prebiotic** materials in space seeded the surface of the planet during the phase of massive meteorite bombardment until about 3.8 Ga. Simple organic compounds such as HCN, formic acid, aldehydes and acetylenes are certainly abundant in a group of meteorites known as **carbonaceous chondrites**, as well as in the 'heads' of comets and in some interstellar dust clouds. An extreme version of this hypothesis is that DNA may also be found in space. Experiments certainly suggest that DNA can tolerate high radiation doses when desiccated and at low temperatures.

The **hydrothermal hypothesis** (Fig. 6.2) argues that amino acid to DNA synthesis took place around hot, alkaline hydrothermal vents, possibly like the 'black smokers' associated with modern mid oceanic ridges (Russell & Hall 1997). Further support for this model is provided by molecular sequence evidence.

Did life originate on Mars?

In August 1996 Dr David McKay and a team from NASA announced to the world that they had found possible microfossils and geochemical evidence consistent with life in ALH 84001, a martian meteorite, confirmed by a distinctive $^{15}N/^{14}N$ isotopic ratio. Full details of this exciting discovery can be found in Treiman (2001). The orthopyroxene minerals in the meteorite crystallized ~4.5 Ga ago, it had suffered

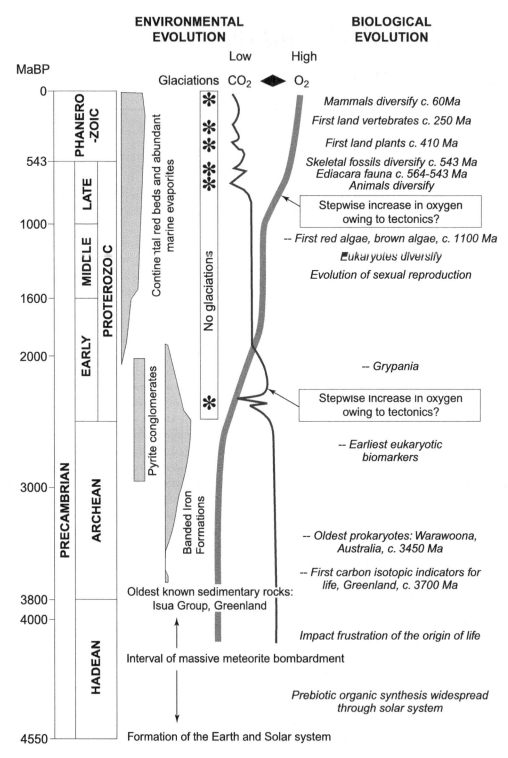

Fig. 6.1 The main succession of events inferred for the evolution of the biosphere alongside geological evidence for changing levels of atmospheric oxygen and carbon dioxide during the Precambrian. (Modified from Brasier *et al.* 2002.)

Chapter 6: The origin of life and the early biosphere

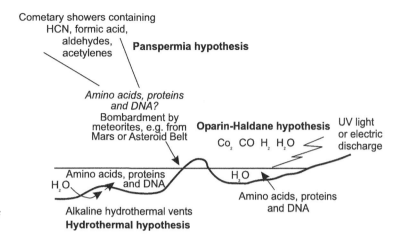

Fig. 6.2 Hypothesis for the origins of life on Earth (from various sources).

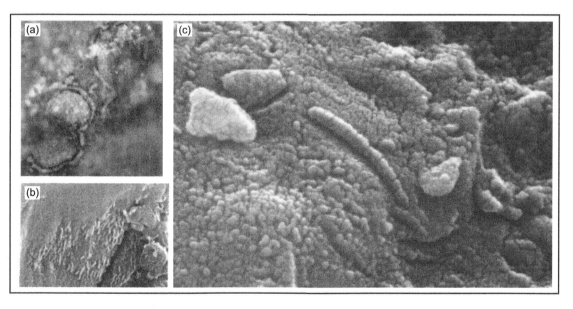

Fig. 6.3 Reported biogenic structures from ALH 84001. (a) Carbonate globule. (b), (c) Scanning electron micrographs of elliptical and rod-like structures. The specimen in (c) is approximately 2 μm long. (Photographs courtesy of the Lunar Planetary Institute.)

impact shocks at 4 Ga and 15 Ma ago and landed in Antarctica 13,000 years ago. In addition to the many lines of evidence proposed by McKay et al. (1996) in support of life, zoned carbonate globules (Fig. 6.3a) were thought to provide evidence for the existence of liquid water essential for life. The geochemistry of these globules suggested bacteria-like metabolism and the presence of organic compounds thought to have been derived from microbial degradation. More provocatively, they described bacteria-like microfossils (e.g. Figs 6.3b,c). Some have suggested that this is direct evidence that life originated on Mars, though others have strongly criticized this interpretation (e.g. Grady et al. 1996; Bradley et al. 1997).

Part 2: The rise of the biosphere

Evidence for the earliest biosphere

The fossil evidence for life on Earth gets increasingly scarce as age increases. This is because older rocks have suffered more exposure to erosion and a greater chance of alteration by metamorphism. The rules for accepting microfossil-like objects as evidence for life include them being demonstrably biogenic and indigenous to the formation of the rocks of known provenance. Biogenicity is the most difficult to demonstrate, as with the martian objects. Whilst the oldest sedimentary rocks on Earth have been too heavily metamorphosed to yield preserved microfossils, molecular and biochemical evidence indicates life may have existed when these rocks were deposited. Evidence no longer indicates that life was already established on the Earth by 3.5 Ga (Brasier *et al.* 2002).

Evidence from molecular sequences and biogeochemistry

Comparisons of the rRNA sequences and ultrastructures of diverse bacteria, protozoa, fungi, plants and animals have recently resulted in two contrasting views on the origin and early evolution of life. The currently widely accepted hypothesis, based largely upon rRNA phylogeny (Woese *et al.* 1990), views life on Earth as three primary domains: the **Archaebacteria** (or **Archaea**), which includes the autotrophic methanogenic and sulphur bacteria; the **Eubacteria** (or **Bacteria**), including **cyanobacteria** and the **Eukaryota** (or **Eukarya**), including all the protozoa, fungi, plants and animals (Fig. 6.4). The following represents the chronological appearance of grades within **autotrophic** prokaryotes (that used carbon dioxide as their sole

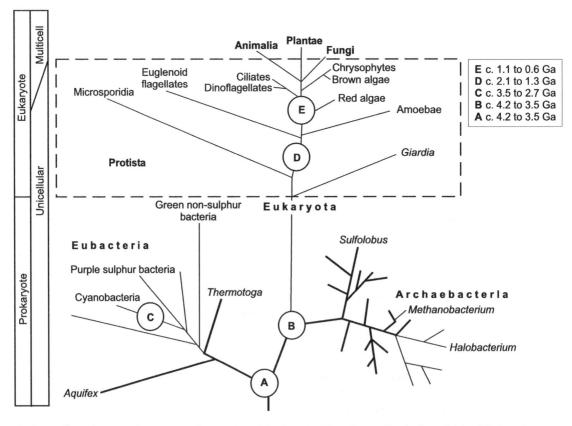

Fig. 6.4 The threefold branches of the tree of life, in which all the deep seated branches are taken by hyperthermophilic bacteria (shown in bold). The approximate times of branching are shown. Time increases along the branches, but not necessarily in a linear fashion nor at the same rate in each branch. Longer branches relate to faster evolution. The times of branching are speculative and are hotly contested. (Adapted from Woese *et al.* 1990, Sogin 1994 and Nisbet & Fowler 1996; Brasier 2000.)

source of carbon) based upon the traditional phylogenetic interpretation:

1 **anaerobic chemolithotrophic** bacteria, which mainly use H_2 produced from inorganic reactions between rock and water as their main electron source;
2 **anaerobic anoxygenic** bacteria such as green and purple sulphur bacteria, which use photosynthesis to reduce CO_2 to form organic matter, with H_2S as the electron source, in the absence of oxygen;
3 **oxygenic cyanobacteria**, use photosynthesis to reduce CO_2 to form organic matter, with H_2O as the electron source, in the presence of oxygen.

In the traditional scenario the deepest roots of the tree of life are occupied by **hyperthermophilic** bacteria (Fig. 6.4). At the present day these are adapted to life in hot hydrothermal springs or life deep in the Earth's crust, at temperatures of >80°C or more and are seldom able to grow below 60°C. This fact has been taken as evidence to suggest that the last common ancestor of all living organisms was a hyperthermophile (Nisbet & Fowler 1996). This proposal is however inconsistent with the fundamental ultrastructural differences to be found within the prokaryotes (i.e. **monoderms** having a single cell membrane and **diderms** with a double cell membrane) and phylogenetic trees based on signature protein sequences or 'indels'. A second hypothesis (Gupta 1998, 2000) recognizes the uniqueness of the Eukaryota and Prokaryota but points to fundamentally different divisions and evolution within prokaryotes. Specifically, a close relationship is envisaged between the Archaebacteria and the gram-positive bacteria (Eubacteria), both of which are monoderm prokaryotes and distinct from the rest of life. This hypothesis postulates the earliest prokaryote was a gram-positive bacterium from which the Archaebacteria and diderm prokaryotes evolved in response to selection pressures exerted by antibiotics produced by some gram-positive bacteria. Accepting this hypothesis (and the underlying phylogenetic methodology) allows for the evolution of the early forms of life from a common ancestor through gram-positive (Low G + C; Archaebacteria), gram-positive (High G + C; Archaebacteria), *Deinococcus* Group, Green non-sulphur bacteria, Cyanobacteria, Spirochetes, *Chlamydia*–Green sulphur bacteria to *Proteobacteria*.

The gram-positive (Low G + C) bacteria appear to be the earliest Eubacteria and include anoxygenetic photosynthetic organisms (e.g. *Heliobacterium*). The phylogenetic inference of this is that the common ancestor of all life on Earth was a photosynthetic anaerobe. If this hypothesis is correct then the major evolutionary changes have, and will continue to have, a linear line of descent.

Geochemical proxies for early life

Biomarkers

The three domains of life contain within their cell walls diagnostic molecules called **lipids** which turn into hydrocarbons in sediments. 2-Methyl-bacteriohopane-polyols (2-methyl-BHP) are characteristic of the cell walls of cyanobacteria and are found in cyanobacterial mats. These are converted to 2α-methylhopanes in sediments found in high abundance in bitumens in the ~2.5 Ga Mt McRae Shale of the Hamersley Basin in Western Australia (Summons *et al.* 1999) and indicate that oxygenic photosynthesis was important by this time.

Stable carbon isotopes

The carbon isotopic record of the Archean is still poorly known and there are large ranges of values for specific time intervals (Fig. 6.5a). Carbonates in the Isua Group rocks, as old as 3.8 Ga, have $\delta^{13}C_{carb}$ signatures close to 0‰ comparable to modern marine bicarbonate (Fig. 6.5b). Organic matter of this age yields $\delta^{13}C_{org}$ values of −15‰ lighter than those of associated carbonates, comparable to the light isotopic values found in modern living organic matter. Some scientists have argued this as evidence for oxygenic photosynthesis during the deposition of the Isua sediments but this is highly contentious. The Isua rocks have both more negative $\delta^{13}C_{carb}$ and less negative organic $\delta^{13}C_{org}$ values than those in later sediments, indicating values from the Isua may be entirely due to metamorphism. Similar data have also led to claims that organic matter within phosphatic grains from metasediments in the Itsaq Group of Greenland (~3.85 Ga) provides evidence for a biological origin (Mojsis *et al.* 1996). This claim is controversial as the sedimentary origin of the phosphate is questionable

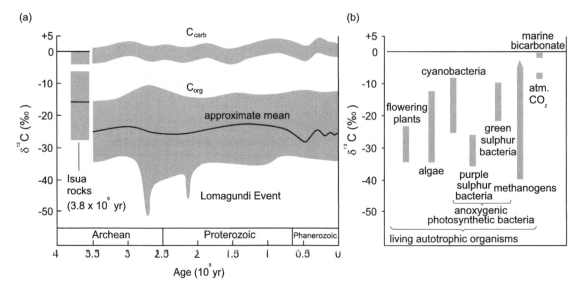

Fig. 6.5 (a) Changes in stable carbon isotope values through the history of life. Values are expressed as $\delta^{13}C_{carb}$ from carbonates and $\delta^{13}C_{org}$ from kerogen samples. (b) $\delta^{13}C$ values of modern autotrophs and recently oxidized inorganic carbon. This figure is commonly described as the Schidlowski diagram. (From Schidlowski 1988, figure 4, permission from *Nature*. Copyright © 1988 Macmillan Magazines Ltd.)

and the age of the phosphate grains may also be significantly younger, around 3.7 Ga (Kamber & Moorbath 1997).

The mean $\delta^{13}C_{org}$ for 3.5-Ga-old sediments is −26‰, falling within the range of $\delta^{13}C_{org}$ values for living anaerobic autotrophic bacteria. A major negative $\delta^{13}C_{org}$ excursion of −50‰ has been found at around 2.7 Ga with a further negative excursion of similar scale at 2.1 Ga. The causes of these excursions are not known. It has been suggested that large amounts of carbon burial during this time brought about a stepwise increase in the oxygen levels in the atmosphere (Karhy & Holland 1996). It may be coincidence that it is only after this event that unequivocal eukaryote organization is found.

Sulphur isotopes

Sulphur isotopes can also be used to trace the history of **sulphate reduction**. In this case, ^{32}S is preferentially taken up by sulphate-reducing bacteria, leaving the water column enriched in the heavier isotope ^{34}S. Studies of $^{34}S/^{32}S$ ($\delta^{34}S$) ratios in sedimentary pyrite and in gypsum and anhydrite have shown that sulphate reduction may not have taken place before 2.8 Ga. This may be because there was insufficient free oxygen in the atmosphere to form the sulphate ions needed for sulphate reduction or that surface water temperatures were too high to produce a measurable fractionation.

Banded iron formations (BIFs)

BIFs are deeper water sediments that show millimetric laminations of Fe_2O_3-rich (hematitic) chert and iron-poor chert (chalcedony). These laminations can be remarkably continuous – some have been traced for up to 300 km. BIFs were particularly common in Archean and Palaeoproterozoic marine basins between about 3.5 and 1.8 Ga. Their presence suggests that some kind of seasonal 'rusting' of the oceans took place, in which oxygen released by blooms of photosynthetic microbes was mopped up by Fe^{2+} ions in solution. These ions were widely available in the water column owing to the reducing chemistry of the early oceans and widespread hydrothermal exhalation. Settling of hematite precipitates through the water column formed laminae on the ocean floor. This interpretation does not require the production of oxygen by photosynthesis, since this oxygen may have a source from photodissociation of water or even from volcanic oxidized mineral species.

After about 1.8 Ga, BIFs seldom appear in the rock record and continental red beds begin to become widespread. This suggests that the ferrous iron **oxygen sink** had become saturated, and that oxygen was now able to accumulate in the atmosphere, leading to the oxygenic weathering of terrestrial rocks.

Archean fossils

Stromatolites (Fig. 6.6)

These sedimentary structures (see Chapter 8) are known to occur in carbonate rocks as old as 3.5 Ga in the Pilbara Supergroup of Western Australia (Fig. 6.6r) and 3.4 Ga in the Swaziland Supergroup of South Africa. Although an origin from the growth of cyanobacterial mats has often been inferred, they do not

Fig. 6.6 Pseudofossils, stromatolites and microfossils from the Archaean and Proterozoic. (a) Corner of the 'microfossiliferous' clast reported by Schopf (1993) from 3.46 billion year old Apex chert, reinterpreted by Brasier *et al.* (2002) as a shard within a subsurface hydrothermal dyke. (b–l). Detailed views of Earth's oldest supposed 'microfossils' (shown at white arrows in (a)), regarded by Brasier *et al.* (2002) as carbonaceous pseudofossils. (k) Pseudofossil *Primaevifilum delicatulum.* (n–o) Similar pseudofossils from resampled cherts. (p) Pseudofossil *Eoleptonema apex* from the 3.46 Gyr Apex chert, showing angular morphology caused by wrapping around crystal margins, shown alongside original interpretation as a beggiatoan bacterium (inset, photographs and drawing at right). (q) Pseudofossil *Archaeoscillatoriopsis disciformis* from the 3.46 Gyr Apex chert, showing branched morphology and proximity to crystal growths (arrowed) alongside original interpretation as an oscillatoriacean cyanobacterium (inset, photographs and drawing at right). (r) 3420 Myr 'stromatolite' from the Strelley Pool chert of Western Australia, controversially claimed as earliest evidence on Earth for microbial entrapment of sediment. (s, t) Proterozoic (1900 Myr) filaments of unquestioned biogenic origin (probable iron bacteria), Gunflint chert of Mink Mountain, Ontario, Canada. Black scale bar: (c) = 10 cm. White scale bar: (a) = 400 μm; (l, t) = 100 μm; (b–k, m–q, s) = 40 μm.

contain microfossils and they have simple rotational symmetry and isopachous sedimentary laminae. It has also been shown that many Archean stromatolites may have formed by the direct precipitation of aragonite from sea water. The evidence from stromatolites is therefore less than conclusive. Even though the size, shape and millimetre scale laminations within these structures are a little bit like those of younger, both fossil and modern, stromatolites, some of which may also be abiogenic.

Silicified microbiotas

Early diagenetic silica has preserved the cells of prokaryotic and even eukaryotic microorganisms at a number of localities with latest Archean and Paleoproterozoic rocks (c. 2.7–1.8 Ga; Fig. 6.6(s, t)). These microfossils, which can be well preserved in three dimensions, are usually studied by means of standard petrographic thin sections at high magnification. Most of these chert microbiotas are associated with stromatolitic carbonates in evaporitic settings.

One of the oldest cherts to have yielded a supposed bacterial microflora is associated with basalt lava flows in the **Warrawoona** Group of Western Australia, dated at 3.465 Ga (Schopf 1992) and from the Barberton mountains of South Africa. Eleven species of bacterial cells and cyanobacterial filaments have been described from the Apex Chert within the Warrawoona Group and were once taken as the oldest morphological evidence for life on Earth. The structures are nearly 1 Ga older than putative cyanobacterial biomarkers. Recent reanalysis of these 'microfossils' has led to questions about their authenticity (Brasier *et al.* 2002), and further work shows they are pseudofossils formed by recrystallization of the chert.

An even more famous microbiota preserved in the **Gunflint Chert** (~1.9 Ga) comprises about 12 species, some of which closely resemble modern coccoid and filamentous cyanobacteria, while others resemble iron bacteria (Schopf & Klein 1992; see Fig. 8.2).

Palynology of shales

The techniques of palynological maceration (see Appendix) have been applied to organic-rich Archean and Proterozoic shales, with interesting results. In rocks older than about 1.8 Ga, the macerations consist largely of very small (10–20 µm) and relatively simple compressed spheres which have been called **cryptarchs**, owing to their uncertain biological affinities. These could be the remains of either benthic or planktonic cyanobacterial spores. After about 1.8 Ga, there was a slow increase in size and complexity, suggestive of the progressive development of morphologies relating to eukaryotic protozoan organization.

REFERENCES

Bradley, J.P., Harvey, R.P. & McSween, H.Y. 1997. No 'nanofossils' in Martian meteorite. *Nature* 390, 454–456.

Brasier, M.D. 2000. The Cambrian explosion and the slow burning fuse. *Science Progress* 83, 77–92.

Brasier, M.D., Green, O.R., Jephcoat, A.P., Kleppe, A.K., Van Kranendonk, M.J., Lindsay, J.F., Steele, A. & Grassineau, N.V. 2002. Questioning the evidence for Earth's oldest fossils. *Nature* 416, 76–81.

Grady, M., Wright, I. & Pillinger, C., 1996. Opening a Martian can of worms? *Nature* 382, 575–576.

Gupta, R.S. 1998. Protein phylogenies and signature sequences: a reappraisal of evolutionary relationships among Archaebacteria, Eubacteria and Eukaryotes. *Microbiology and Molecular Biology Reviews* 62, 1435–1491.

Gupta, R.S. 2000. The natural evolutionary relationships among prokaryotes. *Critical Reviews in Microbiology* 26, 111–131.

Kamber, B.S. & Moorbath, S. 1998. Initial Pb of the Amîtsoq gneiss revisited: implication for the timing of Early Archaean crustal evolution in West Greenland. *Chemical Geology* 150, 19–41.

Karhy, J.A. & Holland, H.D. 1996. Carbon isotopes and the rise of atmospheric oxygen. *Geology* 24, 867–870.

McKay, D.S., Thomas-Keprta, K.L., Romanek, C.S. *et al.*, 1996. Evaluating the evidence for past life on Mars – response. *Science* 274, 2123–2125.

Miller, S.L. 1953. A production of amino acids under possible primitive earth conditions. *Science* 206, 1148–1159.

Mojsis, S.J., Arrenhius, G., McKeegan, K.D., Harrison, T.M., Nutman, A.P. & Friend, C.R. 1996. Evidence for life on Earth before 3800 million years ago. *Nature* 384, 55–59.

Nisbet, E.G. & Fowler, C.M.R. 1996. Early life – some liked it hot. *Nature* 382, 404–405.

Russell, M.J. & Hall, A.J. 1997. The emergence of life from iron monosulphide bubbles at a submarine hydrothermal redox and pH front. *Journal of the Geological Society, London* **154**, 377–402.

Schidlowski, M. 1988. A 3800 million year isotopic record of life from carbon in sedimentary rocks. *Nature* **333**, 316.

Schidlowski, M. & Golubic, S. 1992. *Early Organic Evolution: implications for mineral and energy resources.* Springer-Verlag, Berlin.

Schopf, W.J. 1992. *Major Events in the History of Life.* Jones & Bartlett, Boston.

Schopf, J.W. 1993. Microfossils of the early Archaen Apex Chert. New evidence of the antiquity of life. *Science* **260**, 640–646.

Schopf, J.W. & Klein, C. (eds). 1992. *The Proterozoic Biosphere.* Cambridge University Press, Cambridge.

Sogin, M.L. 1994. The origin of eukaryotes and evolution into major kingdoms. In: Bengtson, S. (ed.) *Early Life on Earth.* Columbia University Press, New York, pp. 181–192.

Summons, R.E., Jahanke, L.L., Hope, J.M. & Logan, G.A. 1999. 2-Methylhopanoids as biomarkers for cyanobacterial oxygenic photosynthesis. *Nature* **400**, 554–557.

Treiman, A.H. 2001. http://cass.jsc.nasa.gov/lpi/meteorites/life.html.

Woese, C.R., Kandler, O. & Wheelis, M.L. 1990. Towards a natural system of organisms: proposals for the domains of Archaea, Bacteria and Eucarya. *Proceedings of the National Academy of Sciences USA* **87**, 4576–4579.

CHAPTER 7

Emergence of eukaryotes to the Cambrian explosion

Emergence of eukaryotes

The divide between prokaryote and eukaryote cells can be regarded as one of the largest discontinuities within the living world. Eukaryotes differ from prokaryotes in the presence of a proper membrane-bound **nucleus** (to contain the DNA in genes arranged on chromosomes) plus a generally larger cell size and presence of cell **organelles** such as mitochondria and chloroplasts. Reproduction in eukaryotes may be asexual, involving strictly controlled cell division by **mitosis**, or sexual, involving strictly controlled cell division by **meiosis**.

The **Serial Endosymbiotic Theory** of cell evolution (e.g. Margulis 1981; Fig. 7.1a) argues that the remarkable complexity of the eukaryotic cell was assembled over a long time period by symbiotic associations between different kinds of prokaryotes and an amitochondriate protozoa host. Purple bacteria were perhaps acquired first to provide the mitochondrial organelles, while photosynthetic prokaryotes such as coccoid cyanobacteria and their relatives were probably acquired last to form chloroplasts.

The **Neomuran Hypothesis** (Cavalier-Smith 2002) modifies the serial endosymbiotic theory and argues for the aggregation of the DNA and the formation of a primitive nuclear membrane in an ancestral gram-negative eubacterium, to form a nucleate pre-eukaryote (Fig. 7.1b). This form had the key evolutionary innovation of a flexible cell wall which separated it from the Archaea and allowed a phagotrophic (feeding by engulfing) mode of life. Through phagocytosis the symbiotic acquisition of mitochondria in ciliate and aciliate pre-eukaryotic forms led to the Amoebozoa. Secondary symbiotic acquisition of chloroplasts in an aciliate amoebozoan produced the common ancestor of all plants. This hypothesis predicts that mitochondria were present in the common ancestors of all living eukaryotes and that anaerobic eukaryotes must have lost their mitochondria. It also allows for the relatively rapid acquisition of the eukaryotic grade of organization.

A billion years of environmental stability

Extremely long periods of nutrient and climatic stability may have been needed for the host–symbiont relationship to become fused into a single eukaryotic organism. This is because the relationship between symbiont and host can be easily destroyed by strong physical perturbations. Indeed, there appears to have been nearly a billion years of environmental stability between at least 2 Ga and 1 Ga ago, when ice ages are unknown and $\delta^{13}C_{org}$ isotopic values barely departed from the mean. It was during this interval that the complex organization of the eukaryotes was evolving (Brasier 2000; Fig. 7.2).

Evidence for the earliest eukaryotes

Controversial evidence for the emergence of eukaryotes is provided by macroscopic carbonaceous compression fossils, interpreted as the remains of an algal megaflora. Spiral ribbons of *Grypania* have been reported from rocks supposedly as old as 2.1 Ga (Fig. 7.3a; Han & Runnegar 1992), although comparable remains do not reappear for another 700 Ma. These structures grade from ribbons to large sack-shaped structures that some regard as the envelopes of cyanobacterial colonies such as *Nostoc*.

The gradual emergence of eukaryote organization by about 1.8 Ga is suggested by **acritarchs** of >60 μm

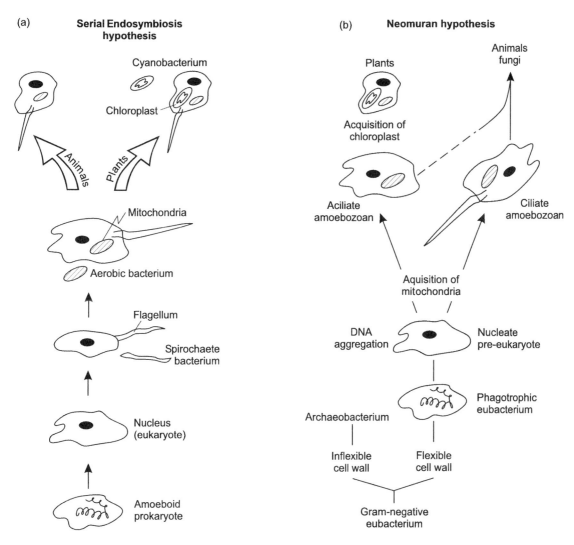

Fig. 7.1 (a) The Serial Endosymbiotic Theory suggests that eukaryote organelles arose from successive endosymbioses between different kinds of prokaryote and an amitochondriate host bacterium such as *Thermoplasma*. (b) The Neomuran Hypothesis indicates a common ancestry in a gram-negative bacterium followed by secondary acquisition of mitochondria and chloroplasts by serial endosymbiosis in different lineages.

diameter (Schopf & Klein 1992; Knoll 1994) while sterane biomarkers typical of eukaryotes have been obtained from the Barney Creek Formation of northern Australia (~1.64 Ga; Summons & Walter 1990). Between 1.3 and 1 Ga, the diversity of acritarchs began to increase rapidly, to include not only simple **sphaeromorphs** but also **megasphaeromorphs** larger than 200 µm and spiny forms known as **acantho-** morphs (e.g. Schopf 1992; Schopf & Klein 1992; Knoll 1994; Chapter 9). The enigmatic tetrad form *Eotetrahedrion* (Fig. 7.2c) also appears in this interval as did the red algae (Fig. 7.2d; Butterfield *et al.* 1990), and according to recent rRNA sequence data this was followed by a major eukaryotic radiation which is thought to have involved ciliates, brown algae, green algae, plants, fungi and animals.

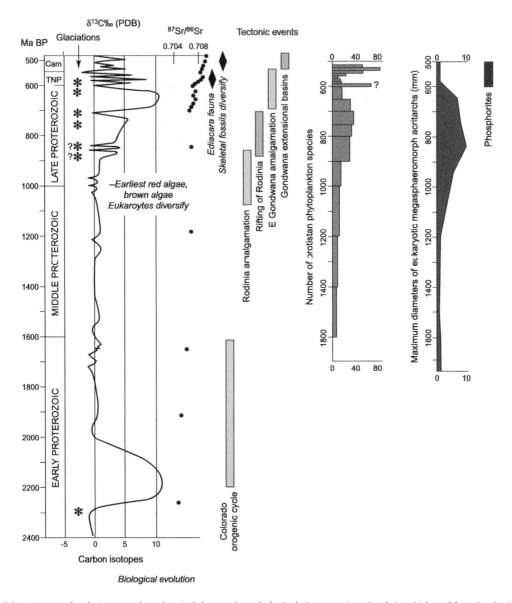

Fig. 7.2 Summary of evolutionary and geochemical changes through the Early Proterozoic to Cambrian. (Adapted from Brasier 2000 and sources therein.)

Cellular differentiation, root-like structures and the presence of nucleus-like spots are arguable indications of eukaryotic organization in Neoproterozoic chert microfloras (Fig. 7.3). There is controversial evidence for vegetative reproduction and sexual reproduction (meiotic spore tetrads, Fig. 7.3b,c) in the Bitter Springs Chert (about 800 Ma). The existence of branched cells like those of siphonalean green algae (Fig. 7.3e) suggests sexual reproduction had evolved by this time.

The sexual revolution

The appearance of sexual reproduction, the exchange of genes to form new genetic recombinations, has many evolutionary advantages over asexual reproduction. For

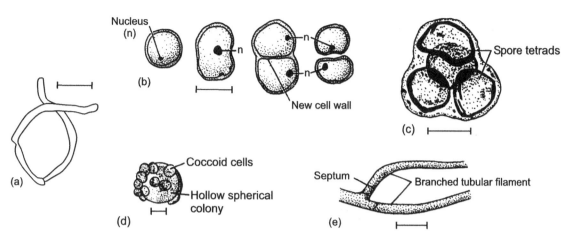

Fig. 7.3 Early fossil 'eukaryotes'. (a) The carbonaceous film '*Helminthoidichnites*' (=*Grypania*) *meeki* from the Greyson Shale, Montana. Scale bar = 2 mm. (b) Sequence of Precambrian fossils claimed to indicate mitosis in *Glenobotrydion*. Scale bar = 10 µm. (c) Tetrad of Precambrian *Eotetrahedrion*. Scale bar = 10 µm. (d) Precambrian *Eosphaera*. Scale bar = 10 µm. (e) Precambrian siphonalean-like filament. Scale bar = 10 µm. ((a) and (b) based on Schopf 1972; (c) and (d) based on Cloud 1976.)

example, 10 genetic mutations in an asexual population can result in only 11 genotypes, the original type plus those of the 10 mutants. The same number of mutations in a primitive diploid sexual population could be combined to produce up to 59,049 distinct genotypes (Schopf *et al.* 1973). Hence, in theory, the evolution of eukaryotic sexuality must have resulted in a prodigiously increased genetic variety of organisms, expressed by an increased rate of biological evolution in the fossil record. A plausible explanation for the explosion in the diversity of microfossils after about 1.3 Ga, therefore, is the evolution of **meiosis** (a reduction division of the cells) and **syngamy** (the fusion of gametic cells). Prior to this, primitive eukaryotes are likely to have reproduced asexually, by means of **mitosis**. If sexual reproduction in modern eukaryotes is a shared character that has been inherited from a single common ancestor, then it may be that the major groups of eukaryotes did not diverge much before about 1.3 Ga.

The Cambrian explosion

The biological revolution: microfossils from the Neoproterozoic–Cambrian transition

From about 600 Ma onward, marked changes began to take place in the fossil record, indicating a major revolution in the biosphere. Changes in the marine phytoplankton are heralded by dramatic changes in the diversity and composition of acritarch assemblages. Large **acanthomorph acritarchs** had appeared by about 1 Ga but experienced extinctions over the Varangian glacial interval (c. 600–560 Ma) and again during the Early Ediacarian, before the appearance of the Ediacara fauna. A new more diverse assemblage of small acanthomorphs, including *Cymatiosphaera*, appeared just above the Precambrian–Cambrian boundary, after which acritarch groups diversified dramatically (Vidal & Moczydlowska 1997). These changes in the marine phytoplankton were coincident with the appearance of the first **phosphatized animal embryos** at c. 580 Ma (Fig. 7.4a), with the **Ediacara biota** (impressions of large, soft-bodied multicellular animals or possibly of giant protists; see Seilacher *et al.* 2003; Brasier & Antcliffe 2004) after c. 575 Ma, with the first unequivocal animal **trace fossils** close to 555 Ma, and the appearance of diverse assemblages of **small shelly fossils** close to the Precambrian–Cambrian boundary at 543 Ma.

Most of the earliest skeletal microfossils are only a few millimetres in diameter and have to be studied using micropalaeontological techniques. *Cloudina* is a small, irregularly curved tube with a double layered $CaCO_3$ wall of stacked half rings. It occurs with Ediacara fauna in Namibia in rocks dated between 550 and 543 Myr and was possibly made by a sedentary, suspension-

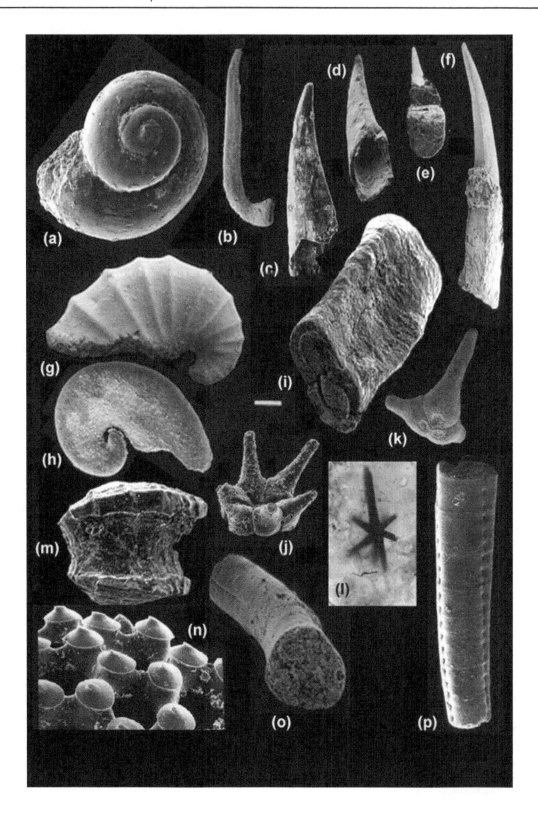

feeding worm living in shallow algal mounds on carbonate platforms. Siliceous biomineralization was also beginning offshore at this time, where the remains of both hexactinellid sponges (Fig. 7.41) and demosponge microfossils have been found in phosphatic and siliceous facies (e.g. Brasier *et al.* 1997).

The base of the Cambrian is marked by the appearance of a complexly branching trace fossil, *Treptichnus* (or *Phycodes*) *pedum* (Brasier *et al.* 1994). This is accompanied in places by small shelly fossils typical of the Nemakit-Daldynian Stage at the base of the Cambrian System. *Platysolenites* is an agglutinated tube, now believed to be the earliest **foraminiferid** test (McIlroy *et al.* 2001). *Anabarites* is a small tapering $CaCO_3$ tube with a three-lobed cross section, commonly preserved as phosphatic iternal moulds (Fig. 7.4 o, p). This was probably the skeleton of a sedentary, suspension-feeding **cnidarian**, related to corals and jellyfish. **Micromolluscs** such as *Latouchella* also made their appearance in this stage. *Latouchella* has a planispirally coiled, flared and bilaterally compressed shell with strong transverse ribs (Fig. 7.4 g). The seven-rayed calcareous spicules of *Chancelloria* (Fig. 7.4 j) differed from those of sponges in being hollow and articulated and were formed by an animal of unknown biology. Later examples, such as *Allonnia*, have a reduced number of rays (Fig. 7.4 k). *Protohertzina* is a small, phosphatic **protoconodont** (Fig. 7.4 b, e, f; see Chapter 21) and is thought to have been part of the feeding apparatus of a predatory, pelagic invertebrate resembling a modern chaetognath worm. Other tooth-shaped objects are also found but are of unknown affinity (e.g. *Maldeotaia*, Fig. 7.4 c, d).

The Tommotian Stage marks a further step in the Cambrian radiation, with the appearance of archaeocyathan sponges, inarticulate brachiopods and a range of possibly related small shelly fossils known as tommotiids. These appear to have had a multielement skeleton of **sclerites**, which may show right- or left-handed symmetry and symmetry transition series (see Qian & Bengtson 1989) as, for example, in the saddle-shaped phosphatic sclerite of *Camenella* (Fig. 7.4 m). Helically coiled **microgastropods** such as *Aldanella* (Fig. 7.4 a) are particularly characteristic of the Tommotian stage, while snails with more rapidly expanding whorls, such as *Pelagiella*, appeared in the following Atdabanian Stage (Fig. 7.4 h).

The Atdabanian is notable for the appearance of arthropod skeletal remains, including not only trilobites but also the first **bradoriids** (see Chapter 20). The elaborately sculptured phosphatic nets of *Microdictyon* (Fig. 7.4 n) appear widely at this time. They appear to have been part of the dorsal skeleton of the caterpillar-like onycophoran arthropod formerly called *Hallucigenia*. A decline took place in the diversity of small shelly fossils during the succeeding Botomian Stage, which also brought about the first well-documented extinction of major reef-building ecosystems. This extinction coincided with a major episode of transgression, which brought anoxic waters onto the shelves (Brasier 1995; Wood & Zhuravlev 1995). The evolutionary trends and stratigraphic utility of these earliest skeletal microfossils are comprehensively reviewed in Brasier (1989).

Rifting of major supercontinents after ~580 Ma was accompanied by an episodic and prolonged rise in sea level through to the end of the Cambrian (Brasier & Lindsay 1998). These changes brought oxygen-depleted and nutrient-enriched oceanic waters over drowning platforms (Brasier 1995; Wood & Zhuravlev 1995). Under these conditions phosphatization was widespread and led to the remarkable preservation of animal embryos from the Duoshantuo Formation of China (Fig. 7.5 a; Ediacarian), phosphatized molluscs from the earliest Cambrian and micro-arthropods (the **Orsten microbiota**) from the Upper Cambrian of Sweden (Fig. 7.5 b–d).

Whilst the first metazoans (multicelled animals) appear abruptly in the fossil record at the end of the Precambrian, some fundamental aspects of this event remain unclear. Are the metazoans a monophyletic

Fig. 7.4 (*opposite*) Representative early skeletal microfossils. All from Lower Cambrian except where stated: (a) from Oxford, UK; (b, g, o from Elburz Mts, Iran; (c–f) from Lesser Himalaya, India; (h, k) from Sichuan, China; (i) from Estonia; (l) from Gobi-Altay, Mongolia; (m) from Nuneaton, UK; (n) from Newfoundland, Canada; (j, p) from Siberia. (a) *Aldanella attleborensis*. (b, e, f) *Protohertzina unguliformis*. (c, d) *Maldeotaia bandalica*. (g) *Latouchella korobkovi*. (h) *Pelagiella emeishanensis*. (i) *Platysolenites antiquissimus*. (j) *Chancellorie lenaica*. (k) *Allonnia erromenosa*. (l) Hexactinellid spicule from latest Proterozoic of Mongolia. (m) *Camenella baltica*. (n) *Microdictyon* cf. *effusum*, width of view 0.3 mm. (o, p) *Anabarites trisculatus*. Scale bar = c. 100 μm unless otherwise stated.

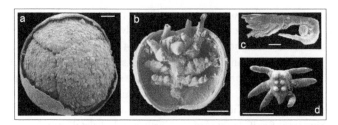

Fig. 7.5 Exceptional Precambrian and Cambrian fossils preserved in calcium phosphate. Scale bars = 100 μm. (a) Fossil embryo from the Doushantuo Formation (570 ± 20 Ma), South China. (b) *Hesslandona* sp. from the Upper Cambrian, Orsten, Vestergötland, Sweden. (c) Microarthropod *Martinssonia elongata* (Müller & Walosseck) from the Upper Cambrian of Sweden. (d) Arthropod larva, dorsal view, from the Upper Cambrian of Sweden. ((a) From Xiao & Knoll 2000, figure 7(2) (with permission of the Paleontological Society); (b) from Müller 1985, plate 1, figure 8 (with permission of the Royal Society, London); (c) from Walosseck & Müller 1990, figure 6 (with permission of the Lethaia Foundation); (d) from Müller & Walosseck 1986, figure 1h (with permission of the Royal Society, Edinburgh).)

group, that is derived from a single unicelled organism (ciliated or aciliate unicells?) or perhaps from multicellular eukaryotes? Were Late Precambrian soft-bodied organisms, so widespread in the Ediacara biota, different from those of the Early Palaeozoic and if so was there a mass extinction in the Late Precambrian? How and when did the major phyla of living animals evolve? 18S ribosomal RNA sequence data suggest the Metazoa are monophyletic whilst other methods have supported a monophyletic ancestry for at least the Eumetazoa (all animals except sponges) and that the coelenterates (Cnidaria and Ctenophora) are the sister group to all other living higher Metazoa (the Bilateria).

The environmental revolution

The early diversification of the Metazoa coincided with a number of exceptional global events including changes in atmospheric and ocean chemistry leading to the transition to a prolonged period of greenhouse climate and the break up of long-lived supercontinents. The end of the Varangian glaciation led to worldwide transgression and the opening of many new shallow-water niches. Is there a connection between the biological revolution and global environmental change? Isotopes can again give us some clues about the context of the biological revolution.

The spread of nutrient-enriched water over low latitude shelves is recorded by fluctuations in $\delta^{13}C_{carb}$. Maximal values of +11‰ are found between 700 and 600 Ma, falling to +8‰ in the Ediacarian and +5‰ in the Cambrian. Minimal values are typical for late glacial to postglacial carbonates. Variations are taken to reflect shifts in the rate of primary productivity and/or the burial of organic matter and imply that rates of carbon burial were maximal between the Late Neoproterozoic glaciations. Increased rates of photosynthesis and carbon burial are stimulated by **eutrophication** of ocean water and led to removal of CO_2 from the atmosphere. This could have resulted in the release of large amounts of O_2 to the atmosphere and led to a negative greenhouse effect.

Nutrient enrichment of the oceans is recorded in the strontium isotopic composition of sea water which reflects the relative input of ^{87}Sr from the weathering of old continental crust, and ^{86}Sr from hydrothermal exchange with younger ocean crust. The largest and longest reported $^{87}Sr/^{86}Sr$ excursion is in the latest Precambrian-Cambrian. Strontium isotopic values show a series of oscillations that broadly parallel the carbon curve in the Neoproterozoic, with a sharp rise in values in the Ediacarian to Late Cambrian. This parallel pattern indicates the close relationship between oceanic nutrient and bioproductivity increase.

The onset of widespread oceanic anoxia (and sulphate reduction) following the Varangian glaciation is recorded in the greatest and most prolonged of all sulphur isotope excursions (Fig. 7.6).

The Deep time versus the Late arrival hypotheses

How reliable is the fossil record as a guide to evolutionary events? Much attention has focused recently upon an apparent mismatch between the evidence

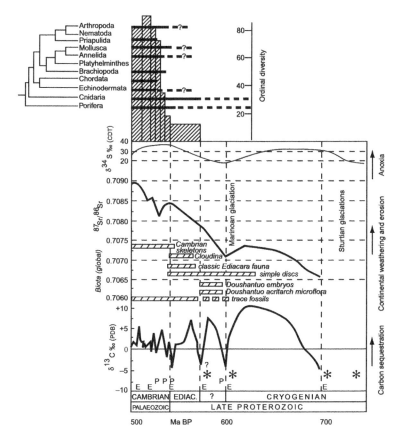

Fig. 7.6 Changes in ocean chemistry as shown by $\delta^{13}C$ and $\delta^{34}S$ and $^{87}Sr/^{86}Sr$, shown alongside evolutionary changes in the fossil record during the Late Proterozoic and Cambrian. P, phosphatized microfossil assemblages (see text); * snowball earth glaciations; E, suspected mass extinctions of acritarch phytoplankton (Redrawn after Brasier 2000.)

provided by fossils and that provided by **molecular clocks**. The oldest unequivocal fossil evidence for metazoans is ~600 million years old but, according to some researchers, this is much younger than evidence suggested by the rRNA of living organisms. Assuming that gene sequences evolve with such regularity that differences can be used as 'molecular clocks', it can be argued that invertebrate lineages began to diverge about 1.2 Ga (Wray et al. 1996). This would mean that the earliest animals are missing from the fossil record because of a low fossilization potential. For example, they may have lived in the water column or as microscopic life in the sediment. The so-called Cambrian 'explosion' after 600 million years would then relate in large part to the acquisition of skeletons.

Molecular clocks are notoriously difficult to calibrate, however, and these 'deep time' estimates for the divergence of major animal phyla have been drastically scaled down to nearer 670 million years, bringing the figures more into line with the fossil record (Ayala & Rzhetsky 1998). A 'late arrival' model would also imply that the evolution of the animal phyla took place both late and rapidly, perhaps in response to the evolution of *Hox* genes (Erwin et al. 1997), or in response to the lifting of some external ecological constraints such as an increase in atmospheric oxygen (Schopf & Klein 1992). Thus while the Cambrian explosion could be viewed as the almost inevitable consequence of the evolution of sexuality and multicellularity between about 1.3 Ga and 600 Ma, its timing appears to have coincided with major geological changes at the Earth's surface.

Biological and evolutionary consequences of the Cambrian explosion

Eukaryotic organisms have the ability to produce proteinaceous membranes capable of mineralization and are able to pump ions through their cell walls. The incorporation of the citric acid cycle into cell metabolism provided the increase in available energy

56 Part 2: The rise of the biosphere

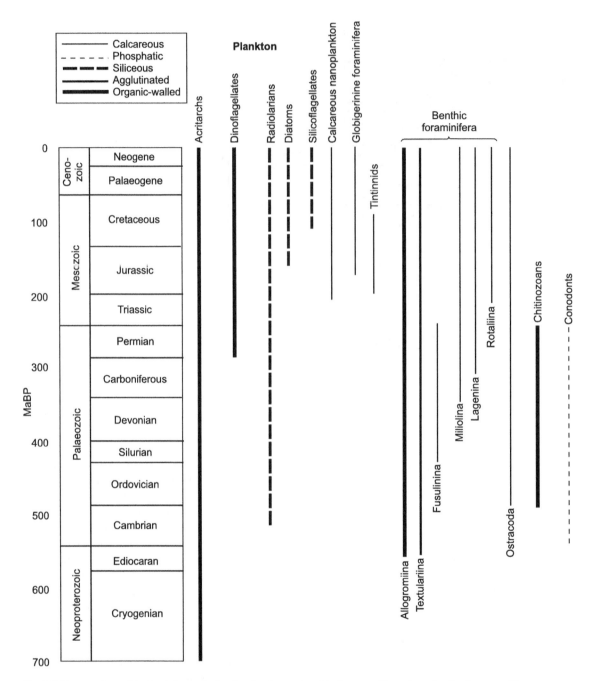

Fig. 7.7 Repercussions of the Cambrian explosion showing the stratigraphical ranges of the major microfossil groups of the Phanerozoic.

required for biomineralization. Eukaryotes are in effect pre-adapted for biomineralization (Simkiss 1989). However, this potential was not realized until metazoan cellular systems were highly differentiated and atmospheric oxygen had risen to a level suitable for the evolution of larger and more complex bodies. The abrupt appearance of shelly fossils can be explained by two factors. Firstly, evolution appears to have been rapid at this time, partly due to intense selection pressures and partly due to the opening of new shallow marine niches. Secondly, the acquisition of a hard skeleton leads to a much better fossilization potential. A number of hypotheses have been suggested for the origination of a hard skeleton. Glaessner (1962) suggested calcareous and phosphatic material may have been excreted onto or accumulated over the skin. Once formed, perhaps accidentally, the hard carapace would provide protection and could provide anchorage for muscles and ligaments; both major evolutionary advantages. Once hard mouth parts had originated (e.g. as preserved in *Protohertzina*) the selection pressure on other organisms to evolve hard protective coverings would be increased initiating the equivalence of an evolutionary 'arms race' between predator and prey.

The widespread appearance of $CaCO_3$ shells after ~600 Ma must have had a dramatic impact upon biogeochemical cycles in the oceans, including the interlinked cycles of carbon, oxygen and the biolimiting nutrients P, N and Fe. Not least the biosphere now moderated and provided for the storage of CO_2 as $CaCO_3$ in the geosphere. In the water column, the expanding and diversifying grazing zooplankton packaged phytoplankton cells as faecal pellets which were able to sink to the sea floor. Faecal pellets can be found in some phosphatized microfossil assemblages. This **faecal pumping** is likely to have lessened the reducing power of slowly settling phytoplankton detritus and brought about an improvement in the oxygenation of the upper water column (Logan *et al.* 1995).

The emergence of a tiered burrowing infauna in the Cambrian which continued to diversify through the Phanerozoic is thought to have improved **porewater irrigation**, displacing the redox boundary downward within the sediment (McIlroy & Logan 1999). Siliciclastic sediment pore waters now had the potential for higher pH and Eh and were potentially less prone to H_2S toxicity. Epifaunal and infaunal niches could therefore be exploited with less risk.

Whilst metazoans have been secreting calcareous skeletons since the beginning of the Cambrian, a different pattern emerges for protozans (Fig. 7.7). Although benthic foraminifera with organic-walled and agglutinated tests appeared in the Early Cambrian, forms with $CaCO_3$ tests did not appear widely until the Devonian to Carboniferous and they have radiated progressively since. There is an absence of carbonate-secreting plankton groups until much later, in the Mesozoic, when calcareous nannoplankton and foraminiferal zooplankton flourished to rock-producing proportions. This delay in the $CaCO_3$ biomineralization of protozoa may be explained by the problems posed by the high surface area to volume ratio of a single cell, and its greater susceptibility to water chemistry than complex metazoans.

Improved ventilation of the upper water column and a subsequent reduction in concentration of dissolved CO_2 and P in the water column over the course of the Phanerozoic may well have reached a threshold level that enabled the secretion of carbonate skeletons among the plankton, such as seen in coccolithophores, calpionellids and foraminifera.

REFERENCES

Ayala, F.J. & Rzhetsky, A. 1998. Origin of the metazoan phyla: molecular clocks confirm paleontological estimates. *Proceedings of the National Academy of Sciences USA* **95**, 606–611.

Bengtson, S. 1992. Proterozoic and earliest Cambrian skeletal metazoans. In: Schopf, J.M. & Klein, C. (eds) *The Proterozoic Biosphere: a multidisciplinary study*. Cambridge University Press, Cambridge, pp. 397–402.

Brasier, M.D. 1995. The basal Cambrian transition and Cambrian bio-events (from terminal Proterozoic extinctions to Cambrian biomeres). In: Walliser, O.H. (ed.) *Global Events and Event Stratigraphy*. Springer Verlag, Berlin, pp. 113–138.

Brasier, M.D. 1989. Towards a biostratigraphy of the earliest skeletal biotas. In: Cowie, J.W. & Brasier, M.D. (eds) (1989) *The Precambrian–Cambrian Boundary*. Oxford Monographs in Geology and Geophysics, No. 12, pp. 117–165.

Brasier, M.D. 2000. The Cambrian Explosion and the Slow Burning Fuse. *Science Progress Millennium Edition* **83**, 77–92.

Brasier, M.D. & Antcliffe, J. 2004. Decoding the Ediacaran enigma. *Science* **305**, 1115–1117.

Brasier, M.D. & Lindsay, J.F. 1998. A billion years of environmental stability and the emergence of eukaryotes: new data from northern Australia. *Geology*, 26, 555–558.

Brasier, M.D., Corfield, R.M., Derry, L.A., Rozanov, A.Yu. & Zhuralev, A.Yu. 1994. Multiple delta-^{13}C excursions spanning the Cambrian explosion to the Botomian crisis in Siberia. *Geology* 22, 455–458.

Brasier, M.D., Green, O.R., Shields, G. 1997. Ediacarian sponge spicule clusters from SW Mongolia and the origins of the Cambrian fauna. *Geology* 25, 303–306.

Briggs, D.E.G. & Crowther, P.R. (eds) 1990. *Palaeobiology: a synthesis*. Blackwell Scientific Publications, Oxford.

Butterfield, N.J., Knoll, A.H., & Swett, K. 1990. A bangiophyte red alga from the Proterozoic of Arctic Canada. *Science* 250, 104–107.

Cavalier-Smith, T. 2002. The phagotrophic origin of eukaryotes and phylogenetic classification of protozoa. *International Journal of Systematic and Evolutionary Microbiology* 52, 297–354.

Cloud, P. 1976. Beginnings of biospheric evolution and their biochemical consequences. *Paleobiology* 2, 351–387.

Erwin, D., Valentine, J. & Jablonski, D. 1997. The origin of animal body plans. *American Scientist*, 85, 126–137.

Glaessner, M.F. 1962. Precambrian fossils. *Biological Reviews* 37, 467–494.

Han, T.-M. & Runnegar, B. 1992. Megascopic eukaryotic algae from the 2.1-billion-year-old Negaunee Iron-Formation, Michigan. *Science* 257, 232–235.

Knoll, A.H. 1994. Proterozoic and Early Cambrian protists: Evidence for accelerating evolutionary tempo. *Proceedings. National Academy of Sciences* 91, 6473–6750.

Logan, G.A., Hayes, J.M., Hieshima, G.B. & Summons, R.G. 1995. Terminal Proterozoic reorganization of biogeochemical cycles. *Nature* 376, 53–56.

McIlroy, D., Green, O.R. & Brasier, M.D. 2001. Palaeobiology and evolution of the earliest agglutinated Foraminifera: *Platysolenites, Spirosolenites* and related forms. *Lethaia* 34, 13–29.

McIlroy, D. & Logan, G.A. 1999. The impact of bioturbation on infaunal ecology and evolution during the Proterozoic–Cambrian transition. *Palaios* 14, 58–72.

Margulis, L. 1981. *Symbiosis in Cell Evolution. Life and its environment on the early Earth*. Freeman, San Francisco.

Moczydlowska, M. 2002. Early Cambrian phytoplankton diversification and appearance of trilobites in the Swedish Caledonides with implications for coupled evolutionary events between primary producers and consumers. *Lethaia* 35, 191–214.

Müller, K.J. 1985. Exceptional preservation in calcareous nodules. *Philosophical Transactions of the Royal Society of London* B311, 67–73.

Müller, K.J. & Walosseck, D. 1986. Arthropod larvae from the Upper Cambrian of Sweden. *Transactions. Royal Society Edinburgh: Earth Sciences* 77, 157–179.

Qian Yi & Bengtson, S. 1989. Palaeontology and biostratigraphy of the Early Cambrian Meishucunian Stage in Yunnan Province, South China. *Fossils and Strata* 24, 156pp.

Schopf, J.W. 1972, Evolutionary significance of the Bitter Springs (Late Precambrian) microflora. *24th International Geological Congress, Montreal* 1, 68–77.

Schopf, W.J. 1992. *Major Events in the History of Life*. Jones & Bartlett, Boston.

Schopf, J.W. & Klein, C. (eds) 1992. *The Proterozoic Biosphere*. Cambridge University Press, Cambridge.

Schopf, J.W., Haugh, B.N., Molnar, R.E. & Satterthwait, D.F. 1973. On the development of metaphytes and metazoans. *Journal of Paleontology* 47, 1–9.

Seilacher, A., Grazhdankin, D. & Legouta, A. 2003. Ediacara biota: the dawn of animal life in the shadow of giant protists. *Paleontological Research* 7, 43–54.

Simkiss, K. 1989. Biomineralization in the context of geological time. *Transactions. Royal Society of Edinburgh: Earth Sciences* 80, 193–199.

Summons, R.E. & Walter, M.R. 1990. Molecular fossils and microfossils of prokaryotes and protists from Proterozoic sediments. *American Journal of Science* 290, 212–244.

Vidal, G. & Moczydlowska, M. 1997. Biodiversity, speciation, and extinction trends of Proterozoic and Cambrian phytoplankton. *Paleobiology* 23, 230–246.

Walosseck, D. & Müller, K.J. 1990. Upper Cambrian stem-lineage crustaceans and their bearing upon the monophyletic origin of Crustacea and the position of *Agnostus*. *Lethaia* 23, 409–427.

Wood, R. & Zhuravlev, A. 1994. IGCP-366 – Ecological aspects of Cambrian radiation. *Episodes* 17, 135.

Wray, G.A. Lewinton, J.S. & Shapiro, L.H. 1996. Molecular evidence for deep Precambrian divergences among metazoan phyla. *Science* 274, 568–573.

Xiao, S. & Knoll, A.H. 2000. Phosphatized animal embryos from the Neoproterozoic Doushantuo Formation at Weng'An, Guizhou, South China. *Journal of Paleontology* 74, 767–788.

CHAPTER 8

Bacterial ecosystems and microbial sediments

Bacteria are the most primitive and oldest kinds of organism on earth. They first appeared in the fossil record some 3.5 Ga ago and they have continued to play a major role in earth surface processes.

Bacterial cells are extremely small cells, generally less than 1 μm in diameter. They may be single or colonial; the latter enclosed within a mucilaginous sheath called a capsule. Many bacterial cells bear a whip-like thread (flagellum) and some contain chlorophyll pigments for photosynthesis.

Bacteria are important in the formation of **microbial** sediments, such as bacterial mats and stromatolites, iron and manganese ores, carbonate concretions, sulphide and sulphate minerals. They also yield important information about early evolution of the cell, and the history of photosynthesis and biogeochemical cycles.

Bacterial habitats

Bacteria are the most successful organisms on the Earth. Although bacteria are almost ubiquitous over the surface of the planet, they are conspicuous today in **bacterial mats** or 'biofilms'. Mats and biofilms can be found forming in many restricted marine and non-marine environments today where there is strong physical stress (e.g. high or variable salinity, high UV light, low oxygen). These stresses discourage grazing and burrowing invertebrates and allow the mat to establish.

Careful examination of modern sediment profiles has revealed marked changes in colour and chemistry within the top metre or so (Fig. 8.1). The top layer comprises an **oxidized zone** of blue-green or reddish-brown colour, owing to the abundance of pigmented

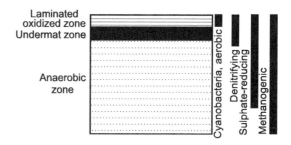

Fig. 8.1 Diagrammatic representation of a cross-section through the upper layers of a sediment profile showing the different domains of major groups of bacteria.

cyanobacteria and other **aerobic bacteria** in the sediment. A cohesive fabric is formed by cyanobacteria and protozoa. Commonly, the organic and sedimentary constituents are more or less diposed in thin **laminae**, each about 1 mm thick (Fig. 8.1). Often the sediment of these laminae is finer grained than in surrounding habitats, being selectively trapped and bound by the mucilaginous sheaths of the cyanobacteria (see below). The laminations may represent a daily growth cycle (e.g. in subtidal conditions) or tidally influenced influxes of sediment (e.g. in intertidal conditions).

A few millimetres beneath the oxidized zone is found the brown-coloured layer of the **undermat zone**. Here are found non-oxygen-producing photosynthethic prokaryotes, such as the purple and denitrifying bacteria. Beneath this lies a black and fetid-smelling **anaerobic zone**. These zones relate to the increasing depletion of oxygen with depth:

nitrate-reducers → sulphate-reducers → methanogens (Fig. 8.1).

The living bacterium

Bacterial cells are small (about 0.25–25 μm) and of spherical, rod or corkscrew shape, and collectively referred to as **cocci**, **bacilli** and **spirilla** (Fig. 8.2a). These cells may be solitary or arranged in filamentous trichomes with or without branching. Most of the bacilli and all of the spirilla possess a whip-like **flagellum** (one or more per cell), but these are very thin and are rarely preserved.

Bacteria may feed either on preformed organic matter (**heterotrophy**) or synthesize organic material from inorganic CO_2 (**autotrophy**). Autotrophic feeding may involve inorganic chemical reactions (**chemoautotrophy**), including minerals in rocks (**chemolithoautrophy**). Others have evolved organic photosynthesis by means of chlorophyll and related pigments in the presence of sunlight, much like green plants (**photoautotrophy**). As a group, the bacteria are relatively unaffected by salinity, and have a temperature tolerance of about 0–125°C. Many dislike a pH outside of the range 6.0–9.0 and will die in bright sunlight. Their habitats range from the deep sea (planktonic and benthic) to terrestrial (including deep subterranean) and aerial.

Bacterial taxonomy

The taxonomy of living bacteria is largely based on staining tests and aspects of biochemistry particularly rRNA sequence data beyond the scope of palaeontology but has fundamental implications for the early evolution of life (see Chapter 6). With the exception of the more highly differentiated cyanobacteria, a morphological classification would prove misleading because similar morphotypes occur in several different orders of bacteria. Similarly bacteria show metabolic versatility and caution has to be applied in using this in classification.

The order Pseudomonadales contains most of the autotrophic bacteria, including the sulphur bacteria which liberate sulphur and sulphates from H_2S. Also included are the stalked bacteria (family Caulobacteraceae) whose fine stalks become encrusted with ferric hydroxide salts from oxidation of dissolved ferrous iron (e.g. Recent *Caulobacter*, Fig. 8.2e). These organisms hence assist in the formation of bog iron ores. Carboniferous iron pyrites nodules have also yielded the somewhat similar genus *Gallionella* (Schopf et al. 1965).

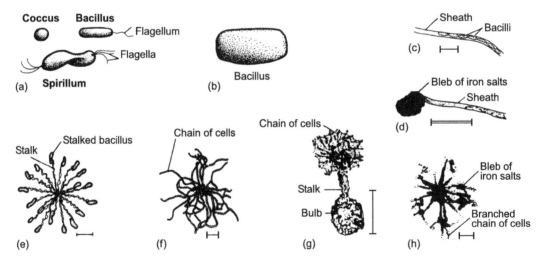

Fig. 8.2 Bacteria. (a) Basic shapes of bacterial cells (schematic). (b) Precambrian *Eobacterium* (length 0.6 μm). (c) Recent sheathed iron bacterium *Sphaerotilus*. (d) Precambrian *Sphaerotilus*-like form. (e) Recent 'stalked' iron bacterium *Caulobacter*. (f) Recent 'budding' bacterium *Metallogenium*. (g) Precambrian *Kakabekia*. (h) Precambrian *Eoastrion*. Single scale bar = 10 μm; double scale bar = 100 μm. ((b) Based on Barghoorn & Schopf 1966; (d) based on Karkhanis 1976: (f) and (h) based on Cloud 1976; (g) based on Barghoorn & Tyler 1965.)

The order Chlamybacteriales, or sheathed bacteria, is also involved in iron ore formation. These have a trichome organization with a sheath that can become encrusted with ferric or manganese oxides much as in the stalked bacteria (e.g. Recent *Sphaerotilus*, Fig. 8.2c). Similar bacteria may have participated in the formation of the world's most extensive iron ores in the early and mid Precambrian banded iron formations (Fig. 8.2d; see Karkhanis 1976) as well as in the formation of iron pyrite (Schopf *et al.* 1965).

The budding bacteria (order Hyphomicrobiales) reproduce by budding; that is, threads grow out either from cells or other threads and themselves produce new cells; these bacteria may also be joined by threads, sometimes in aggregates connected to a common surface by stalks. One such Recent genus, *Metallogenium* (Fig. 8.2f), grows heterotrophically in low-oxygen environments, depositing crusts of manganese oxide around the filaments. Almost identical fossil bacteria, *Eoastrion* and *Kakabekia*, occur in the Gunflint Chert flora in association with banded iron formations (Fig. 8.2g,h; see Cloud 1976).

Examples of possible fossil 'true bacteria' (order Eubacteriales) are reported from the 3.1-Ga-old Fig Tree Chert (*Eobacterium*, Fig. 8.2b). These are tiny bacillus-like structures discovered by electron microscopy of polished chert surfaces (Barghoorn & Schopf 1966), though they may be contaminants. Bacilli may also be involved in the formation of lime mud (Maurin & Noel in Flugel 1977, pp. 136–142) and have been widely reported from various Phanerozoic rocks (Riding 2000).

The Beggiatoales are an order resembling unpigmented filamentous cyanobacteria and thrive in H_2S-rich habitats. Hence, for example, the discovery of *Beggiatoa*-like remains in Carboniferous iron pyrites (Schopf *et al.* 1965). Flexibacteria are an even more cyanobacteria-like group with photosynthetic pigments and they dwell alongside cyanophytes in hot springs to be preserved, eventually, in sinters and stromatolites (Walter 1972). Apart from the biological distinction of their not releasing free oxygen, it would be difficult to differentiate flexibacteria and cyanobacteria except on the tenuous basis of cell diameter – the former rarely exceeding 2 µm in diameter and the latter usually exceeding this.

Cyanobacteria

The cyanobacteria are erroneously called blue-green algae on account of the colour imparted by the photosynthetic pigment **phycocyanin**, but they bear no relationship to algae. Living cyanobacteria may also be olive green or red in colour. They consist of small cells, mostly between 1 and 25 µm in diameter, which may be spherical (coccoid), ovoid, discoidal, cylindrical or pear-shaped (pyriform) in outline. Like other prokaryotes, the cell has a very simple structure, with a nuclear membrane for the chromosomes and without mitochondria. The phycocyanin or chlorophyll pigments are distibuted in lamellae around the edges of the cell where they take part in photosynthesis.

Cyanobacterial cells may be single (unicellular) or arranged in colonies protected by a **mucilaginous sheath** of cellulose fibrils. The arrangement of cells in a colony may be regular to irregular, for example flat, cuboid, spherical, uniseriate filamentous or branched filamentous (Figs 8.3, 8.4). The cells of a filamentous colony comprise the **trichome**.

Cyanobacteria construct organic materials from inorganic materials by **photosynthesis** by means of photosynthic pigments in the presence of sunlight, releasing free oxygen in the process, as in higher plants:

$$CO_2 + H_2O \xrightarrow{\text{sunlight}} CH_2O + O_2$$

The need for light causes them to grow towards the sun. In filamentous forms this may be achieved by gliding upward through the substrate, leaving behind the old sheath in the process. As these sheaths are of resistant cellulose, whilst the cell walls are mostly degradable amino acids and sugars, it is the sheath which has the better chance of preservation in the fossil record.

Cyanobacteria life history

Cyanobacteria are an extremely ancient group which have never developed the controlled division of cells by mitosis or meiosis. Sexual reproduction is therefore unknown, and multiplication is entirely vegetative (asexual), usually brought about by fragmentation, binary fission, or the formation of endospores,

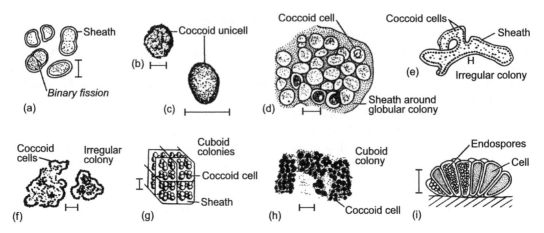

Fig. 8.3 Order Chroococcales. (a) Recent *Synechocystis*. (b) Precambrian *Archaeosphaeroides*. (c) Precambrian *Huroniospora*. (d) Precambrian *Myxococcoides*. (e) Recent *Anacystis*. (f) Fossil *Romaleis*. (g) Recent *Eucapsis* colony. (h) Precambrian *Eucapsis*-like colony. (i) Recent *Entophysalis*. Scale bar = 10 μm. ((b) Based on Schopf & Barghoorn 1967; (c) based on Barghoorn & Tyler 1965; (d) and (h) based on Cloud 1976; (g) based on Fogg *et al.* 1973; (i) from Chapman & Chapman 1973.)

akinetes or hormogonia. Cell division involves the splitting of a cell into two daughter cells by inward growth of the wall (i.e. **binary fission**, Fig. 8.3a). The cell contents are randomly distributed between new cells, unlike the orderly mitotic divisions of eukaryotes. **Fragmentation** simply involves the breaking up of a colony into smaller ones. **Endospores** form by the internal subdivision of cells into two or more spores that are subsequently released to grow into new colonies (Fig. 8.3i). **Akinetes** are also spore cells, but these develop singly from vegetative cells by enlargement and the formation of a thick, often sculptured wall (Fig. 8.4c). After conditions of desiccation or chilling, new filaments germinate from the akinete. **Hormogonia** are characteristic of filamentous forms. These are short detached pieces of the trichome which glide out of their sheath and develop separately (Fig. 8.4a).

Cyanobacterial ecology

Cyanobacteria are very self-sufficient. They can tolerate extremely low oxygen concentrations and some can live anaerobically. They are, with certain other bacteria, the only organisms that can fix their own nitrogen, either with the aid of **heterocyst** cells in aerobic conditions, or without in anaerobic conditions. Cyanobacteria also have a wide resistance to high and low temperatures, ranging from polar climates to hot thermal springs. They are also very resistant to ultraviolet light. Their lack of cell vacuoles gives them great resistance to desiccation and plasmolysis, hence their presence in arid deserts, glacial regions, hypersaline lagoons and freshwater lakes.

Important limitations appear to be pH and light. They prefer neutral and alkaline environments and never more acid than pH 4.0 The blue-green photosynthetic pigment phycocyanin is sensitive to blue light and can work under very low light concentrations, so that cyanobacteria can be found living some 300 mm below the soil surface on land and at depths of 1000 m or more in the oceans.

Where nutrient levels are high enough, certain coccoid and filamentous types thrive as very small **picoplankton** in the water column. Buoyancy is achieved either by the development of pseudovacuoles or by adherence to gas bubbles. Some filamentous forms float in bundles of up to 25 trichomes, forming mats at the surface of the ocean that can extend for many kilometres. In recent years, spring and summer blooms of planktonic cyanobacteria in polluted rivers and lakes have caused poisoning of fish stocks and humans and brought about temporary bottom-water anoxia.

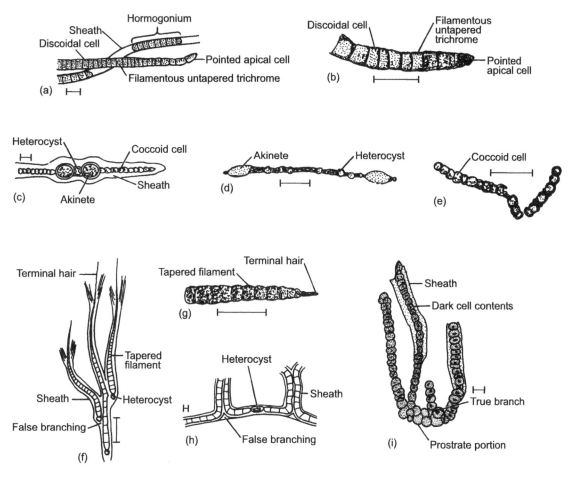

Fig. 8.4 (a)–(h) Order Nostocales. (a) Recent *Oscillatoria*. (b) Precambrian *Oscillatoria*-like filament. (c) Recent *Wollea*. (d) Precambrian *Gunflintia*. (e) Precambrian *Nostoc*-like filament. (f) Recent *Rivularia*. (g) Precambrian *Rivularia*-like filament. (h) Recent *Scytonema*. (i) Order Stigonematales: Devonian, *Kidstonella*. Scale bar = 10 μm. ((b), (e) and (g) based on Schopf 1972; (d) based on Cloud 1976; (c), (f) and (h) redrawn from Fogg *et al.* 1973; (i) based on Croft & George 1959.)

Anaerobic bacteria

Below the level of freely available oxygen in the sediment or water column are found anaerobic bacteria that use other forms of oxygen donor (Fig. 8.1).

Nitrogen-processing bacteria

Nitrogen is an essential component of amino acids and proteins but, because the gas is very inert, its incorporation in the biosphere is largely dependent on **nitrogen fixation** by certain bacteria, including cyanobacteria. These are able to convert gaseous nitrogen into reactive ammonia by means of the enzyme nitrogenase:

$$N_2 + 3H_2 \xrightarrow{\text{nitrogenase}} 2NH_3$$

This ammonia can then be incorporated into proteins. The **nitrifying bacteria** are able to convert ammonium ions into nitrite while others can convert the latter into nitrate ions:

$$NH^{4+} \rightarrow HNO^{2-} \rightarrow HNO^{3-}$$
(nitrite) (nitrate)

Nitrate is extremely important as a biolimiting nutrient for photoautotrophic primary production. Formation of gaseous nitrogen from nitrate is achieved by the anaerobic **denitrifying bacteria** which use the nitrate ion as a hydrogen receptor for the oxidation of sulphur:

$$6KNO_3 + 5S + 2CaCO_3 \rightarrow$$
$$3K_2SO_4 + 2CaSO_3 + 2CO_2 + 3N_2$$

Other bacteria can produce nitrites and later ammonia using the nitrate ion for reduction while some other substance is oxidized:

$$H.COOH + HNO_3 \rightarrow CO_2 + H_2O + HNO_3$$

$$4H + HNO_2 \rightarrow NH_2OH + H_2O$$

$$2H + NH_2OH \rightarrow NH_3 + H_2O$$

This form of **anaerobic respiration** has a lower energy yield than aerobic respiration. Denitrification of this kind is widely found in the **oxygen minimum zone** of the oceans. The removal of nitrate from surface waters can act as a check on photosynthetic algal blooms and thereby prevent runaway anoxia.

Sulphur-processing bacteria

Beneath the denitrifying bacteria may be found the **sulphate-reducing bacteria** (e.g. *Desulfovibrio*) which use sea-water sulphate to oxidize pre-formed organic matter, with a lower net energy yield than found in the zone above:

$$2H_2O + SO_4^{2-} \rightarrow 2HCO_3^- + HS^- + H^+$$

The release of sulphide during this form of anaerobic respiration is highly toxic to most anaerobes, and can lead to the dissolution of carbonate (including fossils) in the sediment. Commonly, iron monosulphides and then pyrite are precipitated:

$$Fe + S \rightarrow FeS \rightarrow FeS + S \rightarrow FeS_2 \text{ (pyrite)}$$

This process can produce pyritic infillings and/or replacements of fossils in the sediment. Because of disequilibrium fractionation processes during sulphate-reduction, this sulphide is depleted in the stable isotope ^{34}S by 4–46‰ compared with standard sea water.

A second group of sulphur-processing bacteria can convert this sulphur and sulphide back into sulphate, using oxygen as the electron acceptor. These **sulphide-** and **sulphur-oxidizing bacteria** may occur as biofilms above the sulphate-reducing zone. They also bloom copiously around hot, sulphide-rich submarine vents known as 'black smokers', where they play an important role in the food chain of the vent community. Their metabolism results in further ^{34}S depletion of the sulphur isotopes (to −60‰), which can be measured in the fossil record.

Methane-processing bacteria

Beyond a certain depth in the sediment, all the pore-water sulphate ions are used up (Fig. 8.1). If there is still a supply of useable organic matter, the anaerobic bacteria here must feed by **fermentation**, much like yeast in the fermentation of beer or wine, producing methane and bicarbonate gas as waste products:

$$H_2O + 2CH_2O \rightarrow CH_4 + HCO_3^- + H^+$$

The energy yield here is even lower than for sulphate reduction. These **methanogenic** bacteria continue the process of organic decomposition so that very little organic matter may remain in the sediment. The hydrogen ions thus released can reduce available Fe^{3+} to Fe^{2+} ions, and the latter may combine with dissolved bicarbonate ions to precipitate out as $FeCO_3$ (siderite) concretions. The methane produced in this way is also important as a 'greenhouse' gas.

Some geologically significant bacteria

Bacteria have seldom been reported as microfossils, perhaps because of their small size and the difficulty of distinguishing them from fossil cyanophytes or fungi, or even from recent contamination, inorganic structures and artifacts formed during preparation of material. They have, however, been reported from a wide range of lithologies including limestones, cherts, phosphorites, iron and manganese ores (including deep sea manganese nodules, pyrite nodules and banded iron ores), tonsteins, bauxites, oil shales, coal seams, plant tissues, coprolites and fossil animal remains. In cherts and phosphorites the cell wall may be preserved but usually the wall, sheath or the whole structure have been replaced by minerals (Riding 2000 is recommended for a review of microbial carbonates). Fossil bacteria are usually identified on shape, their associations and substrate-bound occurrence. Bacteria commonly occur in multispecies colonies, known as **consortia**.

Stromatolites

This is the geological term used for a laminated benthic microbial deposit (Riding 1999), for which an origin from a bacterial mat is often inferred. In fossil stromatolites, the organic and trapped sediment layers are usually seen as alternating or intergrading pale and dark **lamellae** (Fig. 8.5). Preservation of filaments is rare but is known in silicified and fine-grained carbonate stromatolites. More often, only the sheaths or the upward gliding trails are preserved. In most cases, however, stromatolites preserve no relics of an organic origin.

The gross form of a stromatolite is controlled by a combination of factors including mat viscosity (e.g. from mucilage, which may reflect the biological components of the mat) and surface roughness of the mat (e.g. from grain size, which may reflect current energy and sediment supply). Precambrian palaeontologists often employ a binomial system of nomenclature in the description of stromatolites. **Groups** (= genera) are based on a general shape (planar, domed, columnar, oncolitic), mode of branching (straight, digitate), morphology of the 'wall' (i.e. marginal zone) and

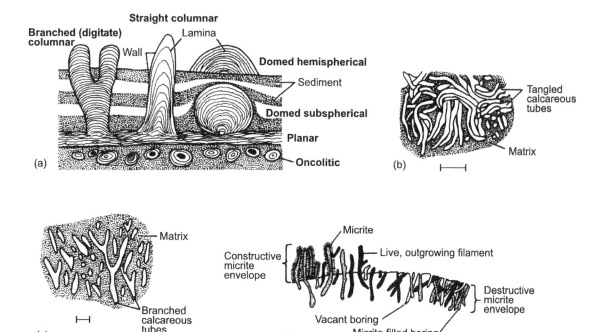

Fig. 8.5 Cyanophyte sedimentary structures. (a) Stromatolite types in vertical section, ×1. (b) *Girvanella* tubes in skeletal oncolite. (c) *Ortonella* tubes in skeletal oncolite. (d) Section through endolithic cyanophyte borings and skeletal envelopes (diagrammatic). Scale bars = 100 μm. ((d) Based on Kobluck & Risk 1977.)

geometry of the laminae (Fig. 8.5a). **Forms** (= species) are distinguished by microscopic textures and lamina geometry.

Skeletal stromatolites differ from the **non-skeletal** stromatolites in that the form of the cells or sheaths has been moulded in $CaCO_3$, giving rise to micritic tubes within a micritic or sparry calcite internal filling. The tubes appear to represent calcification of the sheath, such as can occur during life in certain autotrophs because of CO_2 uptake during photosynthesis. Post-mortem calcification is also possible however, and this possibility is suggested by stratigraphical intervals during which such calcification seems to have been more widespread (Kazmierczak 1976). Such skeletal stromatolites are known from both freshwater and marine waters. In *Girvanella* (Fig. 8.5b, L. Camb.-Rec.) the tubes are tangled and unbranched and occur in both oncolites and thrombolites. *Ortonella* (Fig. 8.5c, L. Carb.-Perm.) is an oncolitic form with branched tubes.

Thrombolites differ from stromatolites in lacking the internal laminations, having instead a mottled or clotted microtexture. They were probably built by coccoid cyanobacteria (e.g. *Renalcis*, Fig. 8.3f) or by filamentous forms with wispy, tufted, branched or tangled growth rather than vertical growth. Thrombolites are typically found in sublittoral, often calcareous facies in association with reef-dwelling invertebrates.

Travertine develops in $CaCO_3$-supersaturated waters in which coccoid and filamentous cyanobacteria may become encrusted by physicochemical precipitation of $CaCO_3$, forming hollow tubes. In this case, however, the crystals greatly exceed the diameter of the original organic sheath. The moulds left by such fossilization are of little taxonomic value. During the Phanerozoic, travertine stromatolites have largely been confined to fresh or extremely hypersaline waters. Many Early Precambrian (Archean) stromatolites are actually marine travertine. This suggests that there have been long-term changes in the chemistry of the oceans and atmosphere.

Endolithic cyanobacteria

A variety of marine cyanobacteria bore into the surface of hard calcareous substrates such as shells and limestone by chemical dissolution (Fig. 8.5d). This **endolithic** boring is for protection rather than for food. Under conditions of $CaCO_3$ supersaturation the vacated borings are filled with micritic carbonate and the substrate thereby acquires an outer **micrite envelope**. Eventually, boring may lead to the destruction of the substrate and the formation of lime mud. If the filaments extend outwards from their borings and become calcified after death, however, a **skeletal envelope** may form (Fig. 8.5d). This constructive process requires relatively quiet conditions. Cyanobacterial microborings are not easily distinguished from those of algae or fungi, but they are generally narrower than the former and broader than the latter (i.e. about 4–25 μm wide). The depth of water in which such borings may be found varies with water clarity and latitude, but is mostly shallower than 75 m.

Precambrian organic-walled microfossils of uncertain affinity, **cryptarchs**, could be the resting cysts of cyanobacteria. They have been obtained from Precambrian rocks up to 2.0 Ga old. Filaments and coccoid cells have been widely reported from restricted, often hypersaline, bacterial mat facies of the Precambrian where they form an important component of chert microbiotas (see Chapter 6).

Unfortunately the fossil record is as yet too incomplete to comment on the history or applications of the group. It seems probable that bacteria are older than cyanobacteria, their ability to live in anaerobic conditions being a legacy from early and mid-Precambrian times. The parasitic and saprozoic bacteria may, in part, be Phanerozoic developments, and their evolution could have had important consequences for ecosystem evolution in general.

Hints for collection and study

Living bacteria are easily cultured. Take a sample of water from decaying pond vegetation and place a drop on a glass slide with water and a cover slip. Allow the slide to dry out in a warm, dark place. When viewed with transmitted light at over 400× magnification the slide will often be seen to contain clusters of minute bacilli. Fossil bacteria may well be encountered in thin sections of sedimentary ironstones, phosphatized faecal pellets, bauxites and evaporites, but the observer must be wary of the likelihood of more recent 'contamination'.

REFERENCES

Barghoorn, E.S. & Schopf, J.W. 1966. Micro-organisms three billion years old from the Precambrian of South Africa. *Science* 152, 758–763.

Barghoorn, E.S. & Tyler, S.A. 1965. Microorganisms from the Gunflint chert. *Science* 147, 563–577.

Chapman, V.J. & Chapman, D.J. 1973. *The Algae*. Macmillan, London.

Cloud, P. 1976. Beginnings of biospheric evolution and their biochemical consequences. *Paleobiology* 2, 351–387.

Croft, W.N. & George, E.A. 1959. Blue-green algae from the Middle Devonian of Rhynie, Aberdeenshire. *Bulletin. British Museum Natural History (Geology)* 3, 341–353.

Flugel, E. 1977. *Fossil Algae. Recent results and developments*. Springer-Verlag, Berlin.

Fogg, G.E., Stewart, W.D.P., Fay, P. & Walsby, A.E. 1973. *The Blue-green Algae*. Academic Press, London.

Karkhanis, S.N. 1976. Fossil iron bacteria may be preserved in Precambrian ferroan carbonate. *Nature* 261, 406–407.

Kazmierczak, J. 1976. Devonian and modern relatives of the Precambrian *Eosphaera*: possible significance for the early eukaryotes. *Lethaia* 9, 39–50.

Kobluk, D.R. & Risk, M.J. 1977. Microtization and carbonate-grain binding by endolithic algae. *Bulletin of the American Association of Petroleum Geology* 61, 1069–1083.

Kutznetsov, S.I., Ivanov, M.V. & Lyalikova, N.N. 1963. *Introduction to Geological Microbiology*. McGraw Hill, New York.

Riding, R. 1999. The term stromatolite: towards an essential definition. *Lethaia* 32, 321–330.

Riding, R. 2000. Microbial carbonates: the geological record of calcified bacterial-algal mats and biofilms. *Sedimentology* 47, 179–214.

Schopf, J.W. 1972. Evolutionary significance of the Bitter Springs (Late Precambrian) microflora. *24th International Geological Congress, Montreal* 1, 68–77.

Schopf, J.W. & Barghoorn, E.S. 1967. Alga-like fossils from the Early Precambrian of South Africa. *Science* 156, 508–512.

Schopf, J.M., Ehlers, E.G., Stiles, D.V. & Birle, J.D. 1965. Fossil iron bacteria preserved in pyrite. *Proceedings. American Philosophical Society* 109, 288–308.

Walter, M.W. 1972. Stromatolites and the biostratigraphy of the Australian Precambrian and Cambrian. *Special Papers in Palaeontology*, no. II.

PART 3
Organic-walled microfossils

PART 3

Organic-walled microfossils

CHAPTER 9

Acritarchs and prasinophytes

Acritarchs are hollow, organic-walled, eukaryotic unicells of unknown biological affinity. Many are probably the resting stage (cyst) in the life cycle of marine phytoplanktonic algae. Since their discovery about 150 years ago many have been assigned to the green algae. In particular some acritarchs bear a close similarity to the non-motile stage (**phycoma**) in the life cycle of modern **prasinophytes**, a group of well-known primitive green algae included here with the acritarchs for convenience.

Ranging from mid-Precambrian to Recent times, acritarchs reached their acme in the Palaeozoic. Like dinoflagellates, they are useful for biostratigraphical correlation and palaeoenvironmental analysis. Perhaps more importantly acritarchs probably represent the remains of the phytoplankton, the primary producers of the Proterozoic and Palaeozoic.

Morphology

The vesicle

The acritarch wall consists of a complex of polymers known as **sporopollenin**. Most acritarchs are 20–150 µm across consisting of a **vesicle** enclosing a **central cavity** from which may project spine-like **processes** and **crests**. The shape of the vesicle, presence or absence of processes and of ornamentation are important criteria for defining species and genera. Compression, pyrite growth and other diagenetic processes and extraction techniques can considerably modify the original shape.

Many acritarchs have a wall composed of a single layer, whilst double and complex wall ultrastructures are not uncommon. Wall thickness can also vary considerably from <0.5 µm in *Leiosphaeridium* (Fig. 9.1a) through 2–3 µm in *Baltisphaeridium* (Fig. 9.1b) and up to 7 mm in *Tasmanites* (Fig. 9.3d, a prasinophyte). Wall ultrastructure is very poorly known. In the prasinophytes and *Baltisphaeridium* the vesicle ultrastructure comprises a single layer penetrated by narrow canals, usually only discernible by SEM and TEM. A thin two-layer wall structure separated by a zone comprising 0.5–2-µm-diameter pores has been described in *Acanthodiacrodium* (Fig. 9.1c) and is similar to that found in some dinoflagellates. A double wall occurs in *Visbysphaera* (Fig. 9.1d), which develops processes from the outer layer.

The exterior surface of the vesicle may be smooth, granulate, or may bear a variety of spinose or reticulate ornaments, indentations or micropores. These processes may be hollow and connected with the central cavity open (e.g. *Diexallophasis*, Figs 9.1e, 9.3a) or closed at the base, or solid. The tips of the processes can be simple, bifurcated, branched or connected by a thin membrane, the **trabeculum** (e.g. *Tunisphaeridium*, Fig. 9.1f). Processes on an individual vesicle are termed **homomorphic** if all are similar or **heteromorphic** if more than one type is developed. Processes may be variously branched, smooth or bear a secondary ornament of granules.

Excystment structures

If some acritarchs were resting cysts, comparable to those produced by the dinoflagellates, then the contents must have escaped through an opening, the excystment structure. Excystment structures are not found in all acritarchs but sufficient are known to

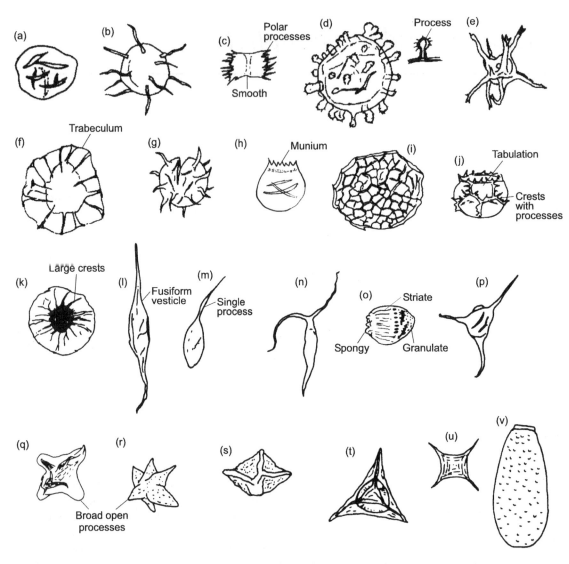

Fig. 9.1 Acritarchs. (a) *Leiosphaeridium*, ×400. (b) *Baltisphaeridium*, ×250. (c) *Acanthodiacrodium*, ×400. (d) *Visbysphaera*, ×700. (e) *Diexallophasis*, ×250. (f) *Tunisphaeridium*, ×345. (g) *Micrhystridium*, ×1200. (h) *Ammonidium*, ×390. (i) *Cymatiosphaera*, ×400. (j) *Cymatiogalea*, ×600. (k) *Pterospermella*, ×330. (l) *Leiofusa*, ×400. (m) *Deunffia*, ×400. (n) *Domasia*, ×400. (o) *Ooidium*, ×450. (p) *Veryhachium*, ×300. (q) *Pulvinosphaeridium*, ×300. (r) *Estiastra*, ×300. (s) *Octoedryxium*, ×300. (t) *Polyodryxium*, ×350. (u) *Neoveryhachium*, ×600. (v) *Melanocyrillium*, ×300. ((a), (i), (j), (l) Redrawn from Mendelson *in* Lipps 1993; (g) redrawn from Tappan 1990; (c) redrawn from Evitt in Tschundy & Scott 1969; (u) redrawn from Molyneux *et al.* in Jansonius & McGregor 1996.)

suggest that the style of opening is taxon specific. Excystment openings form six major types. A simple **lateral rupture** (or cryptosuture) is the most common and comprises a simple, more or less straight suture that does not divide the vesicle completely (e.g. *Micrhystridium*, Fig. 9.1g). The lateral rupture is similar but has an ornamented border or thickening (e.g. *Diexallophasis*, Figs 9.1e, 9.3a). A **median split** divides the vesicle into two roughly equal halves (e.g. *Ammonidium*, Fig. 9.1h). The **trochospiral suture** traces a lateral split found in the spindle-shaped acritarchs (e.g. *Leiofusa*, Figs 9.1l, 9.3c). The **epityche** opening forms a hemispherical flap of wall and is characteristic of *Veryhachium* (Fig. 9.1p). A **pylome** is a circular opening situated above the equator (e.g. *Cymatiogalea*, Fig. 9.1j). A **circinate suture** is only found in *Circinatisphaera* and defines a circular suture coiling in a levorotary direction, often with an attached lid-like **operculum**. **Munium** and **munitium** are apical apertures with denticulate margins commonly found in vesicles fossilized prior to excycstment.

Classification

Group ACRITARCHA

Downie (1973, 1974), Fensome *et al.* (1990) and Dorning (in Benton 1993, pp. 33–34) have reviewed the classification of the acritarchs. Informal groupings have been established on overall morphology, wall structure and type of excystment opening. None of the published schemes reflects biological affinity or evolution and some workers prefer to list taxa alphabetically. Biometrical studies (e.g. Servais *et al.* 1996) and the chemistry of the vesicle (e.g. Colbath & Grenfell 1995) may offer the potential for a more natural classification to be developed. Alete spores can be distinguished from acritarchs with difficulty based on their thicker walls and colour variation within an individual specimen. Most acritarchs fall into three morphological groups, each of which includes one or more acritarch subgroups.

1 Acritarchs lacking processes or crests

Subgroup Sphaeromorphitae (Precamb.-Rec.; Fig. 9.2) This includes acritarchs with a spherical or ellipsoidal vesicle which may be variously ornamented. The often thin, simple, imperforate wall may develop an irregular or cyclopyle opening. Many of the large Neoprotoerozoic acritarchs, such as *Chuaria* (U. Precamb.), may belong here, although this genus is exceptionally large (<5 mm diameter) and may

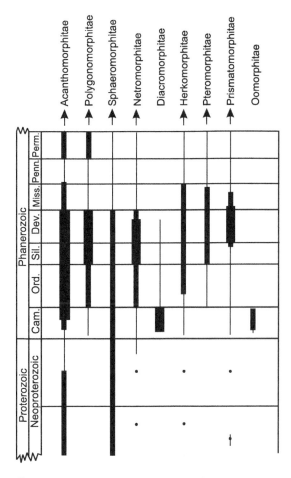

Fig. 9.2 Generalized ranges of major acritarch groups. Arrows indicate groups with a post-Permian record, width of line approximates to number of species. (After Mendelson in Lipps 1993.)

represent the carbonaceous impression of a nostocales cyanophyte alga (Martin 1993). *Leiosphaeridia* (U. Precamb., Palaeozoic, Fig. 9.1a) may have had green algal affinities.

2 Acritarchs with crests but lacking processes

Subgroup Herkomorphitae (Camb.-Rec.; Fig. 9.2) These have spherical to subpolygonal acritarchs in which the vesicle is divided into polygonal fields by crests, for example *Cymatiosphaera* (Camb.-Rec., Fig. 9.1i). The

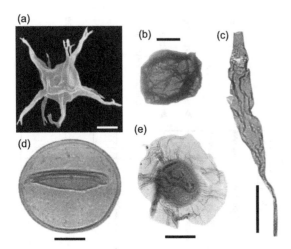

Fig. 9.3 Photomicrographs of selected acritarchs and a prasinophyte. Scale bars = 20 μm. (a) SEM image of *Diexallophasis* sp. (b) *Leiosphaeridium* sp., from the Upper Proterozoic. (c) *Leiofusa* sp., from the Whitcliffe Formation, Ludlow. (d) *Tasmanites pradus*, a prasinophyte. (e) *Pterospermella* sp., Silurian. ((a) Reproduced from Lipps 1993, figure 6.12, D2); (d) reproduced from Traverse 1988, figure 6.9l; both with permission.)

wall is perforate and without known excystment openings. The vesicle, originally spherical or polygonal, is divided into fields by crests. Members of this subgroup are now considered along with the Prasinophyta as green algae. In *Cymatiogalea* (M. Camb.-Tremadoc, Fig. 9.1j) the vesicle is divided into polygonal fields by crests, somewhat resembling a proximate dinoflagellate cyst, but it has a large cyclopyle opening. Some species of *Cymatiogalea* bear processes and these may belong in the Acanthomorphitae.

Subgroup Pteromorphitae (Ord.-Rec.; Fig. 9.2) These are similar to the herkomorph acritarchs in overall shape, but are distinguished by possessing an equatorial flange (**ala**), for example *Pterospermella* (Figs 9.1k, 9.3e).

3 Acritarchs with processes, with or without crests

Subgroup Acanthomorphitae (Precamb.-Rec.; Fig. 9.2) These spherical acritarchs lack an inner body and crests; processes are simple or branching, for example *Baltisphaeridium* (L. Camb.-L. Sil., Fig. 9.1b), has a spherical central body, over 20 μm in diameter, with simple hollow or solid processes with closed tips. *Micrhystridium* (L. Camb.-Rec., Fig. 9.1g) has a spherical central body <20 μm in diameter with simple processes. *Visbysphaera* (L. Sil.-L. Dev., Fig. 9.1d) is spherical, characterized by a double-layered wall and a lateral rupture and bears processes that are produced from the outer wall.

Subgroup Diacromorphitae (Camb.-Dev.; Fig. 9.2) This subgroup comprises acritarchs with a spherical or elliptical vesicle in which the equatorial zone is smooth and the polar areas are ornamented. The simple wall tends to split up into angular plates when damaged. The openings are of varying kinds but the vesicles are typically elongate with the sculpture concentrated at one or both poles. *Acanthodiacrodium* (M. Camb.-M. Ord., Fig. 9.1c) has small processes at both poles and an equatorial constriction.

Subgroup Netromorphitae (Precamb.-?Triassic; Fig. 9.2) Long elongate, fusiform acritarchs in which one or both poles may be extended as processes, with a median- or lateral-split or a C-shaped epityche opening, for example *Leiofusa* (U. Camb.-U. Carb., Figs 9.1l, 9.3c). *Deunffia* (Sil., Fig. 9.1m) bears a single process whilst *Domasia* has three processes (Sil., Fig. 9.1n).

Subgroup Oomorphitae (Camb.) This subgroup has an egg-shaped, polarized, vesicle, generally smooth at one end and ornamented at the other. *Ooidium* (Camb., Fig. 9.1o) is ovate with a granular sculpture at one pole and a spongy sculpture at the other, with fine striae between.

Subgroup Polygonomorphitae (Camb.-Rec.; Fig. 9.2) Acritrachs in this subgroup bear a polygonal vesicle with simple processes. *Veryhachium* (U. Camb.-Mioc., Fig. 9.1p) has a polygonal central body with from three to eight hollow pointed spines with closed tips. *Pulvinosphaeridium* (Camb.-Ord., Fig. 9.1q) and *Estiastra* (M. Ord.-U. Sil., Fig. 9.1r) may be included in this group. The vesicle is star shaped with wide processes.

Subgroup Prismatomorphitae (Camb.-Rec.; Fig. 9.2) These acritrachs have a prismatic or polygonal vesicle, the edges of which may be extended into a flange, for example *Octoedryxium* (Fig. 9.1s) and *Polyodryxium* (Fig. 9.1t).

Acritarch affinities and biology

Acritarchs are considered to be the resting cysts of phytoplanktonic eukaryotic algae. The presence of dinosterane and 4α-methyl-24-ethylcholestane, two biomarkers characteristic of dinoflagellates, in samples of Neoproterozoic (including the Bitter Springs Chert) and Cambrian age indicate acritarchs are likely ancestors of the dinoflagellates (Moldowan & Talyzina 1999). This conclusion is supported by RNA sequence data that indicate the dinoflagellates diverged before the Foraminifera and the Radiolaria which both have a known Cambrian fossil record.

By comparison with the dinoflagellates, the cyst was formed to protect the cell during binary fission or to survive adverse environmental conditions. Monospecific clusters of acritarchs have been found, especially in Precambrian and Cambrian rocks. Although it has been suggested that these are the spores of multicellular algae this need not follow. Dinoflagellates are also known to aggregate in clusters. Chemically the wall most closely resembles the sporopollenin wall of vascular plant spores, algal spores and dinoflagellate cysts.

Most herkomorphs, pteromorphs and prismatomorphs are now considered to be prasinophytes or other green algae. The sphaeromorphs compare with the spores of multicellular algae. The remainder may have affinities with the naked dinoflagellates (Gymnodiniales) which are known to develop non-tabulate cysts. Acritarchs differ from most peridinalean dinoflagellate cysts in the absence of both reflected tabulation and of pre-formed excystment openings of definite form. However, at least one living peridinialean dinoflagellate is known to produce an acritarch-like cyst (Dale 1976).

Acritarch ecology

Poor understanding of taxonomy, biological affinity and rarity in modern environments hinder palaeoecological interpretations. Acritarchs have mostly been found in marine strata, especially in shales and mudstones, but also occur in sandstones and limestones. Non-marine examples are first reported from Recent strata.

Lagoonal facies are characterized by low diversity and monospecific assemblages of sphaeromorph and netromorph acritarchs and prasinophytes. The boundary between nutrient-rich coastal waters and nutrient-poor oceanic water is reflected in inshore-offshore trends in plankton communities, abundance and diversity. Dorning (1981, 1997) documented acritarch distribution across the Ludlow (Silurian) shelf in Wales and the Welsh Borderland. Based on the relative abundances of 17 genera he concluded that acritarch assemblages in nearshore and deep offshore environments had low diversity, dominated by sphaeromorphs. He found a much higher diversity in mid-shelf environments. Inshore facies contained abundant *Micrhystridium*, whereas quieter offshore facies are reflected in assemblages with longer, more delicate and elaborate processes and crests. A similar pattern was described for Middle Ordovician acritarchs (Wright & Meyers 1991). This simple palaeoecological model has been widely accepted (e.g. Hill & Molyneux 1988; Wicander *et al.* 1999) though probably belies much more complex physico-chemical and relative sea-level changes (e.g. Jacobsen 1979; Colbath 1990). Vecoli (2000) reported some Early Ordovician, high latitude acritarchs, including *Acanthodiachrodium*, may have been facies controlled implying a benthonic mode of life. A number of other acritarchs are known to be facies controlled. *Neoveryhachium* (Fig. 9.1u) occurs in turbid environments whereas *Pulvinosphaeridium* (Fig. 9.1q) and *Estiastra* (Fig. 9.1r) are most common in warm-water carbonate facies. In the Late Devonian reefs of western Canada sphaeromorphs predominate in near-reef facies further away from the reef thin-spined acanthomorphs and finally thick-spined acanthomorphs and polygonomorphs came to dominate assemblages (Staplin 1961). Salinity control on acritarchs has not yet been widely demonstrated, though Servais *et al.* (1996) did suggest that process length may vary with salinity, a feature found in some dinoflagellate cysts. Acritarchs along with dinoflagellates track transgressive and regressive sequences in the Jurassic of Britain and France. In the Mesozoic, acanthomorphs appear to have favoured inshore environments whilst polygonomorphs and some netromorphs favoured open marine environments.

Though temperature range had a primary control on acritarch distribution, evidence for provinciality is patchy in the Palaeozoic. It appears that acritarchs did not show marked provinciality until the Cambrian–Ordovician boundary. By the Tremadoc-Arenig two well-developed provinces were established, in warm tropical and cool temperate-boreal latitudes. Ordovician acritarchs appear following a pattern similar to that of modern dinoflagellate cysts, primarily controlled by latitude but also following continental margins and modified by surface ocean currents (Li & Servais 2002; Servais et al. 2003). Later in the Palaeozoic, geographically restricted assemblages have been reported from the Silurian (Le Hérissé & Gourvennec 1995). Cramer & Diez (1974) suggested Silurian acritarchs were provincial assemblages paralleling palaeolatitude. The *Deunffia-Domasia* Assemblage is thought to be characteristic of low latitudes during the mid-Silurian and has been shown to be associated with outer-shelf environments independent of temperature control. However, it is now known the group had a wide overall tolerance, being found from periglacial to tropical palaeoenvironments. Late in the Devonian provinciality broke down with the appearance of many cosmopolitan forms (Le Hérissé et al. 1997).

General history of acritarchs

The oldest-known unequivocal acritarchs occur in the mid-Proterozoic Belt Supergroup in Montana, USA (~1400 Ma). These are smooth, spherical sphaeromorphs some tens of micrometres across; similar forms may range back into the Early Proterozoic (~1900 Ma, Mendelson & Schopf 1992). The majority of Proterozoic acritarchs are sphaeromorphs which first became abundant in marine sediments about 1000 Ma. The first radiation in the Late Precambrian (900–600 Ma) is characterized by the appearance of large sphaeromorphs (up to 400 µm), acanthomorphs and polygonomorphs. This radiation predates the Ediacara Fauna but was short lived and many of these forms became extinct during the Vendian glaciation. The earliest prismatomorphs (e.g. *Octoedryxium*, Fig. 9.1s) and curious vase-shaped forms called **melanocyrillids** (Fig. 9.1v) appeared at this time.

During the second major radiation in the Early Cambrian many small, spinose acanthomorphs (e.g. *Micrhystridium*, *Baltisphaeridium*), herkomorphs (e.g. *Cymatiosphaera*), netromorphs and diacromorphs appeared. These last, plus *Cymatiogalea* and similar forms, were at their acme in Late Cambrian and Early Ordovician times. The radiation of ornamented and densely spinose acritarchs in the Early Cambrian coincides with the major radiation of invertebrate suspension feeders. It is therefore possible that acritarch evolution has played an important role in the Cambrian explosion (see Brasier 1979). The acanthomorphs flourished throughout the Ordovician but declined in the Early Silurian, a period dominated by *Micrhystridium*, *Veryhachium* and similar netromorph genera. Rich Early Devonian assemblages containing these plus diverse prismatomorphs were followed by a general decline in acritarch diversity and abundance (Fig. 9.2). Acritarchs then became scarce throughout the Carboniferous, Permian and Triassic.

Late Triassic and Jurassic acanthomorphs, polygonomorphs and herkomorphs have been documented. Certain genera made a limited come back in the Jurassic, Cretaceous and Tertiary, for example the prasinophyte genus *Tasmanites* and the acritarchs *Cymatiosphaera* and *Micrhystridium* but in the Mesozoic and Cenozoic palynologists have largely concentrated on dinoflagellate cysts.

Applications of acritarchs

Acritarchs have been used largely to correlate upper Precambrian and Palaeozoic rocks. Papers by Martin (1993) and Vidal & Knoll (1993) illustrate their potential in Precambrian rocks and by Molyneux et al. (1996) in the Palaeozoic, whilst Wall (1965) examined their value in some Mesozoic strata.

Geographically distinct acritarch provinces in the Ordovician, Silurian and Devonian may assist the reconstruction of ancient ocean currents or climatic belts. However generalized interpretations are more criticized than followed. Dinoflagellates and acritarchs have been shown to track climatic changes associated with the Northern Hemisphere glaciation (2.9 Ma to 2.2 Ma) (Versteegh 1997). Reworked acritarchs are useful in detecting uplift and erosion of basin margins (Turner 1992) and in sedimentary provenance studies (e.g. McCaffrey et al. 1992). The acritarch alteration index (AAI, see Fig. 5.1) can be related to the burial/thermal history of the enclosing sedimentary rock (e.g. Dorning 1996). Colour alteration is indispensable in recognizing reworked material from older/deeper strata.

Phylum Prasinophyta

The prasinophytes are a group of non-cellulosic, green, flagellate algae. Modern species are characterized by a scaly, quadriflagellate (or biflagellate) motile stage. Some encyst, producing a resistant phycoma that resembles pteromorph acritarchs in morphology. There is considerable disagreement over the taxonomy of the prasinophytes. Some believe this group may be the ancestral to all the green algae, however cladistic analysis fails to separate the prasinophytes as a natural group. The most recent summaries of the group can be found in Tappan (1990) and Mendelson (in Lipps 1993, pp. 77–105); four orders are recognized by Hart (in Benton 1993, pp. 24–25).

Fossil prasinophytes are exclusively marine and generally much larger than acritarchs. They have a perforate wall with a cyclopyle or median-split opening. The vesicles were spherical and lacked spines or crests. Such forms (e.g. *Tasmanites*, Fig. 9.3d) range from Ordovician to Recent times.

Further reading

Introductory reviews can be found in Traverse (1988), Mendelson (in Lipps 1993) and Martin (1993), whilst Tappan (1990) provides a comprehensive treatment of the acritarchs and fossil prasinophytes. Dorning (in Benton 1993, pp. 33–34) outlines the classification of the acritarchs and Hart (in Benton 1993, p. 25) the prasinophytes. Fensome et al. (1990, 1991) provide an index and lists of acritarch and prasinophyte genera and species. Le Hérissé & Gourvennec (1995) described the palaeoenvironmental and biogeographical controls of upper Llandovery and Wenlock acritarchs, and Richardson & Rasul (1990) documented acritarch palynofacies for the mid-Silurian of Wales. Servais et al. (2003) provides a comprehensive review of Ordovician acritarch palaeoecology and palaeobiogeography.

Hints for collection and study

Acritarchs can often be obtained from dark carbonaceous shales, mudstones and clays disaggregated by methods A to E (see Appendix). Those occurring with dinoflagellate cysts in Mesozoic and Cenozoic rocks are usually easier to extract. Acritarchs can be sorted and concentrated by methods H and K. Temporary and permanent mounts on glass slides should be scanned with well-condensed transmitted light at 400× magnification. For a fuller treatment of techniques see Martin (1993).

REFERENCES

Benton, M.J. (ed.) 1993. *The Fossil Record 2*. Chapman & Hall, London.

Brasier, M.D. 1979. The Cambrian radiation event. In: M.R. House (ed.), *The Origin of Major Invertebrate Groups*. Academic Press, London.

Colbath, G.K. 1990. Palaeobiogeography of Middle Palaeozoic organic-walled phytoplankton. In: McKerrow, W.S., Scoteses, C.R. (eds), Palaeozoic Palaeogeography and Biogeography. *Memoir. Geological Society of London* 12, 207–213.

Colbath, G.K. & Grenfell, H.R. 1995. Review of the biological affinities of Paleozoic acid-resistant, organic walled eukaryotic algal microfossils (including 'acritarchs'). *Review of Palaeobotany and Palynology* 96, 297–314.

Cramer, F.H. & Diez, M. del, C.R. 1974. Silurian acritarchs, distribution and trends. *Review of Palaeobotany and Palynology* 19, 137–54.

Dale, B. 1976. Cyst formation, sedimentation and preservation: factors affecting dinoflagellate assemblages in Recent sediments from Trondheimsfjord, Norway. *Review of Palaeobotany and Palynology* 22, 39–60.

Dorning, K.J. 1981. Silurian acritarch distribution in the Ludlow shelf sea of South Wales and the Welsh Borderland. In: Neale, J.W. & Brasier, M.D. (eds), *Microfossils from Recent and Fossil Shelf Seas*. Ellis Horwood, Chichester, pp. 31–36.

Dorning, K.J. 1996. Organic microfossil geothermal alteration and interpretation of regional tectonic provinces. *Journal of the Geological Society, London* 143, 219–220.

Dorning, K.J. 1997. The organic palaeontology of Palaeozoic carbonate environments. In: M.B. Hart (ed.), *Micropalaeontology of Carbonate Environments*. British Micropalaeontological Society, Chichester, pp. 256–265.

Downie, C. 1973. Observations on the nature of the acritarchs. *Palaeontology* 16, 239–59.

Downie, C. 1974. Acritarchs from near the Pre-Cambrian/Cambrian boundary – a preliminary account. *Review of Palaeobotany and Palynology* 19, 57–60.

Fensome, R.A., Williams, G.L., Barss, M.S. et al. 1990. Acritarchs and fossil prasinophytes: an index to genera, species and intraspecific taxa. *AASP Contributions Series* 25, 1–771.

Fensome, R.A., Williams, G.L., Barss, M.S. et al. 1991. Alphabetical listing of acritarch and fossil prasinophyte species. *AASP Contributions Series* 26, 1–111.

Hill, P.J. & Molyneux, S.G. 1988. Biostratigraphy, palynofacies and provincialism of Late Ordovician-Early Silurian acritarchs from northeast Libya. In: El-Arnauti, A., Owens, B. & Thusu, B. (eds), *Subsurface Palynostratigraphy of Northeast Libya*. Garyounis University Publications, Benghazi, pp. 27–43.

Jacobsen, S.R. 1979. Acritarchs as palaeoenvironmental indicators in Middle and Upper Ordovician rocks from Kentucky, Ohio and New York. *Journal of Paleontolology* 53, 1197–1212.

Jansonius, J. & McGregor, D.C. (eds) 1996. *Palynology: principles and applications*. American Association of Stratigraphic Palynologists, Salt Lake City, pp. 81–107.

Le Hérissé, A. & Gourvennec, R. 1995. Biogeography of Upper Llandovery and Wenlock acritarchs. *Review of Palaeobotany Palynology* 96, 111–133.

Le Hérissé, A., Gourvennec, R. & Wicander, R. 1997. Biogeography of Late Silurian and Devonian acritarchs and prasionphytes. *Review of Palaeobotany and Palynology* 98, 105–124.

Li, J. & Servais, T. 2002. Ordovician acritarchs of China and their utility for global palaeobiogeography. *Buletin. Societe Geologique de France* 173, 399–406.

Lipps, J.H. (ed.) 1993. *Fossil Prokaryotes and Protists*. Blackwell Scientific, Oxford.

McCaffrey, W.D., Barron, H.F., Molyneux, S.G. & Kneller, B.C. 1992. Recycled acritarchs as provenance indicators-implications for Caledonian Terrane reconstruction. *Geological Magazine* 129, 457–464.

Martin, F. 1993. Acritarchs – a review. *Biological Reviews* 69, 475–539.

Mendelson, C.V. & Schopf, J.W. 1992. Proterozoic and Early Cambrian acritarchs. In: Schopf, J.W. & Klein, C. (eds) *The Proterozoic Biosphere. A multidisciplinary study*. Cambridge University Press, Cambridge, pp. 219–232.

Moldowan, J-M. & Talyzina, N.M. 1999. Biogeochemical evidence for the dinoflagellate ancestors in the Early Cambrian. *Science* 281, 1168–1170.

Molyneux, S.G., Le Hérissé, A. & Wicander, R. 1996. Paleozoic phytoplankton. In: Jansonious, J. & McGregor, D.C. (ed.) *Palynology: principles and applications*, vol. 2. American Association of Stratigraphic Palynologists Foundation, pp. 493–529.

Richardson, J.B. & Rasul, S.M. 1990. Palnofacies in a Late Silurian regressive sequence in the Welsh Borderland and Wales. *Journal of the Geological Society, London* 147, 675–696.

Servais, T., Brocke, R., Fatka, O. 1996. Variability in the Ordovician acritarch *Dicrodiacrodium*. *Palaeontology* 39, 389–405.

Servais, T., Li, J., Molyneux, S. & Raevsaya, E. 2003. Ordovician organic-walled microphytoplankton (acritarch) distribution: the global scenario. *Palaeogeography, Palaeoclimatology, Palaeoecology* 195, 149–172.

Staplin, F.L. 1961. Reef-controlled distribution of Devonian microplankton in Alberta. *Palaeontology* 4, 392–424.

Tappan, H. 1990. *The Palaeobiology of Plant Protists*. Freeman, San Francisco.

Traverse, A. 1988. *Paleopalynology*. Unwin Hyman, Boston.

Tschundy, R.H. & Scott, R.A. (eds) 1969. *Aspects of Palynology*. Wiley Interscience, New York.

Turner, R.E. 1992. Reworked acritarchs from the type section of the Ordovician Caradoc Series, Shropshire. *Palaeontology* 25, 119–143.

Vecoli, M. 2000. Palaeoenvironmental interpretation of microphytoplankton diversity trends in the Cambrian-Ordovician of the northern Sahara Platform. *Palaeogeography, Palaeoclimatology Palaeoecology* 160, 329–346.

Versteegh, G.J.M. 1997. The onset of major Northern Hemisphere glaciations and their impact on dinoflagellate cysts and acritarchs from the Singa section, Calabria (southern Italy) and DSDP Holes 607/607A. *Marine Micropalaeontology* 30, 319–343.

Vidal, G. & Knoll, A.H. 1993. Proterozoic plankton. *Memoir. Geological Society of America* 161, 265–267.

Wall, D. 1965. Microplankton, pollen and spores from the Lower Jurassic in Britain. *Micropalaeontology* 11, 151–190.

Wicander, R., Playford, G. & Robertson, E.B. 1999. Stratigraphic and palaeogeographic significance of an upper Ordovician acritarch flora from the Maquoketa Shale, northeastern Missouri, USA. *Journal of Paleontology* 73, supplement 6, 1–38.

Wright, R.P. & Meyers, W.C. 1981. Organic walled microplankton in the subsurface Ordovician of northeastern Kansas. *Kansas Geological Survey, Subsurface Geology Series* 4, 1–53.

CHAPTER 10

Dinoflagellates and ebridians

Dinoflagellates (meaning whirling whips) are second only to the diatoms as primary producers in the world's oceans. They are single-celled organisms generally between 20 and 150 μm in maximum diameter, with both plant and animal characteristics. Most dinoflagellates are distinguished by a **dinokaryon**, a special form of eukaryote nucleus. Their **carotenoid** pigments dinoxanthin and peridinin give to these organisms flame-like colours and produce 'red tides' when populations bloom. Many living dinoflagellates are also bioluminescent.

The majority of dinoflagellates exhibit alternation of generations in the life cycle and bear two flagella for propulsion. Motile (**theca**) cells are equipped with one longitudinal whip-like and one transverse ribbon-like flagellum for propulsion, have a prominent nucleus and a sculptured cell wall (Fig. 10.1). Both **heterotrophic** and **autotrophic** modes of nutrition occur, although the latter predominate. Dinoflagellates have formed an important part of oceanic phytoplankton since at least mid-Mesozoic times. Although motile cells are abundant and wide ranging, it is the resistant resting **cyst** which leaves a fossil record. Dinoflagellate cysts have proved to be valuable tools in biostratigraphy and are also important in palaeoecology, palaeoclimatology and evolutionary palaeontology.

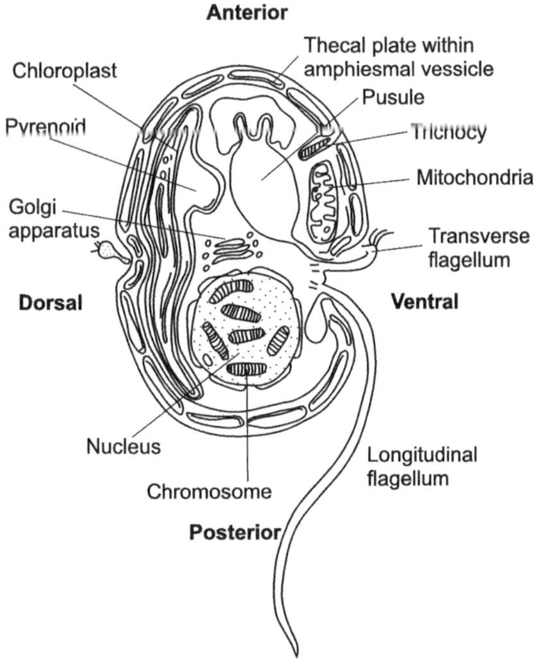

Fig. 10.1 The dinoflagellate cell. (After Edwards in Lipps 1993, pp. 105–127.)

The living dinoflagellate

Motile stage

Dinoflagellate cells range in size from 5 to 2000 μm (Fig. 10.1). These organisms are amongst the most primitive of the eukaryotes and have been regarded as intermediates between prokaryotes and eukaryotes.

Fig. 10.2 (*opposite*) Dinoflagellate motile stage. (a) Schematic section through the wall of an unarmoured dinoflagellate. (b) Schematic section through the wall of an armoured dinoflagellate. (c), (d) Tabulation of a hypothetical peridinialean motile stage: (c) ventral side; (d) dorsal side. (e), (f) Motile cell of a Recent *Peridinium* approx. ×505. (g), (h) Motile cell of a Recent *Gonyaulax*, approx. ×750. ((e)–(h) Based on Sarjeant 1974.)

Chapter 10: Dinoflagellates and ebridians 81

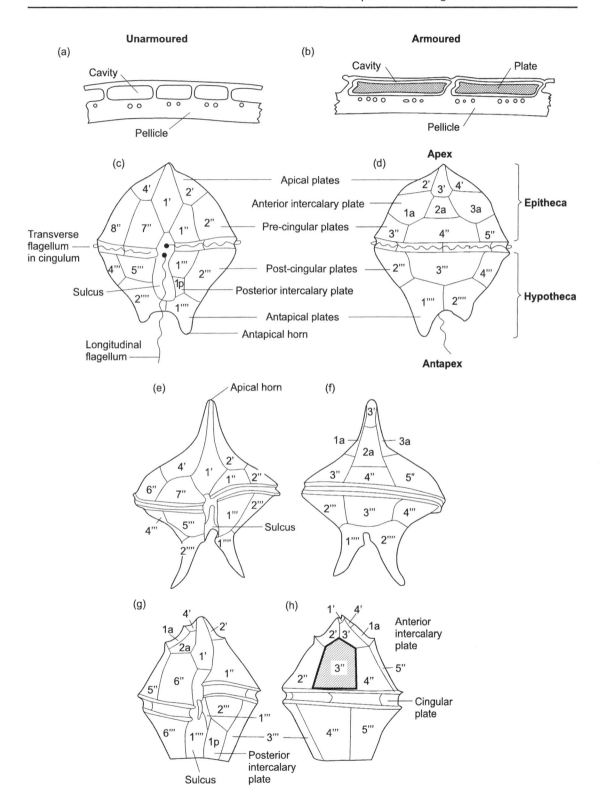

The cell wall may be either, flexible and **unarmoured** or rigid and **armoured** (Fig. 10.2). In the former case it comprises a proteinaceous envelope (**pellicle**) containing flattened cavities near the surface (Fig. 10.2a). In the armoured cell wall these cavities are filled by plates of fibrous cellulose to form a closely fitting **theca** (Fig. 10.2b). The mode of arrangement of these plates, known as **tabulation**, is consistent within a species.

The cell contains eukaryote organelles such as a single large nucleus, the endoplasmic reticulum, Golgi apparatus and mitochondria (Fig. 10.1). But, as in the prokaryotes, the chromosomes remain condensed throughout life and the nuclear spindle which forms during meiosis lies external to the nuclear membrane. Within the cell, several fluid-filled vessels (**pusules**) are connected to the exterior via canals. Photosynthetic pigments, where present, are contained in round chloroplasts at the cell margins. Light sensory eye spots may also be present.

The two flagella arise either from pores at the anterior end or from the ventral surface (Fig. 10.2c,d). Two furrows, each of which bears a flagellum, generally traverse the cell surface. One occupies a more or less equatorial position in a transverse furrow called the **cingulum**, the other lies in a longitudinal furrow called the **sulcus**. That half of the cell anterior to the cingulum is called the **epitheca** and that posterior to it is termed the **hypotheca** (Fig. 10.2c,d). The side bearing the sulcus is **ventral** (Fig. 10.2c,e,g), whilst the opposite side is **dorsal** (Fig. 10.2d,f,h). Many cells and cysts are dorso-ventrally compressed so that these two views are the ones usually illustrated.

The sulcus extends in a posterior direction and may terminate in a depression flanked by one or two **antapical horns** (Fig. 10.2c,d). The other anterior, or **apical**, end is often rounded, pointed or produced into an **apical horn** (Fig. 10.2e). Overall cell shape can be very varied even within a single genus, but includes spherical, subspherical, ovoid, biconical, fusiform, rod-shaped, rectangular, polygonal, discoidal and peridinioid outlines.

Tabulation refers to the arrangement of plates in the armoured motile cells of the class Peridinea. In these, five plate series are found to encircle each cell, each plate being numbered for reference in a counter-clockwise direction using the Kofoidian System (Fig. 10.2c–h). This system of nomenclature is objective and purely descriptive and does not normally imply homology between plates in different taxa. Around the epitheca occur the **apical** and **precingular** series. In the cingulum lie the **cingular** series whilst the **postcingular** and **antapical** series occur on the hypotheca. Additional anterior and posterior **intercalary** plates may also develop at sites between the series, and the sulcus bears small **sulcal** plates that can also be of taxonomic value.

The functional significance of cell shape and tabulation is little understood. As the planktonic forms maintain their position in the water by active flagellar propulsion rather than by passive floating, the cells tend to be streamlined. Nevertheless, the long horns of certain genera may serve to retard sinking.

Cyst stage

Only about 10–20% of living species are known to encyst following sexual reproduction, yet almost all fossil dinoflagellates are preserved as cysts. Three basic kinds of cyst are recognized, termed proximate, proximochorate and chorate, depending upon the relative length of any ornament, although intergradations between these exist. **Proximate** cysts resemble the theca in both size and shape and presumably formed in close contact with the thecal wall (Fig. 10.3a,b). The tabulation, cingulum and sulcus are all reflected in the surface sculpture of proximate cysts. Proximochorate cysts are an intermediate type between proximate and chorate cysts. They have **processes** that are between 10 and 30% of the overall diameter (Fig. 10.3d,e) and an elaborate ornament. The tips of the processes were in contact with the thecal wall and in some species were plate centred and can be numbered in a similar fashion to proximate cysts. The tips of the processes may be joined by thin, filamentous **trabeculae** giving the impression of an additional layer. **Chorate** cysts (Fig. 10.3c,d) usually exhibit no traces of a reflected cingulum or sulcus.

The cyst is formed within the motile cell and contains the same organelles. The cyst wall (**phragma**), built of organic material resistant to bacterial decay and called **dinosporin**, may consist of one or multiple layers (Fig. 10.3a,f). An **autocyst** has a single layer and

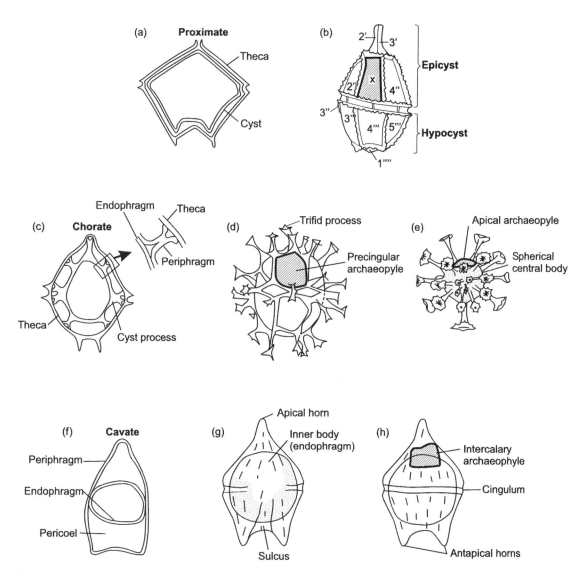

Fig. 10.3 Dinoflagellate cyst stage. (a) Proximate cyst of *Peridinium* (axial section), approx. ×250. (b) Proximate cyst of fossil *Gonyaulacysta*, approx. ×450. (c) Chorate cyst of *Gonyaulax* with detail of wall (axial section), approx. ×250. (d) Chorate cyst of *Hystrichosphaeridium*, approx. ×400. (e) Proximochorate cyst of fossil *Spiniferites*, approx. ×465. (f) Proximate cyst of *Deflandrea* (axial section), approx. ×250. (g), (h) Proximate cyst of *Deflandrea*: (g) ventral; (h) dorsal, approx. ×360. ((b), (c), (f) Based on Sarjeant (1974); (e) based on Evitt 1969.)

its wall is an **autophragm**. A two-layered cyst with connections between the walls has an inner layer, the **autophragm**, and an **ectophragm**. This condition is termed **holocavate**. If the two layers are not connected, the cyst is known as **cavate** and the inner layer is called the **endophragm** (Fig. 10.3f–h) and the outer layer the **periphragm** which are partially separated, usually at the poles (Fig. 10.3f–h). The cavities thus formed (**pericoels**) may promote buoyancy in the cyst. Traces of the tabulation, the cingulum and the sulcus may also be seen on the periphragm, so it is probable that this type of cyst formed just below the thecal cell wall.

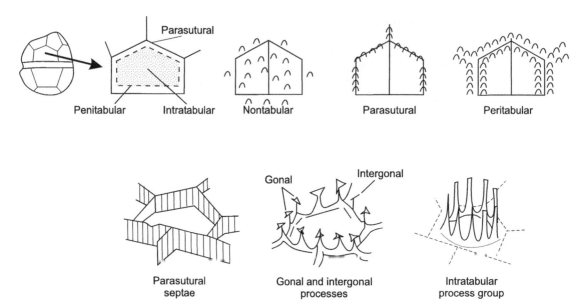

Fig. 10.4 Surface ornament. (Redrawn after Evitt 1985.)

Cyst surface features

Dinoflagellate cysts can be smooth or bear granules, ridges, indentations, raised crests or develop short spines, processes or horns (Fig. 10.4). Processes can be plate centred or form groups. **Tabular** ornament is **sutural** if it defines plate boundaries or **intertabular** if it defines the central parts of plates. Processes that are situated at the intersection of paraplate boundaries are **gonal** and those along boundaries are **intergonal**.

Where a reflected cingulum is present, that portion apical to it is called the **epicyst** and the antapical portion is called the **hypocyst** (Fig. 10.3b).

The function of the cyst is demonstrated by the presence of an escape hole, called an **archaeopyle**. This is formed by the removal of one or several plates (thereby comprising an **operculum**) normally from the apical series, the precingular series, an anterior intercalary plate or a combination of these. The form and position of the archaeopyle is constant within a genus.

Further investigation is needed into the functional and ecological significance of cyst morphology. The processes of chorate cysts and the pericoels of cavate forms may both be mechanisms to minimize the downward sinking of oceanic species. If the cysts of such forms were to sink far below the photic zone before excystment, their chances of survival would be reduced. In laboratory cultures, similar morphotypes of cyst are known to produce markedly different motile cells and apparently identical motile cells can produce very different cysts. In Recent sediments cyst morphology may be directly related to salinity. *Tectatodinium* and *Spiniferites* (Figs 10.3e, 10.9b) are round in outline in normal marine conditions yet cruciform in low-salinity conditions. *Operculodinium* (Fig. 10.7c) species are known to have reduced numbers of processes in low-salinity environments. For these reasons, in fossil assemblages, it is unlikely that there is a simple relationship between the cysts preserved and the motile cells originally present.

Dinoflagellate life history

Sexual reproduction is known to occur in very few living dinoflagellates. Asexual (vegetative) reproduction predominates and involves a division of the cell into two halves by binary fission. Details of the life

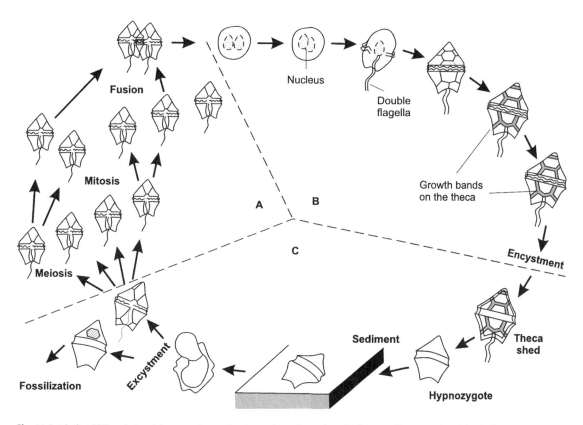

Fig. 10.5 Idealized life cycle involving sexual reproduction and cyst formation. Section A, cells are motile and haploid; section B, cells are motile and diploid; section C, cells are non-motile (except excysted cell on left) and diploid. (Reproduced from Stover *et al.* in Jansonius & McGregor 1996, vol. 2, pp. 641–750 (with the permission of the ASSP Foundation).)

cycle can vary considerably, particularly in the sexual part of the cycle, and a generalized life cycle is described here (Fig. 10.5). With the exception of *Polykrikos* (Fig. 10.6d) and *Noctiluca* (Fig. 10.6e), the vegetative motile stage the **schizont** has a full chromosomal compliment and is **haploid**. Once formed, the zygote enlarges, the cell wall thickens (at this stage it is known as the **planozygote**) and it looses motility. In the **hypnozygote** stage the cytoplasm contracts, the cyst forms and the flagella are lost. The cytoplasm may remain dormant in the cyst for hours to years during which time the first and occasionally second meiotic divisions occur. The resultant cells will emerge through the archeopyle and growth and vegetative division are initiated. In laboratory culture withholding nutrients or reducing temperature and light levels can induce sexual reproduction. Though cysts can form in the schizont, the majority of fossil dinoflagellate cysts are believed to be hypnozygote cysts. Many dinoflagellate cysts remain dormant on the sea floor through the winter. During this period the surrounding thecal plates may drop away and begin to decay. With the amelioration of conditions in spring, the motile stage excysts through the archeopyle to leave a resistant cyst for the fossil record.

Dinoflagellate ecology

Dinoflagellates currently form a major part of the ocean plankton, especially the armoured and autotrophic forms, and they play a prominent rôle in the

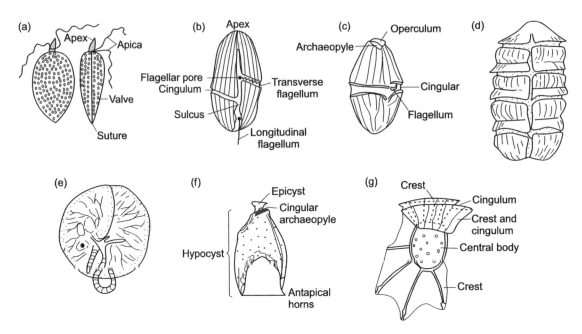

Fig. 10.6 Examples of cyst from the Prorocentroidia and Bilidinea. (a) Recent *Prorocentrum* (Prorocentroidia), about ×350. (b) Recent *Gymnodinium* (Gymnodinoidia), about ×350. (c) Fossil *Dinogymnodinium* (Gymnodinoidia). (d) Recent *Polykrikos* (Bilidinea), about ×500. (e) Recent *Noctiluca* (Noctilucea), about ×180. (f) Fossil *Nannoceratopsis* (Bilidinea), about ×680. (g) Recent *Ornithocercus* (Bilidinea), about ×275. ((a) From Chapman & Chapman 1973; (b) from Kofoid & Swezy 1921; (d) after Dodge 1985; (f) based on Sarjeant 1974; (g) based on Barnes 1968.)

food chains of the marine realm. The autotrophic forms thrive in areas of upwelling currents that are rich in nutrients such as nitrates and phosphates, whilst they are rarely found alive below 50 m depth because of their need for light. Flagella locomotion is employed in bringing them to the surface at night and withdrawing them to greater depths in the day because they must avoid harmful ultraviolet light. In Mesozoic, Cenozoic and Recent dinoflagellate assemblages only a few taxa appear to have palaeoecological or palaeobiogeographical significance. Dale (1976) described dinoflagellate cyst ecology and discussed the geological implications. Of the primary ecological factors one of the most important for controlling cyst assemblages is sea surface temperature. As a whole, the group has a wide temperature tolerance (1–35°C) with an optimum for most species of 18–25°C. *Ceratium* (Fig.10.7d) shows temperature-related morphological variability particularly in the length and angle between the antapical horns.

Dale (1976) noted a change of only a few degrees might be sufficient to cause differentiation into biogeographical provinces. One of the most important temperature boundaries controlling the distribution of dinoflagellate cysts in the Northern Hemisphere occurs between the main bodies of cooler and warmer water in the North Atlantic Ocean. This boundary lies between Cape Cod and Nova Scotia (42–43°N) and between the English Channel and southwestern Norway (Dale 1983; Taylor 1987). Dale (in Jansonius & McGregor 1996, vol. 3, pp. 1249–1275) described the distribution of selected cyst assemblages compared to the modern biogeographical zones (polar, subpolar, temperate and equatorial) for the Atlantic Ocean. In this some species range from pole to pole, whilst others are restricted to the zones and have obvious applications in biogeographical and climate studies. On a global scale modern dinoflagellates occupy broad latitudinal low-, middle- and high-latitude zones (Taylor 1987).

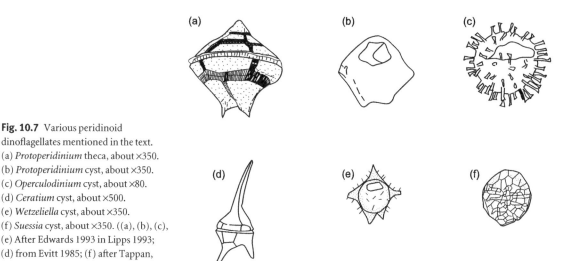

Fig. 10.7 Various peridinoid dinoflagellates mentioned in the text. (a) *Protoperidinium* theca, about ×350. (b) *Protoperidinium* cyst, about ×350. (c) *Operculodinium* cyst, about ×80. (d) *Ceratium* cyst, about ×500. (e) *Wetzeliella* cyst, about ×350. (f) *Suessia* cyst, about ×350. ((a), (b), (c), (e) After Edwards 1993 in Lipps 1993; (d) from Evitt 1985; (f) after Tappan, 1980.)

Dinoflagellates can tolerate a wide range of salinities and are found in lakes, ponds and rivers. Certain genera, such as *Gymnodinium* (Fig. 10.6b) and *Peridinium* (Fig. 10.2e,f), are found in both fresh and salt water, although the majority of species are marine and show optimal growth at salinities of 10–20‰. Recent experiments on dinoflagellate cultures show that, for a single species, the size and morphology of the cyst may vary considerably with salinity. The greatest variation lies in the number, density and structure of the processes. Dale (1983) described similar effects in the morphology of the resting cyst of *Lingulodinium* from the Black Sea. Other examples can be found in Ellegaard (2000) and Hallett & Lewis (2001).

Autotrophic species live in the photic zone where trace element availability limits their productivity. Cyst-forming species live almost exclusively in marine environments, particularly in shallow coastal waters. Sudden blooms of dinoflagellates, called **red tides**, may occur under optimal conditions and the build up of toxins can kill great numbers of fish and invertebrates.

Planktonic forms with a predatory or parasitic mode of life are usually unarmoured and belong mostly to the Subclass Gymnodinoidia. Others of limited palaeontological interest contain immobile, benthic, colonial forms and the zooxanthellae that live symbiotically in the tissues of reef-building corals and larger foraminifera.

At present, dinoflagellate cysts are most abundant in sediments from coastal to continental slope and rise settings, with 1000–3000 cysts per gram. There is also a tendency for specific diversity to increase with distance from shore and to be greatest in tropical waters, a pattern reflected in many groups of marine plankton. In modern sediments specific assemblages of dinoflagellate cysts are known from estuarine, nearshore, neritic and oceanic environments. Ocean currents can be traced in cyst distribution patterns. Mudie (in Head & Wrenn 1992, pp. 347–390) documented inshore–offshore trends in transects across the temperate, subarctic and arctic margins of eastern Canada and mapped the distribution of selected dinoflagellate cysts in the northwestern Atlantic Ocean.

Modern ocean currents influence the distribution of dinoflagellate cysts, and the marine microplankton as a whole. Matthiessen (1995) reported the transport of cysts by currents in the Norwegian-Greenland Sea. Mudie & Harland (in Jansonius & McGregor 1996, vol. 2, pp. 843–877) noted the warm water of the North Atlantic Drift into the eastern Arctic was responsible for the mixing of dinoflagellate assemblages.

There are several inherent problems in interpreting the palaeoecology of fossil dinoflagellates. Firstly, those of pre-Quaternary age are not easy to relate to taxa of known habit, although lineages can be traced in a few cases. Secondly, many dinoflagellates do not encyst and therefore leave no fossil record. Thirdly, dinoflagellate cysts may sink and drift to be preserved at depths and conditions beyond the tolerance of the species. Some studies however suggest a strong correlation between cyst assemblages from the sea floor and the overlying water-mass, suggesting little post-mortem transport; however many contradictory examples are also known.

The distribution and ecology of Recent and Quaternary dinoflagellates are reviewed more fully by Williams (in Funnell & Reidel 1971, pp. 91–95, 231–243; in Ramsay 1977, pp. 1288–1292), Wall *et al.* (1977) and Harland (in Powell 1992, pp. 253–274). Additional useful modern syntheses can be found in Fensome *et al.* (in Jansonius & McGregor 1996, pp. 107–171) and Stover *et al.* (in Jansonius & McGregor 1996, pp. 641–787).

Classification

Kingdom PROTOZOA
Subkingdom DICTYOZOA
Phylum DINOZOA
Subphylum DINOFLAGELLATA

At one time many dinoflagellate cysts were classed with the problematic hystrichospheres. Evitt (1961, 1963) demonstrated that some of these were true dinoflagellate cysts, designating the remaining problematica to the group Acritarcha.

The classification of dinoflagellates commonly represented in the fossil record is outlined in Box 10.1 and follows that proposed by Cavalier-Smith (1998). A major reclassification of the dinoflagellates by Fensome *et al.* (1993b) independently created the taxon Dinokaryota but differs in that it includes all dinoflagellates that have histones in at least one stage of their life cycle. Since six of the eight classes of dinoflagellates are totally non-photosynthetic, plus about half the species in the remaining two classes, it seems more appropriate to treat the whole phylum under the Zoological rather than Botanical Code of Nomenclature. The classification of living forms takes account of molecular sequence data, position of flagellar insertion, predominant habit (e.g. mobile and flagellate, mobile amoeboid, immobile solitary or immobile colonial), presence of armour, tabulation, shape and sculpture of the motile cell. Fossil dinoflagellate cysts are classified according to cyst type, reflected tabulation, archaeopyle position, shape and sculpture (see Fensome *et al.* 1993a).

Class Peridinea

Subclass Prorocentroidia These are thought to be the most primitive dinoflagellates. They have two flagella of equal length inserted at the anterior end of the motile cell, which is unarmoured. Dinoflagellate cysts are unknown but may be included amongst the acritarchs. *Prorocentrum* (Fig. 10.6a) is a living genus

---

Box 10.1 Higher level classification of dinoflagellates (based on Cavalier-Smith 1998)

Kingdom Protozoa	Superclass Dinokaryota
Subkingdom Dictyozoa	Class Peridinea
Infrakingdom Neozoa	Subclass Gymnodinoidia*
Parvkingdom Alveolata	Subclass Peridinoidia*
Phylum Dinozoa	Subclass Prorocentroidia*
	Subclass Desmocapsoidia
Subphylum Dinoflagellata	Subclass Thoracospaeroidia
(=Dinophyta)	Class Bilidinea
Superclass Hemidinia	Order Dinophysida*
Class Noctilucea*	Order Nannoceratopsida*

*Contain fossil representatives.

involved in red tides. The theca is divided into two equal valves by a longitudinal suture.

Subclass Gymnodinoidia The gymnodinoids are predatory and parasitic forms lacking armour but having a flexible pellicle. The theca is commonly spherical, traversed by a deep equatorial cingulum and a shallow longitudinal sulcus. Although cysts are known, the lack of tabulation-related features makes it difficult to infer biological affinities. Such uncertain forms may therefore become classified as acritarchs.

The Recent genus *Gymnodinium* (Fig. 10.6b) has a motile cell with an equatorial cingulum. The common Late Cretaceous genus *Dinogymnodinium* (Fig. 10.6c) is probably a proximate cyst and has longitudinal folds, a cingulum and an apical archaeopyle.

Subclass Peridinoidia This subclass includes forms with an armoured motile stage. In these the cingulum is equatorial with a slight spiral offset and there is a longitudinal sulcus. The plates are arranged into apical, precingular, cingular, postcingular and antapical series, with additional intercalary and sulcal plates.

Inevitably, classification of the peridinoids (Peridiniales) has proceeded along two independent lines, one for the fossil dinoflagellate cysts (the bulk of which belong here) and one for the living motile cells. In principle it would be best to combine this divergent information into a single, natural, classification. Unfortunately, however, cyst genera and motile genera do not always correspond, evolution having proceeded at different rates for these different stages of the life cycle (**mosaic evolution**).

The motile cell of Recent *Peridinium* (Fig. 10.2e,f) is laterally compressed and almost bilaterally symmetrical. The cyst stage is proximate with a peridinioid shape, clearly reflected tabulation and furrows. Both theca and cysts may bear two antapical horns. *Deflandrea* (L. Cret.-U. Olig., Fig. 10.3f) is a fossil cavate cyst of ellipsoidal shape, commonly with horns. The reflected tabulation is rarely visible but of *Peridinium*-type with an anterior intercalary archaeopyle.

In Recent *Gonyaulax* (Fig. 10.2g,h) the motile stage usually lacks horns and the tabulation is relatively asymmetrical. Its cyst is chorate or intermediate proximochorate with a precingular archaeopyle, being of the type once called *Hystrichosphaera* but now called *Spiniferites* (U. Jur.-Rec., Figs 10.3e, 10.9b). The fossil proximate cyst *Gonyaulacysta* (M. Jur.-M. Mioc., Figs 10.3b, 10.9a) also has a reflected tabulation *of Gonyaulax*-type with a precingular archaeopyle and sutures that are marked by crests and it bears an apical horn.

Hystrichosphaeridium (U. Jur.-M. Mioc., Fig. 10.3d) is a fossil chorate cyst with a spherical body and radiating hollow processes, often with trumpet-like openings at the distal end. Each process corresponds to the centre of a plate on the once enclosing theca. The archaeopyle is apical.

Class Bilidinea

This class includes the orders **Dinophysida** and **Nannoceratopsida**. Although armoured these dinoflagellates lack a distinctive tabulation. The cingulum is anterior in position and less spiralled than in the peridinoids (Peridiniales), uniting with the sulcus in a T- or Y-shaped junction. Both furrows are bordered by flange-like crests, as in the Recent *Ornithocercus* (Fig. 10.6g). The cysts are proximate, the archaeopyle and operculum **epicystal**, comprising the whole of the epicyst. Fossil examples are few but may include *Nannoceratopsis* from the Jurassic (Fig. 10.6f). In this there are normally two prominent antapical horns and a cingular archaeopyle.

General history of dinoflagellates

Although the apparent primitive organization implies great age for the group, the acme of peridinalean history appears to have been reached in the Mesozoic and Cenozoic Eras. Dinosterane and 4α-methyl-24-ethylcholestane, two dinoflagellate biomarkers in samples of Upper Proterozoic and Cambrian age (Moldowan & Talyzina 1999), and RNA sequence data indicate the dinoflagellates diverged before the Foraminifera and the Radiolaria which both have a Cambrian fossil record. It is evident that Late Precambrian and Palaeozoic radiations of acritarchs may represent an earlier stage in dinoflagellate history, when non-tabulate forms thrived. The subsequent evolutionary history of the dinoflagellates

90 Part 3: Organic-walled microfossils

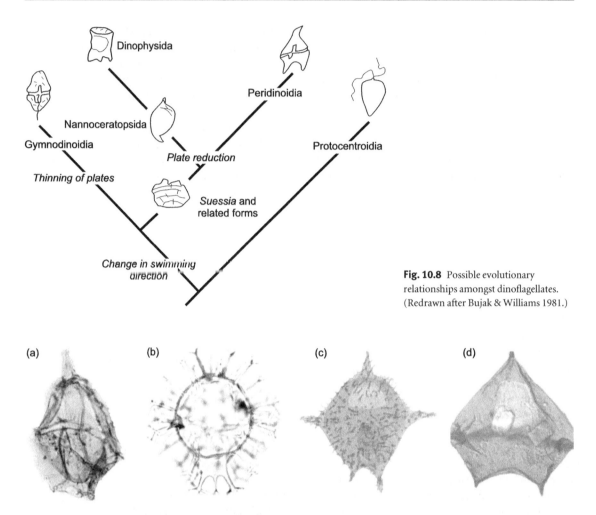

Fig. 10.8 Possible evolutionary relationships amongst dinoflagellates. (Redrawn after Bujak & Williams 1981.)

Fig. 10.9 Photomicrographs of selected dinoflagellate cysts. (a) *Gonyaulacysta jurassica* (Deflandre 1938), ×528. A proximate cyst ranging from the early Oxfordian to late Kimmeridgian. (b) *Spiniferites mirabilis* (Rossignol 1964), ×611. Chorate cysts ranging from the Quaternary to Recent. (c) *Wetzeliella articulata* Eisenack 1939, ×294. A proximochorate cyst ranging from the Ypresian to Rupelian. (d) *Protoperidinium communis* Biffe & Grignani 1983, ×515. This is a proximate cyst with a cavate wall structure and shows the archeopyle. Range: Early Pleistocene. ((a) Reproduced from Riding & Thomas in Powell 1992, plate 2.17.3; (b) reproduced from Harland in Powell 1992, plate 5.3.2; (c) reproduced from Powell in Powell 1992, plate 4.6.10; (d) reproduced from Harland in Powell 1992, plate 5.1.6. All photomicrographs reproduced with permission from Kluwer Academic Publishers.)

has been reviewed by Bujak & Williams (1981; Fig. 10.8).

The earliest record of an equivocal peridinoid cyst is *Arpylorus*, from the Silurian. This has been interpreted as possessing tabulation, a cingulum and a precingular archaeopyle. The main dinoflagellate radiation, however, began in the mid- to Late Triassic with the appearance of genera such as *Suessia* (Figs 10.7f, 10.8). Proximate cysts were common throughout the Jurassic (e.g. *Gonyaulacysta jurassica*, Figs 10.3b, 10.9a), although chorate and proximochorate cyst types had all appeared by the Middle Jurassic.

Many Cretaceous forms are chorate (e.g. *Hystrichosphaeridium*, Fig. 10.3d) or proximochorate (e.g.

Spiniferites ramosus, Figs 10.3e, 10.9b), and it was at this time that the greatest diversity of dinoflagellate cysts was reached.

Cavate peridinoid dinoflagellate cysts began to flourish in Aptian-Albian (e.g. *Deflandrea*, Fig. 10.3f, *Wetzeliella*, Figs 10.7e, 10.9c) and dominated many Tertiary assemblages until the Oligocene, almost dying out in the Pliocene. Proximate and chorate dinoflagellate cysts with complex processes occur in the Eocene and Oligocene, but simpler forms have prevailed since then. Dinoflagellate cysts first appeared in freshwater sediments during the Tertiary.

Applications of dinoflagellate cysts

Dinoflagellate cysts are ideal biostratigraphical indices. Williams & Bujak (1985) provide a detailed review of cyst biozones with subsequent modifications to be found in Stover et al. (in Jansonius & McGregor 1996, vol. 2, pp. 641–750). Powell (1992) provides an accessible laboratory manual for identifications and biozonations.

Late Triassic dinoflagellate assemblages are known from Alaska, Arctic Canada, Australia, England and Austria; only one biozone has been recognized, the *Rhaetogonyaulax rhaetica* Interval Biozone. Early Mesozoic assemblages are low in species diversity but by the mid-Jurassic dinoflagellates were an important part of the phytoplankon. Provincialism means different biozonations have been erected for the Arctic, Boreal, Tethyan and Southern Hemisphere realms, corresponding approximately to molluscan faunal provinces (Davies & Norris 1980; Stancliffe & Sarjeant 1988). Similar provinciality has been recognized at least for the Early Cretaceous (Lentin & Williams 1980; Williams et al. 1990) and Tertiary (Williams & Bujak 1977, 1985; Williams et al. 1990). Dinoflagellates did not show appreciable levels of increased extinction at the K-T boundary, although the character of assemblages did change. The usefulness of dinoflagellate cysts in sequence stratigraphy was recognized by Haq et al. (1987), whose correlation chart includes last and first appearance datums for selected cyst species. Habib et al. (1992) reported the relationship between cyst species diversity and system tracts in the K-T boundary beds of Alabama, with a minimum diversity in the lowstand, whilst Monteil (1993) were able to define third-order sequences in the Berriasian-type section using cyst first and last appearance datums. Stover & Hardenbol (1994) demonstrated the sensitivity of cyst assemblages to transgressive and regressive cycles in the Lower Oligocene, Boom Clay of Belgium.

The palaeoecological utility of dinoflagellate cysts was reviewed by Williams (in Ramsay 1977, pp. 1292–1302) and the taphonomic factors affecting the preservation of dinoflagellate assemblages have been considered by Dale (1976). The gonyaulacean to peridinialean ratio (or variations of this) have been used extensively as a palaeoshoreline indicator (e.g. Bint 1986). Powell et al. (1992) also used this ratio as a proxy for upwelling strength and by inference palaeotemperature in post-Paleogene sediments; an increasing ratio indicating cooler water. A similar study in the Coniacian-Maastrichtian of Israel showed this ratio also varied with upwelling intensity and suggests a much wider utility (Eshet et al. 1994).

Lower Tertiary deposits of southern England contain four distinct dinoflagellate cyst-acritarch assemblages, characterized by the dominance of a single species. These assemblages are also lithology specific. Assemblages dominated by the gonyaulacean genera *Areoliera* and *Spiniferites* indicate open water, *Micrhystridium* (an acritarch) dominates in inner neritic environments and marks the initial and closing stages of marine transgressions. *Wetzeliella* (a peridinoid genus) dominates in estuarine environments. The use of dinoflagellate cysts as palaeosalinity indicators is however poorly developed, although it appears living dinoflagellates follow a similar distribution to living molluscs (Wall et al. 1977).

Dinoflagellates are increasingly being used in palaeoclimate research. Studies include Tertiary sections in cores from the North Atlantic and the Eocene-Oligocene of the Gulf of Mexico and western Atlantic, as part of the ODP, where regional distribution patterns agree with inferred palaeocurrent configurations (e.g. Damassa et al. 1990). Mudie et al. (1990) summarized the climatic control on Neogene dinoflagellate cysts and acritarch distribution in polar oceans. Head (1993) used dinoflagellate cysts to indicate a subtropical to tropical climate in the Pliocene of southwest England. High winter sea surface temperatures (of around 15°C) were thought to indicate the presence of the Gulf Stream and global warming. Edwards et al. (1991) and Edwards (in Lipps 1993, pp. 105–127; in Head & Wrenn 1992, pp. 69–87) have attempted to use cyst abundance data to predict ocean temperatures, whilst Jarvis et al. (1988) and Palliani & Riding (1999) focused on deciphering the temporal variations in cyst assemblages through ocean anoxic events.

The Quaternary deposits of the North Sea glacials are marked by low-abundance and low-diversity assemblages with predominantly *Spiniferites elongatus* and round species of *Protoperidinium* (Fig. 10.9d). Interglacials or more temperate conditions are represented by high abundances of *Operculodinium centrocarpum* (Fig. 10.7c), *Spiniferites mirabilis* (Fig. 10.9b and pentagonal species of *Protoperidinium* (Fig. 10.7a,b) (Harland, in Powell 1992, pp. 253–274).

Dinoflagellate cysts and acritarchs reworked into younger sediments can be used to indicate the provenance of sediments and the directions of transport (Stanley 1966; Riding *et al.* 1997, 2000).

Further reading

Useful introductions to the group may be found in Edwards (in Lipps 1993, pp. 105–127), Fensome *et al.* (in Jansonius & McGregor 1996, vol. 1, pp. 107–171) and Stover *et al.* (in Jansonius & McGregor 1996, vol. 2, pp. 641–750). Many dinoflagellate cysts can be identified with the assistance of the catalogues by Fensome *et al.* (1991, 1993a), Williams & Bujak (1985) and Powell (1992). Significant reviews on living dinoflagellates include Spector (1984) and Taylor (1987); Popovsky & Pfiester (1990) reviewed non-marine forms.

Hints for collection and study

Dinoflagellate cysts are common in dark grey and black argillaceous rocks of post-Triassic age. They can be disaggregated by methods A to E (especially D) and sorted and concentrated by methods H or K (see Appendix). Temporary or permanent mounts on glass slides should be scanned with well-condensed transmitted light at over 400× magnification. More sophisticated methods of preparation and concentration are reviewed by Wood *et al.* (in Jansonius & McGregor 1996, vol. 1, pp. 29–51). Dodge (1985) provides a compendium of SEM images of living motile cells and cysts.

Ebridians

The ebridians are unicellular, marine and planktonic with an endoskeleton of silica, but unlike that of the similar silicoflagellates, they are solid with a tetraxial or triaxial symmetry. Ebridians possess two flagella of unequal length and lack photosynthetic pigments, surviving instead by the ingestion of food (especially diatoms) with the aid of pseudopodia. Reproduction is mostly by asexual division.

Classification of ebridians is complicated by their uncertain biological status, resembling algal groups such as the silicoflagellates and dinoflagellates as much as animal groups like radiolarians. Generally regarded as algae they are placed by some in the division Chrysophyta and by others in the Pyrrhophyta as a distinct class, the Ebriophyceae.

Genera and species are distinguished on the basis of endoskeleton morphology (see Loeblich *et al.* 1968). For example, *Ebria* (Mioc.-Rec., Fig. 10.10c) has three

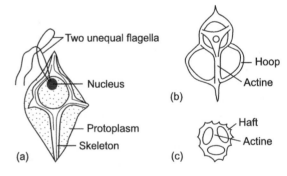

Fig. 10.10 Ebridians. (a) A living cell and skeleton of *Hermesinium*. (b) *Hermesinum* skeleton, ×500. (c) *Ebria* skeleton, ×533. ((a) Based on Hovasse 1934.)

or four radiating bars (**actines**) with the ends joined by curved hoops called **hafts**. *Hermesinum* (Palaeoc.-Rec., Fig. 10.10a,b) consists essentially of four actines resembling a sponge spicule in their tetraxial arrangement, the ends of which are joined by a series of subcircular **hoops**.

Ebridians are known in rocks of Palaeocene age, the majority of genera thriving until the Pliocene when their diversity dropped sharply (Tappan & Loeblich 1972). The geological value of ebridians has been little exploited as yet, largely because they are neither abundant nor uniformly distributed and preserved. None the less, they have been used successfully with silicoflagellates in Cenozoic biozonal schemes, such as those in the north Pacific area (see Ling 1972, 1975). A comprehensive review can be found in Ernisse (in Lipps 1993, pp. 131–141).

REFERENCES

Barnes, R.D. 1968. *Invertebrate Zoology*. W.B. Saunders, Philadelphia.

Bint, A.N. 1986. Fossil Ceratiaceae: a restudy and new taxa from the mid-Cretaceous of the Western Interior, USA. *Palynology* 10, 135–180.

Bujak, J.P. & Williams, G.L. 1981. The evolution of dinoflagellates. *Canadian Journal of Botany* 59, 2077–2087.

Cavalier-Smith, T. 1998. A revised six-kingdom system of life. *Biological Reviews of the Cambridge Philosophical Society* 73, 203–266.

Chapman, V.J. & Chapman, D.J. 1973. *The Algae*. Macmillan, London.

Dale, B. 1976. Cyst formation, sedimentation and preservation: factors affecting dinoflagellate assemblages in Recent sediments from Trondheimsfjord. Norway. *Review of Palaeobotany and Palynology* 22, 39–60.

Dale, B. 1983. Dinoflagellate resting cysts 'benthic plankton'. In: Fryxell, G.A. (ed.) *Survival Strategies of the Algae*. Cambridge University Press, Cambridge, pp. 68–136.

Damassa, S.P., Goodman, D.K. & Kidson, E.J. 1990. Correlation of Paleogene dinoflagellate assemblages to standard nannofossil zonation in North Atlantic DSDP sites. *Review of Palaeobotany and Palynology* 65, 331–339.

Davies, E.H. & Norris, G. 1980. Latitudinal variations in encystment modes and species diversity in Jurassic dinoflagellates. In: Strangway, D.W. (ed.) *The Continental Crust and its Mineral Deposits. Special Paper. Geological Association of Canada* 20, 361–373.

Dodge, J.D. 1985. *Atlas of Dinoflagellates*. Farrand Press, London.

Edwards, L.E., Mudie, P.J. & Devernal, A. 1991. Pliocene paleoclimatic reconstruction using dinoflagellate cysts – comparison of methods. *Quaternary Science Reviews* 10, 259–274.

Ellegaard, M. 2000. Variation in dinoflagellate cyst morphology under conditions of changing salinity during the last 2000 years in the Limfjord Denmark. *Review of Palaeobotany and Palynology* 109, 65–81.

Eshet, Y., Almogi-Labin, A. & Bein, A. 1994. Dinoflagellate cysts, paleoproductivity and upwelling system: a Late Cretaceous exampe from Israel. *Marine Micropalaeontology* 23, 231–240.

Evitt, W.R. 1961. Observations on the morphology of fossil dinoflagellates. *Micropalaeontology* 7, 385–420.

Evitt, W.R. 1963. A discussion and proposals concerning fossil dinoflagellates, hystrichospheres and acritarchs. *Proceedings of the National Academy of Sciences USA* 49, 158–164, 298–302.

Evitt, W.R. 1969. Dinoflagellates and other organisms in palynological preparations. In: Tschundy, R.H. & Scott, R.A. (eds) *Aspects of Palynology*. Wiley Interscience, New York, pp. 439–481.

Evitt, W.R. 1985. *Sporopollenin Dinoflagellate Cysts. Their morphology and interpretation*. American Association of Stratigraphic Palynologists Foundation, Salt Lake City.

Fensome, R.A., Gocht, H., Stover, L.E. & Williams, G.L. 1991. *The Eisenack Catalog of Fossil Dinoflagellates*, new series, vol. 1. E. Schweizerbart'sche Verlagsbuchhandlung, Stuttgart, pp. 1–828.

Fensome, R.A., Gocht, H., Stover, L.E. & Williams, G.L. 1993a. *The Eisenack Catalog of Fossil Dinoflagellates*, new series, vol. 2. E. Schweizerbart'sche Verlagsbuchhandlung, Stuttgart, pp. 829–1461.

Fensome, R.A., Taylor, F.J.R., Norris, G., Sarjeant, W.A.S., Wharton, D.I. & Williams, G.L. 1993b. A classification of living and fossil dinoflagellates. *Micropalaeontology*, special publication, no. 7.

Funnell, B.M. & Riedel, W.R. (eds) 1971. *The Micropalaeontology of Oceans*. Cambridge University Press, Cambridge.

Habib, D., Moshkovitz, S. & Kramer, C. 1992. Dinoflagellate and calcareous nannofossil response to sea level changes in the Cretaceous–Tertiary boundary sections. *Geology* 20, 165–168.

Hallett, R. & Lewis, J. 2001. Salinity, dinoflagellate cyst growth and cell biochemistry. *34th Annual Meeting of the American Association of Stratigraphic Palynologists*, abstracts.

Haq, B.U., Hardenbol, J. & Vail, P.R. 1987. Chronology of fluctuating sea levels since the Triassic. *Science* 235, 1156–1167.

Head, M.J. 1993. Dinoflagellates, sporomorphs and other palynomorphs from the Upper Pliocene, St Erth Beds of Cornwall, southwestern England. *Palaeontological Society, Memoir* 31, 1–62.

Head, M.J. & Wrenn, J.H. (eds) 1992. *Neogene and Quaternary Dinoflagellate Cysts and Acritarchs*. American Association of Stratigraphic Palynologists Foundation, Salt Lake City.

Jansonius, J. & McGregor, D.C. (eds) 1996. *Palynology: Principles and Applications*. American Association of Stratigraphic Palynologists, Salt Lake City.

Jarvis, I., Carson, G.A., Cooper, M.K.E., Hart, M.B., Leary, P.N., Tocher, B.A., Horne, D. & Rosenfeld, A. 1988. Microfossil assemblages at the Cenomanian-Turonian (Late Cretaceous) Oceanic Anoxic Event. *Cretaceous Research* 9, 3–103.

Kofoid, C.A. & Swezy, O. 1921. *The Fossilizing Unarmoured Dinoflagellata*. California Press, Berkeley.

Lentin, J.K. & Williams, G.L. 1981. Fossil dinoflagellates: index to genera and species, 1981 edition. *Report Series. Bedford Institute of Oceanography* B1-R-81-12, 1–345.

Lipps, J.H. (ed.) 1993. *Fossil Prokaryotes and Protists*. Blackwell Scientific, Boston.

Matthiessen, J. 1995. Distribution patterns of dinoflagellate cycsts and other organic-walled microfossils in recent Norwegian–Greenland Sea sediments. *Marine Micropalaeontology* 24, 307–334.

Moldowan, J-M. & Talyzina, N.M. 1999. Biogeochemical evidence for the dinoflagellate ancestors in the Early Cambrian. *Science* 281, 1168–1170.

Monteil, E. 1993. Dinoflagellate cyst biozonation of the Tithonian and Berriasian of southeast France correlation with seismic stratigraphy. *Bulletin des Centres de Reseches Exploration-Production Elf-Aquitaine* 17, 249–275.

Mudie, P.J., De Vernal, A. & Head, M.J. 1990. Neogene to Recent palynostratigraphy of circum-arctic basins: results of ODP leg 104, Norwegian Sea, leg 105, Baffin Bay, and DSDP site 611, Irminer Sea. In: Bleil, U. & Thiede, J. (eds) *Geological History of Polar Oceans: Arctic versus Antarctic*. Kluwer Academic Publishers, Dordecht, pp. 609–646.

Palliani, R.B. & Riding, J.B. 1999. Relationships between the Early Toarcian anoxic event and organic-walled phytoplankton in central Italy. *Marine Micropalaeontology* 37, 101–116.

Popovsky, J. & Pfeister, L.A. 1990. Dinophyceae (Dinoflagellida). In: Ettl, H., Gerloff, J., Heynig, H. & Mollenhauer, D. (eds) *Süsswasserflora von Mitteleuropa; begründet von A. Pascher*, vol. 6. Gustav Fischer Verlag, Jena.

Powell, A.J. (ed.) 1992. *A Stratigraphical Index of Dinoflagellate Cysts*. British Micropalaeontological Publication Series. Chapman & Hall, London.

Powell, A.J., Lewis, J. & Dodge, J.D. 1992. The palynological expressions of post-Palaeogene upwelling: a review. In: Summerhayes, C.P., Prell, W.I. & Emeis, K.C. (eds) Upwelling systems: evolution since the Early Miocene. *Geological Society of London, Special Publication* 64, 215–226.

Ramsay, A.T.S. (ed.) 1977. *Oceanic Micropalaeontology*. Academic Press, London.

Riding, J.B., Moorlock, B.S.P., Jeffery, D.M., *et al.* 1997. Reworked and indigenous palynomorphs from the Norwich Crag Formation (Pleistocene) of eastern Suffolk: implications for provenance, palaeogeography and climate. *Proceedings. Geological Association* 108, 25–38.

Riding, J.B., Head, M.J. & Moorlock, B.S.P. 2000. Reworked palynomorphs from the Red Crag and Norwich Crag formations (Early Pleistocene) of the Ludham Borehole, Norfolk. *Proceedings. Geological Association* 111, 161–171.

Sarjeant, W.A.S. 1974. *Fossil and Living Dinoflagellates*. Academic Press, London.

Spector, D.L. (ed.) 1984. *Dinoflagellates*. Academic Press, Orlando.

Stancliffe, R.P.W. & Sarjeant, W.A.S. 1988. Oxfordian dinoflagellate cysts and provincialism. In: Rocha, R.B. & Soares, A.F. (eds) *Second International Symposium on Jurassic Stratigraphy, Lisbon, 1987. Centro de Estratigrahia e Paleobiologia da Universidade Nova de Lisboa e Centra de geosciencias da Universidade de Coimbra, Lisbon* 2, 763–798.

Stanley, E.A. 1966. The problem of reworked pollen and spores in marine sediments. *Marine Geology* 4, 397–408.

Stover, L.E. & Hardenbol, J. 1994. Dinoflagellates and depositional sequences in the Lower Oligocene (Rupelian) Boom Clay Formation, Belgium. *Bulletin de la Société belge de Géologie* 102, 5–77.

Stover, L.E. & Williams, G.L. 1982. Dinoflagellates. *Proceedings of the Third North American Palaeontological Convention* 2, 525–533.

Tappan, H. 1980. *The Paleobiology of Plant Protists*. W.H. Freeman, San Fransisco.

Taylor, F.G.R. (ed.) 1987. Ecology of dinoflagellates. In: The Biology of Dinoflagellates. *Botanical Monographs*, vol. 21. Oliver and Boyd, Edinburgh, pp. 399–501.

Wall, D.B., Dale, B., Lohmann, G.P. & Smith, W.K. 1977. The environmental and climatic distribution of dinoflagellate cysts in modern marine sediments from regions in the

North and South Atlantic Oceans and adjoining seas. *Marine Micropalaeontology* 2, 121–200.

Williams, G.L. & Bujak, J.P. 1977. Distribution patterns of some North Atlantic Cenozoic dinoflagellate ccysts. *Marine Micropalaeontology* 2, 223–233.

Williams, G.L. & Bujak, J.P. 1985 Mesozoic and Cenozoic dinoflagellates. In: Bolli, H.M., Saunders, J.B. & Perch-Nielsen, K. (eds) *Plankton Stratigraphy*. Cambridge University Press, Cambridge, pp. 847–965.

Williams, G.L., Ascoli, P., Barss, M.S., Bujk, J.P., Davies, E.H., Fensome, R.A. & Williamson, M.A. 1990. Biostratigraphy and related studies. In: Keen, M.J. & Williams, G.L. (eds) *Geology of the Continental Margin of Eastern Canada; Geology of Canada* (also: *The Geology of North America, Geological Society of America* 1–1, 87–137), pp. 87–137.

Ebridians

Hovasse, R. 1934. Ebriacees, Dinoflagellés et Radiolaires. *Comptes Rendus Hebdonmadaires des Seances* **198**, 402–404.

Ling, H.Y. 1972. Upper Cretaceous and Cenozoic silicoflagellates and ebridians. *Bulletin of American Paleontology* **62**, 135–229.

Ling, H.Y. 1975. Silicoflagellates antlebridians from Leg 31. *Initial Reports of the Deep Sea Drilling Project* 31, 763–773.

Loeblich III, L.A., Tappan, H. & Loeblich Jr, A.R. 1968. Annotated index of fossil and Recent silicoflagellates and ebridians with descriptions and illustrations of validly proposed taxa. *Memoir Geological Society America* no. 106.

Tappan, H. & Loeblich Jr, A.R. 1972. Fluctuating rates of protistan evolution, diversification and extinction. *24th International Geological Congress, Montreal* 7, 205–213.

CHAPTER 11

Chitinozoa

The Chitinozoa are flask- or bottle-shaped, hollow organic vesicles of uncertain affinity. Appearing first in the Early Ordovician, they evolved rapidly through the Palaeozoic Era. The majority became extinct at the end of the Devonian Period. Although Chitinozoa have been reported from the Carboniferous and Permian, these records are considered suspect or may include reworked specimens. Chitinozoa are commonly associated with acritarchs, scolecodonts and graptolites in mudstones and siltstones. Chitinozoan walls are resistant to oxidation, thermal alteration, tectonism and recrystallization of the rock matrix. Indeed, chitinozoans may be the only organic fossils recognizable in some rocks (such as slates) and from this derives their particular value to biostratigraphy and thermal maturity studies.

Morphology

The vesicle

The chitinozoan **vesicle** ranges from 30 to 1500 μm, but most are 150–300 μm long. The vesicle has a longitudinal axis of symmetry, sections taken at right angles to this being radially symmetrical. The wall (Figs 11.1, 11.2) is two-layered and of a dark brown or black chitin-like substance (**pseudochitin**). It encloses an empty **body chamber** that once housed the organism. The **oral** end, which bears the **aperture**, is usually produced into a neck, whilst the **aboral** end is broader and closed. The aperture is occluded by a separate **operculum**, whose form and position is of taxonomic value.

The outer wall of the vesicle may be smooth, striate, tuberculate, hispid (i.e. hairy), folded into hollow spines or extended into a tubular **sleeve**. The inner wall can also give rise to spines that penetrate through the outer wall. Many chitinozoans are found united in long chains or clusters, the vesicles welded together at the operculum (i.e. the **oral pole**) and at the base (i.e. the **aboral pole**, Fig. 11.1a). In certain genera the operculum is deeply recessed within the neck so that adhesion of the adjacent vesicle must be achieved by a basal, tubular appendage called a **copula** (Fig. 11.2c).

Distribution and ecology of chitinozoans

Though this group of microfossils is extinct it is possible to infer their palaeoecology indirectly with reference to associated benthic and trace fossil groups (e.g. Bergström & Grahn 1985; Miller in Jansonius & McGregor 1996, vol. 1, pp. 307–337), enclosing sediments (Laufeld 1974) and morphology (Grahn 1978). Miller (in Jansonius & McGregor 1996, vol. 1, pp. 307–337) reviewed the recurrent species associations through time. Chitinozoa were exclusively marine and can be found in a wide range of shelf environments but are undoubtedly most abundant in outer shelf slope and basinal settings. The presence of abundant Chitinozoa in anoxic black shales indicates the majority were planktic. Environmental controls are very poorly documented. The highest abundances of chitinozoans are in high-latitude waters; reefs were apparently unfavourable habitats (Laufeld 1974).

Classification

Group CHITINOZOA

Because Chitinozoa are invariably opaque with their internal structure obscured, classification has concentrated

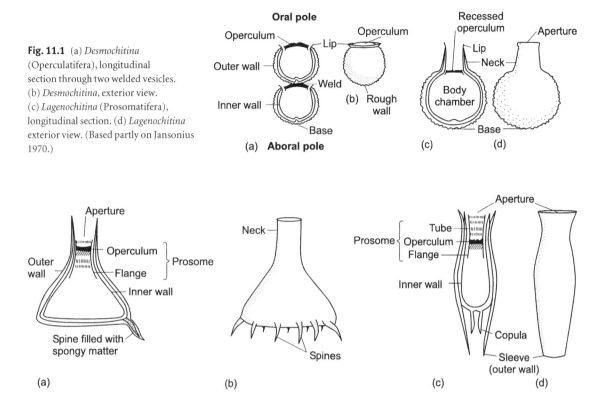

Fig. 11.1 (a) *Desmochitina* (Operculatifera), longitudinal section through two welded vesicles. (b) *Desmochitina*, exterior view. (c) *Lagenochitina* (Prosomatifera), longitudinal section. (d) *Lagenochitina* exterior view. (Based partly on Jansonius 1970.)

Fig. 11.2 Prosomatifera (diagrammatic). (a) *Ancyrochitina*, longitudinal section. (b) *Ancyrochitina*, exterior view. (c) *Velatachitina*, longitudinal section. (d) *Velatachitina*, exterior view. (Based partly on Jansonius 1970.)

largely on form genera defined on their outline shape (i.e. silhouette). Chitinozoa are generally arranged alphabetically under a version of the suprageneric classification proposed by Eisenack (1972). Orders are the highest taxonomic category used. The Operculatifera are characterized by an operculum, reduced oral tubes (usually with a collarette, but lacking a neck). This order contains one family, the Desmochitinidae, and six subfamilies; for example *Desmochitina* (L. Ord.- U. Sil; Fig. 11.1a,b) had a relatively small subspherical vesicle with short **lips** but no neck and was commonly united in chains.

The Prosomatifera have a prosome and well-developed necks. This order contains two families, the Conochitinidae and the Lagenochitinidae, distinguished by the relationship between the chamber and the neck. The operculum is recessed within the neck of *Lagenochitina* (L. Ord.-L. Sil., Fig. 11.1c,d) which had a relatively large vesicle with a cylindrical neck. Twelve subfamilies are distinguished by the type and distribution of the ornament and the morphology of basal edge structures. The Complexoperculati bear a recessed operculum provided with a sleeve-like extension, the **flange**, which together are called the **prosome** (Fig. 11.2). This prosome is simple in the Sphaerochitinidae, whose vesicles lack aboral sleeves and copulae, as for example in *Ancyrochitina* (Ord.-Dev., Fig. 11.2a,b) which has a flask-shaped vesicle with a ring of spines around the base. The Tanuchitinidae display elaborate differentiation at the aboral end and their vesicles are often tubular. *Velatachitina* (L. Ord.-L. Sil., Fig. 11.2c,d) is subcylindrical with a sleeve at either end, formed from the outer wall. The inner wall is produced aborally into a copula, whilst

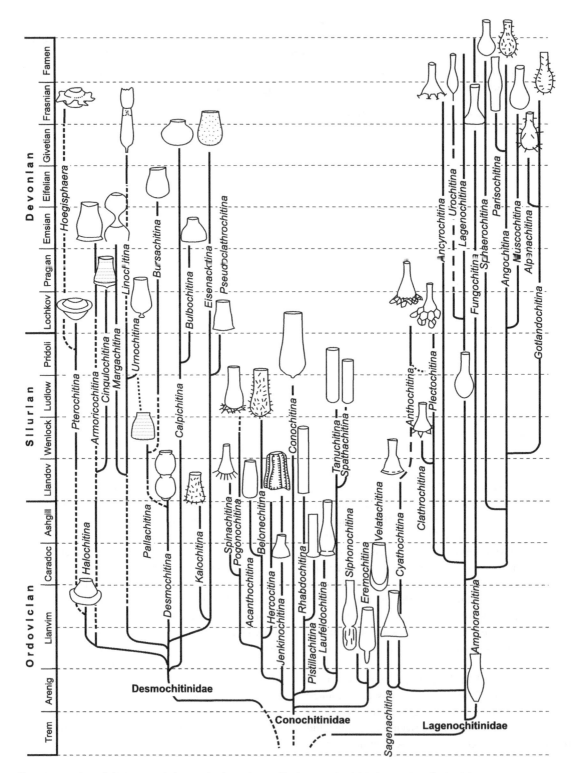

Fig. 11.3 Provisional chitinozoan phylogeny. (Redrawn from Miller in Jansonius & McGregor 1996 after Paris.)

the prosome has an orally extended tube with ring-like markings (**annulations**).

Affinities of chitinozoans

The pseudochitin wall of the Chitinozoa suggests animal affinities, but whether metazoan or protistan is still uncertain. Comparison may be made with the egg-cases of worms and Recent gastropods, an analogy further strengthened by discoveries of fossil cocoons filled with chitinozoan chains and clusters. It is unlikely that these represent the eggs of graptolites, as has been suggested (Jenkins 1970), because the geological ranges are dissimilar. Evidence of asexual reproduction (i.e. budding) and of secondary thickening of the wall are more compatible with a protozoan origin (Cramer & Diez 1970). Similar pseudochitinous shells are built by testacean and foraminiferid rhizopods and by ciliated protozoa like the tintinnids. The chains and clusters may also be compared with those formed by dinoflagellates and acritarchs. Cashman (1990, 1991) reinterpreted a variety of morphological features and suggested an affinity with the Rhizopoda, whereas Jaglin & Paris (1992) have argued that chitinozoans were the egg-cases of some extinct planktonic metazoan.

General history of chitinozoans

Figure 11.3 shows a provisional phylogeny for Chitinozoa. The oldest chitinozoans may be the *Desmochitina*-like sacs from the upper Precambrian Chuar group of Arizona (~750 Ma, Bloeser *et al.* 1977), but this affinity has not yet been demonstrated conclusively. During the Ordovician, morphologically smooth vesicles of desmochitinid and conochitinid type were characteristic, the latter stock gradually dwindling through the Silurian and dying out at the end of it. There is a general evolutionary trend towards smaller size. The more complex tanuchitinids also appeared in the Arenig, whilst the sphaerochitinids appeared in the Caradoc and came to characterize Silurian and Devonian assemblages. Chitinozoa with stout basal horns and appendix-bearing forms are typical of many Silurian and Early Devonian assemblages, although Late Devonian forms are more often covered with short spines. Chains of Chitinozoa are also prominent in Silurian and Devonian assemblages. Carboniferous chitinozoans are rare, following extinctions from the mid-Silurian onwards, but specimens have been reported from Permian sediments (Tasch 1973, p. 826) though these may be fungal spores.

Applications

Because of their rapid evolution and widespread nature Chitinozoa can be useful both for local and global stratigraphical correlations (e.g. Keegan *et al.* 1990; Grahn 1992; Al Hajri 1995; Verniers *et al.* 1995), particularly in the subsurface. Their resistant nature allows dating of metamorphosed and deformed rocks, such as phyllites from the Black Forest, Germany (Montenari *et al.* 2000).

Although the majority of chitinozoan genera appear to be cosmopolitan, others show latitudinal provinciality. Separate biozonations are required for Baltic (Nôlvak & Grahn 1993), Gondwana (Paris 1990) and Laurentia (Achab 1989). Paris (1993) applied the distribution of chitinozoans as a test of a palaeogeographic reconstruction of Europe during the Early Palaeozoic. Chitinozoa reflectance studies are becoming important in the elucidation of the thermal histories of Palaeozoic sedimentary basins (Tricker 1992; Tricker *et al.* 1992; Obermayer *et al.* 1996).

Further reading

Useful introductions to the Chitinozoa may be found in Miller (in Jansonius & McGregor 1996, vol. 1, pp. 307–337). Paris (in Jansonius & McGregor 1996, vol. 2, pp. 531–553) provides an excellent review of chitinozoan biostratigraphy and palaeoecology.

Hints for collection and study

Fossil Chitinozoa may be extracted from Palaeozoic argillaceous rocks by the same techniques recommended for acritarchs and other organic-walled microfossils (q.v.). Make permanent mounts from strews of the organic residue on glass slides and view with transmitted light.

REFERENCES

Achab, A. 1989. Ordovician chitinozoan zonation of Québec and western Newfoundland. *Journal of Paleontology* 63, 14–24.

Al Hajri, S. 1995. Biostratigraphy of the Ordovician Chitinozoa of Northwestern Saudi-Arabia. *Review of Palaeobotany and Palynology* 89, 27–48.

Bergström, S.M. & Grahn, Y. 1985. Biostratigraphy and paleoecology of chitinozoans in the lower Middle Ordovician of the Southern Appalachians. In: Shumaker, R.C. (ed.). *Appalachian Basin Industrial Associates Program – Spring Meeting* 8, 6–31.

Bloeser, B., Schopf, J.W., Horodyski, R.J. & Breed, J.W. 1977. Chitinozoans from the Late Precambrian Chuar Group of the Grand Canyon, Arizona. *Science* 195, 67–69.

Cashman, P.B. 1990. The affinity of the chitinozoans: new evidence. *Modern Geology* 5, 59–69.

Cashman, P.B. 1991. Lower Devonian chitinozoan juveniles – oldest fossil evidence of a juvenile stage in protists, with an interpretation of their ontogeny and relationship to allogromiid Foraminifera. *Journal of Foraminiferal Research* 21, 269–281.

Cramer, F.H. & Diez, M. del C.R. 1970. Rejuvenation of Silurian chitinozoans from Florida. *Revista Espanola de Micropalaeontologia* 2, 45–54.

Eisenack, A. 1972. Chitinozoen und andere Mikrofossilien aus der Bohrung Leba, Pommern. *Palaeontographica, Abteilung A* 139, 64–87.

Grahn, Y. 1978. Chitinozoan stratigraphy and palaeoecology at the Ordovician–Silurian boundary in Skåne, southernmost Sweden. *Sveriges Geologiska Undersökning, Series C* 744, 1–16.

Grahn, Y. 1992. Ordovician chitinozoa and biostratigraphy of Brazil. *Geobios* 25, 703–723.

Jaglin, J.C. & Paris, F. 1992. Examples of Teratology in the Chitinozoa from the Pridoli of Libya and implications for biological significance of this group. *Lethaia* 25, 151–164.

Jansonius, J. 1970. Classification and stratigraphic application of Chitinozoa. *Proceedings. North American Paleontological Convention 1969, Part G* 789–808.

Jansonius, J. & McGregor, D.C. (eds) 1996. *Palynology, Principles and Applications*, vols 1–3. American Association of Stratigraphic Palynologists Foundation, Salt Lake City.

Jenkins, W.A.M. 1970. Chitinozoa. *Geoscience and Man* 1, 1–20.

Keegan, J.B., Rasul, S.M. & Shaheen, Y. 1990. Palynostratigraphy of the Lower Palaeozoic, Cambrian to Silurian of the Hashemite Kingdom of Jordan. *Review of Palaeobotany and Palynology* 66, 167–180.

Laufeld, S. 1974. Silurian Chitinozoa from Gotland. *Fossils and Strata* 5, 130pp.

Montenari, M., Sevais, T. & Paris, F. 2000. Palynological dating (acritarchs and chitinozoans) of Lower Palaeozoic phyllites from the Black Forest/southwestern Germany. *Comptes Rendus de l'Academie des Sciences Paris, Sciences de la Terre et des Planets* 330, 493–499.

Nôlvak, J. & Grahn, Y. 1993. Ordovician chitinozoan zones from Baltoscandinavia. *Review of Palaeobotany and Palynology* 79, 245–269.

Obermayer, M., Fowler, M.G., Goodarzi, F. & Snowdon, L.R. 1996. Assessing thermal maturity of Palaeozoic rocks from reflectance of chitinozoa as constrained by geochemical indicators – an example from southern Ontario, Canada. *Marine and Petroleum Geology* 13, 907–919.

Paris, F. 1990. The Ordovician chitinozoan biozones of the Northern Gondwana Domain. *Review of Palaeobotany and Palynology* 66, 181–209.

Paris, F. 1993. Palaeogeographic evolution of Europe during the Early Palaeozoic – the Chitinozoa test. *Comptes Rendus de l'Academie des Sciences Serie II* 316, 273–280.

Tasch, P. 1973. *Paleobiology of the Invertebrates.* John Wiley, New York.

Tricker, P.M. 1992. Chitinozoan reflectance in the Lower Palaeozoic of the Welsh Basin. *Terra Nova* 4, 231–237.

Tricker, P.M., Marshall, J.E.A. & Badman, T.D. 1992. Chitinozoan reflectance – a Lower Palaeozoic thermal maturity indicator. *Marine and Petroleum Geology* 9, 302–307.

Verniers, J., Nestor, V., Paris, F., Dufka, P., Sutherland, S. & Vangrootel, G. 1995. Global chitinozoa biozonation for the Silurian. *Geological Magazine* 132, 651–666.

CHAPTER 12

Scolecodonts

Scolecodonts are the chitinous mouth-parts of marine polychaetous worms. They are organic and are commonly found as disassociated elements in association with acritarchs and chitinozoans in marine shales. They have a patchy geological record from the Early Ordovician to Recent. They are most diverse in the Upper Ordovician-Devonian shallow marine limestones and shales. Though the biostratigraphical utility of scolecodonts has yet to be realized, attempts to use them as geothermometers have proved successful.

Morphology and classification

They vary in size from around 100–200 μm and have a variable morphology, but most are elongated double-walled plates, denticulated along one margin (Fig. 12.1). Colbath & Larson (1980) showed the chitinous layer covers an inner layer of calcium carbonate that dissolves during palynological preparation; in life the elements were filled with soft tissue. Edgar (1984) described a typical scolecodont jaw apparatus as comprising three groups of elements, antero-ventral **maxillae** (Fig. 12.1c–e), antero-dorsal **mandibles** (Fig. 12.1f) and posterior **carriers** (Fig. 12.1g,h). The mandibles are used for muscle attachment and chiselling. The morphological terms used for the description of fossil scolecodonts are based on direct comparison with modern forms (Clarke 1969). The elements of the maxillary apparatus work as a unit but independently from the mandibles and are used for grasping, biting and rarely poisoning the prey. The carriers are used for muscle attachment and support the first maxillae.

The MI elements of the posterior maxillae (Fig. 12.1a) are the most diagnostic elements and are used to define fossil species. Taxonomic difficulties occur due to the disassociated nature of fossil finds and the fact that some element types in different species are morphologically very similar. Only a few complete fossil scolecodont apparatuses have been found (e.g. Tasch & Stude 1965). All Recent and fossil polychaetes with this type of apparatus belong to the order Eunicida. Only four different types of Palaeozoic apparatuses are known.

Geological history and applications

Scolecodonts first appeared in the Lower Ordovician and diversified rapidly (Underhay & Williams 1995) and had their acme in the Palaeozoic. Most work on fossils has been conducted on material from glacial erratics in Poland (Kielan-Jaworowska 1966; Szaniawski 1968) and on outcrop and borehole material from the Baltic region (e.g. Nakrem *et al.* 2001; Erikson 2002). They are uncommon in the Mesozoic and Cenozoic (Jansonius & Craig 1971; Schäfer 1972; Szaniawski 1974; Germeraad 1980; Courtinat *et al.* 1990; Head 1993) and persist into the Recent. Bergman (1995) and Baudu & Paris (1995) have documented some facies restriction in Silurian and Devonian scolecodonts.

Scolecodonts are primarily used in biostratigraphy from the Ordovician to Permian and in thermal maturity studies (e.g. Goodarzi & Higgins 1988; Bertrand 1990; Bertrand & Malo 2001 for a case study). A useful introduction is to be found in Szaniawski (in Jansonius and McGregor 1996, vol. 1, pp. 337–355).

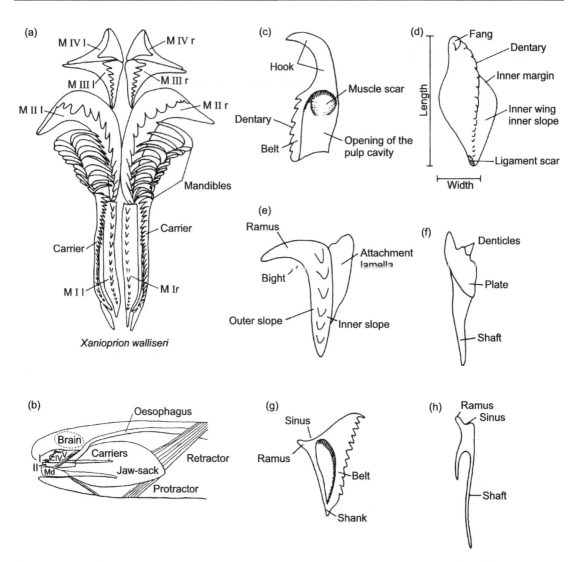

Fig. 12.1 Descriptive terminology of scolecodonts. (a) Diagrammatic representation of the complete apparatus of *Xanioprion walliseri*. (b) Diagrammatic sagittal section through *Eunice siciliensis* showing the relationship of the mandible (Md), maxillae (I–IV) and carriers. (c)–(e) Morphological terminology applied to maxillae. (f) Morphological terminology applied to mandible. (g), (h) Morphological terminology applied to carriers. l, Left; r, right. ((a) After Kielan-Jaworowska 1966; (b) modified from Traverse 1988; (c)–(h) redrawn after Szaniawski in Jansonius and McGregor 1996, vol. 1.)

REFERENCES

Baudu, V. & Paris, F. 1995. Relationships between organic-walled microfossils and paleoenvironments – examples of two Devonian formations from the Armorican Massif and Acquitaine. *Review of Palaeobotany and Palynology* 87, 1–14.

Bergman, C.F. 1995. *Symmetroprion spatiosus* (Hinde), a jawed polychaete showing preference for reef environments in the Silurian of Gotland. *Geologiska Foereningens i Stockholm Foerhandlingar* 127, 143–150.

Bertrand, R. 1990. Correlations among the reflectances of vitrinite, chitinozoans, graptolites and scolecodonts. *Organic Geochemistry* 15, 565–574.

Bertrand, R. & Malo, M. 2001. Source rock analysis, thermal maturation and hydrocarbon generation in the Siluro-Devonian rocks of the Gaspe Belt basin, Canada. *Bulletin. Canadian Petroleum Geology* 49, 238–261.

Clarke, R.B. 1969. Systematics and phylogeny: Annelida, Echiura, Sipuncula. In: Florkin, M. & Scheer, T. (eds) *Chemical Zoology IV, Annelida, Echiura, Sipuncula*. Academic Press, New York, pp. 1–68.

Colbath, G.K. & Larson, S.K. 1980. On the chemical composition of fossil polychaete jaws. *Journal Paleontology* 54, 485–488.

Courtinat, B., Crumiere, J.P. & Meon, H. 1990. Upper Cenomanian organoclasts from the Vocontian Basin (France) – Scolecodonts. *Geobios* 23, 387–397.

Edgar, D.R. 1984. Polychaetes of the lower and middle Paleozoic: a multi-element analysis and phylogenetic outline. *6th International Palynology Conference, Calgary*, abstracts, 39.

Erikson, M. 2002. The palaeobiogeography of Silurian ramphoprionid Polychaete annelids. *Palaeontology* 45, 985–996.

Germeraad, J.H. 1980. Dispersed scolecodonts from Cenozoic strata of Jamaica. *Scripta Geologica* 54, 1–24.

Goodarzi, F. & Higgins, A.C. 1988. Optical properties of scolecodonts and their use as indicators of thermal maturity. *Marine and Petroleum Geology* 4, 353–359.

Head, M.J. 1993. Dinoflagellates, sporomorphs and other palynomorphs from the Upper Pliocene St Erth Beds of Cornwall, southwestern England. *Journal of Paleontology* 67, 1–62.

Jansonius, J. & Craig, J.H. 1971. Scolecodonts: I. Descriptive terminology and revision of systematic nomenclature: II. Lectotypes, new names for homonyms, index of species. *Bulletin. Canadian Petroleum Geology* 19, 251–302.

Jansonius, J. & McGregor, D.C. (eds) 1996. *Palynology, Principles and Applications*, vols. 1–3. American Association of Stratigraphic Palynologists Foundation, pp. 337–355.

Kielan-Jaworowska, Z. 1966. Polychaete jaw apparatuses from the Ordovician and Silurian of Poland and a comparison with modern forms. *Palaeontologia Polonica* 16, 1–152.

Nakrem, H.A., Szaniawski, H. & Mork, A. 2001. Permian-Triassic scolecodonts and conodonts from the Svalis Dome, central Barents Sea, Norway. *Acta Palaeontologica Polonica* 46, 69–86.

Schäfer, W. 1972. *Ecology and Palaeoecology of Marine Environments*. University of Chicago Press, Chicago.

Szaniawski, H. 1968. Three new polychaete jaw apparatuses from the Upper Permian of Poland. *Acta Palaeontologica Polonica* 13, 255–280.

Szaniawski, H. 1974. Some Mesozoic scolecodonts congeneric with Recent forms. *Acta Palaeontologica Polonica* 19, 179–195.

Tasch, P. & Stude, J.R. 1965. A scolecodont natural assemblage from the Kansas Permian. *Transaction. Kansas Academy of Science* 67, 4.

Traverse, A. 1988. *Paleopalynology*. Unwin Hyman, Boston.

Underhay, N.K. & Williams, S.H. 1995. Lower Ordovician scolecodonts from the Cow-Head Group, Western Newfoundland. *Canadian Journal of Earth Sciences* 32, 895–901.

CHAPTER 13

Spores and pollen

Spores and pollen are produced during the life cycle of plants – spores by the lowly bryophytes and ferns, and pollen by the 'higher plants', the conifers and angiosperms. Both types of grain possess a wall that is remarkably resistant to microbial attack and to the effects of temperature and pressure after burial. Produced in vast numbers, these microscopic grains can travel widely and rapidly in wind or water, eventually settling on the bottom of ponds, lakes, rivers and oceans. Such features make them valuable to biostratigraphy, particularly when correlating continental and nearshore marine deposits of Silurian or younger age. Where the ecology of the parent plant is known, spores and pollen can be used for palaeoecological and palaeoenvironmental studies.

Life cycles of 'lower' land plants

Primitive vascular land plants differ from their algal ancestors in their development of special conducting **vascular tissues**. Nonetheless, the **alternation of generations** found in the life cycle of the algae was inherited by the vascular plants. This comprises a life cycle alternating between a spore-producing **sporophyte** generation (reproducing asexually with spores) and a gamete-producing **gametophyte** generation that reproduces sexually with male and female gametes (Fig. 13.1).

Bryophyta (mosses, liverworts and hornworts) appear to have an organization intermediate between the green algae and vascular plants, the **Tracheophyta**. The sporophyte generation is small and totally dependent upon the much larger leaf-bearing gametophyte. The **haploid** gametophyte contains half the chromosomal compliment (1n) of the **diploid** (2n) sporophyte, and is the typical moss or thallose liverwort that is commonly found in damp habitats; it bears apical male (**antheridia**) and female (**archegonia**) reproductive organs. The motile biflagellate sperm swims through a film of water to the archegonia to fertilize the egg. The zygote is the first stage in the sporophyte generation and grows by mitosis into a slender stalk capped by a terminal fruiting body, the **sporangium**. **Spore mother cells** produced in the sporangium divide by meiosis to produce tetrads of four spores; thus, each spore is again haploid. The spores are ejected explosively from the ripe sporangium and germinate in damp habitats, growing to form the prostrate **protonema**, thus completing the life cycle. Bryophytes adapted to dry environments have spores with thick walls capable of long periods of dormancy.

The term **pteridophyta** has no natural classificatory significance but is used here to define the ferns (the **Pterophyta**) and fern allies (the **Psilophyta, Lycopodophyta** and **Sphenophyta**). In these plants, and also the higher pollen and seed-producing Tracheophyta, the sporophyte is much larger and predominates over the gametophyte. Some pteridophytes (e.g. most ferns and some lycopsids) are **homosporous**, producing one type of spore. Other pteridophytes are **heterosporous** producing a male **microspore** and a much larger female **megaspore** (e.g. *Tuberculatisporites*, Fig. 13.13c). In fossil assemblages it can sometimes be difficult to distinguish the two spore types, particularly in Devonian assemblages (e.g. Scott & Hemsley in Jansonius & McGregor 1996, vol. 2, pp. 629–641). The term **miospore** is used to include all spores less than 200 μm in diameter. Heterosporous plants (Fig. 13.2) include the extant fern orders Marsileales and Salvinales

Chapter 13: Spores and pollen   **105**

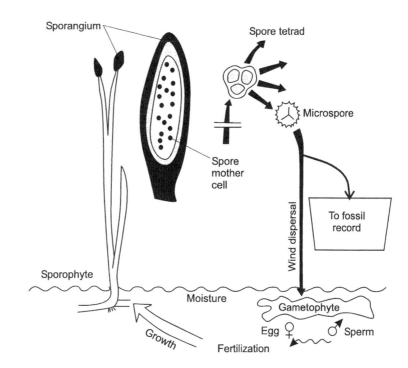

Fig. 13.1 Reconstructed life cycle of a homosporous plant, the Devonian psilopsid *Rhynia*.

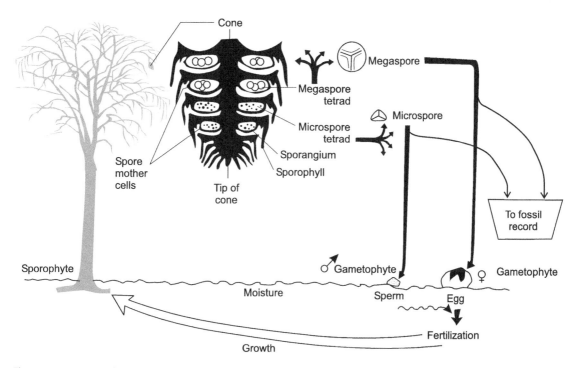

Fig. 13.2 Reconstructed life cycle of a heterosporous plant, the Carboniferous lycopsid *Lepidodendron*.

and four orders of the Division Lycopodophyta (including the extinct Lepidodendrales and the living 'club-mosses', the Isoetales and Selaginellales).

Spore morphology

The morphology of spores can be described according to their shape, apertures, wall structure and size. The shape of a spore owes much to the nature of the meiotic divisions of the spore mother cell. In simultaneous meiosis, the mother cell splits into a **tetrad** consisting of four smaller cells. In tetrahedral tetrads each of the four spores is in contact with all three of its neighbours on the proximal face (Figs 13.3–13.5). The proximal face is characterized by three contact areas that are defined by a Y mark or trilete mark centred on the proximal pole. The arms of the trilete mark may extend to the equator and can take the form of raised ridges or fissures in the surface, **laesurae** (Fig. 13.4). The exterior surface of the spore in the tetrad is the **distal polar face**. In successive stages of meiosis, the mother cell divides at first into two cells, these subdivide further along a single plane at right angles to the first division, or along two planes at right angles (Fig. 13.3). The tetrads here are **tetragonal** and may resemble the segments of an orange in shape; each spore is only in contact with two of its neighbours and only has two contact areas and a single scar. These spores are often bean-shaped. Spores are most commonly compressed proximo-distally in fossil material. The equatorial contour is called the **amb**.

The spores of vascular plants are characterized by well-formed and consistently placed **germinal apertures**. These allow ready germination of the prothallus and accommodate size changes caused by fluctuations in humidity. The form and position of these apertures are important in describing and classifying fossil spores (and pollen).

Trilete spores have three laesurae, which radiate 120 degrees from the proximal pole (Fig. 13.4). The symmetry of trilete spores is therefore **radial**, but heteropolar, i.e. with differently formed polar faces. **Monolete** spores tend to be less common, although they may be abundant in Palaeogene-Recent assemblages and only have one proximal laesura (the **monolete mark**) which separates the contact areas (Fig. 13.5). The symmetry of monolete spores is therefore **bilateral** and heteropolar. Some spores that bear tetrad scars but lack laesurae possess a **hilum**. This can be developed on either the proximal or distal faces and

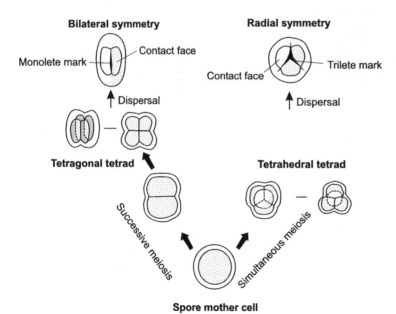

Fig. 13.3 Meiosis and the production of bilaterally or radially symmetrical spores.

Chapter 13: Spores and pollen   107

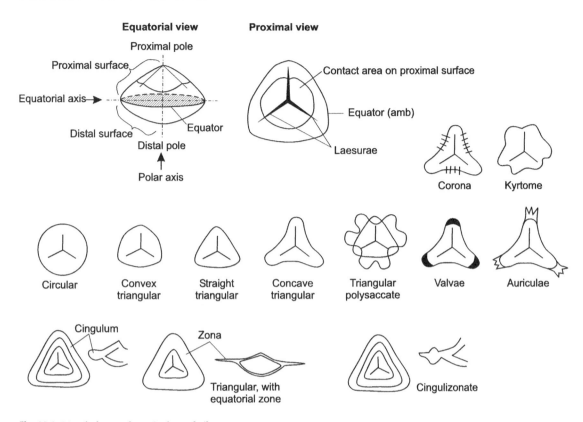

Fig. 13.4 Morphology and terminology of trilete spores.

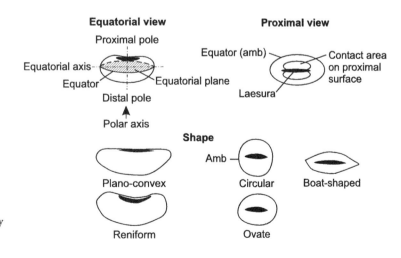

Fig. 13.5 Morphology and terminology of monolete spores.

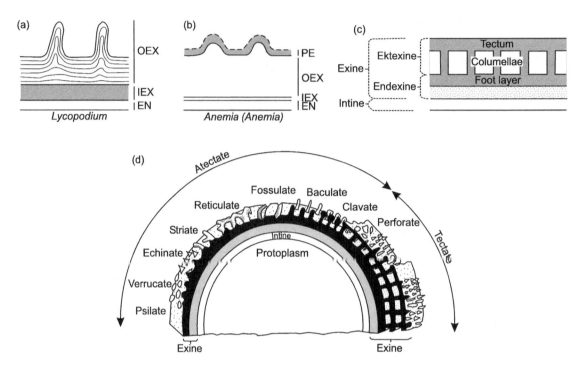

Fig. 13.6 Wall structures and surface ornament found in spores and pollen grains (diagrammatic). (a) The wall structure in the extant *Lycopodium*. (b) Wall structure of the extant genus *Anemia (Anemia)*. (c) Tectate wall of angiosperm pollen. (d) Wall structure and surface ornament of angiosperm pollen. The latter terms are also applied to the surface ornament of spores. Abbreviations of the sporoderm layers: EN, endospore; PE, perispore; OEX, outer exospore; IEX, inner exospore. ((a) After Uehara *et al.* 1991; (b) after Schraudolf 1984.)

functions as the germinal exit in many bryophytes. Spores lacking any apparent dehiscence structures are termed **alete**.

The development of a multilayered wall structure of spores and pollen is markedly different and the two may not be homologous (Fig. 13.6). The inner cellulose layer, or the **endospore**, rarely survives fossilization, the **exospore** is either a single layer or multilayered and consists largely of sporopollenin. The **perispore** is external to the exospore and is composed of sporopollenin material that is more electron dense than the exospore. The wall of many fossil spores (the **sporoderm**) has only one exine layer. Where two layers are present they can be in contact (**acavate**) or are separated to varying degrees (**cavate**). The cavum is most commonly developed in a distal or equatorial position. The layers may be homogeneous or finely lamellate. The layers can be uniform in thickness or variably thickened. A continuous equatorial thickening is known as a **cingulum**; a continuous equatorial flange is a **zona**. Spores with composite equatorial features are termed **cingulizonate**. Discontinuous equatorial features usually developed in the radial areas are **valvae** (smooth) and **auriculae** (ear-like thickenings commonly fluted). The inter-radial areas can also develop **flanges**, **coronae** or **kyrtomes**.

Spore surface sculpture is equally diverse with the descriptive terms also applied to pollen grains. The superficial sculpture of the exine is of considerable importance in the description and classification of spores and pollen grains (Fig. 13.6d). In atectate spores and pollen the surface may be smooth (**psilate** or **laevigate**), covered with small grains (**verucate** or **granulate**), grooved (**fossulate**, Fig. 13.13g), with mesh-like sculpture (**reticulate**), with fine parallel grooves (**striate**), warty (**verrucate**), with rod-like projections (**baculate**), with pointed projections (**echinate**) or with club-shaped projections (**clavate**).

Cryptospores

Spore-like bodies or **cryptospores** have been described from Middle Ordovician, Silurian and Lower Devonian rocks of continental and nearshore settings. They include 'permanent' alete monads, dyads and tetrads (e.g. *Tetrahedrales*, Fig. 13.12a). The botanical and evolutionary affinities of these microfossils are highly controversial. Some resemble the spores of modern bryophytes (Gray 1985; Richardson 1992) and some are referable to the Turma Hilates. A detailed description of this group can be found in Richardson (in Jansonius & McGregor 1996, vol. 2, pp. 555–575).

Life cycle of the 'higher plants'

In gymnosperms (Fig. 13.7) and angiosperms (Fig. 13.8) the gametophyte generation is reduced to a few cells represented by the **ovum** (or **ovule**; female) and the pollen grain (male). In gymnosperms the **megasporangium** (ovule) produces an exposed egg that is fertilized by a free-swimming sperm or one introduced through a **pollen tube**. Modern angiosperm pollen grains contain a tube cell nucleus (which controls the development of the pollen tube) and a generative cell nucleus, which divides before fertilization. The two generative nuclei effect **double fertilization**, in which one nucleus unites with the ovum to produce the zygote and the second unites with two subsidiary sexual nuclei of the female gametophyte to produce a **triploid** (3n chromosomal compliment) endosperm nucleus. This develops into the **endosperm** that nourishes the zygote within the seed. The flowering plants with which we are all familiar are the sporophyte generation.

The vast array of form in flowers and pollen grains reflects adaptations to the many types of pollination mechanism used by the angiosperms. The most common of these is transfer of pollen by insects (**entomophily**). Wind pollination (**anemophily**) is important to the palynologist because of the large quantities of pollen produced by these plants. Much of

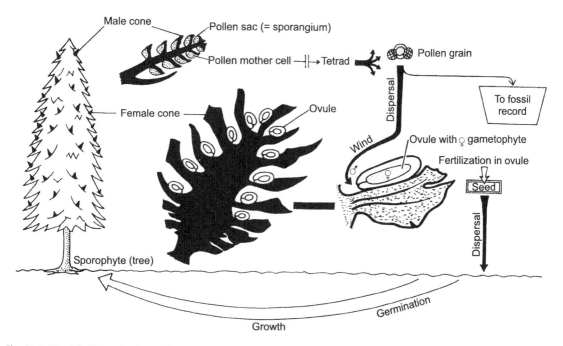

Fig. 13.7 Simplified life cycle of a coniferous gymnosperm.

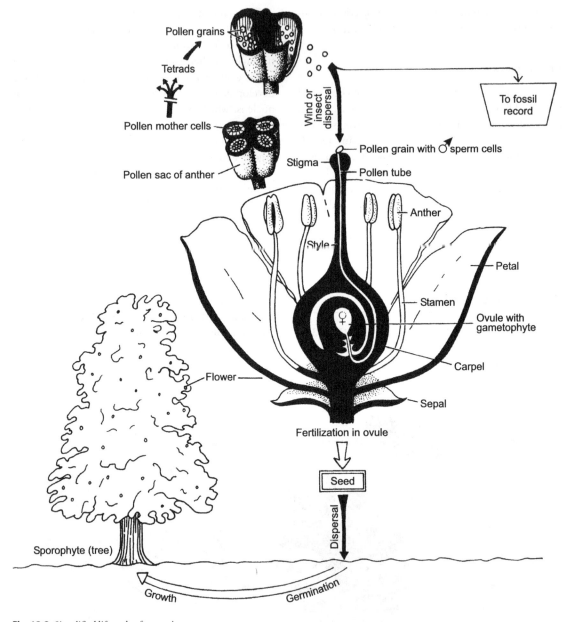

Fig. 13.8 Simplified life cycle of an angiosperm.

this contributes to 'pollen rain' and is preserved in sediments. Most wind-pollinated angiosperms produce small oval and smooth pollen grains ranging from 20 to 40 μm in diameter. The air sacs of many gymnosperms are known to have functioned to increase buoyancy for long-distance wind transport. However, most anemophilous gymnosperm pollen (produced by the Taxaceae, Taxodiaceae, Cupressaceae, cycads and other groups) is oval to spherical, smooth or weakly sculptured, and lacks air sacs.

Pollen morphology

Gymnosperm pollen varies from small, simple, spherical and inaperturate (e.g. modern *Juniperus* and *Cupressus* pollen) to large bisaccate and ornamented grains (e.g. *Abies* (Fig. 13.11a) and *Pinus* (Fig. 13.11b)) and polyplicate forms (e.g. *Ephedra*; Fig. 13.12q). **Saccate** pollen is characteristic of the gymnosperms and grains can bear one (**monosaccate**, e.g. *Tsuga*; Fig. 13.11c), two (**bisaccate**, e.g. *Abies, Picea, Pinus*; Fig. 13.11b; *Striatopodocarpites*, Fig. 13.13d) or rarely three sacs (**trisaccate**, e.g. *Podocarpus*; Fig. 13.11d). Some modern and fossil cycadophytes and ginkgophytes have produced monosulcate pollen (Fig. 13.9). *Ginkgo* and cycad pollen grains are typically subspherical to ellipsoidal, have a single distal furrow (**sulcus**) and smooth to scabrate outer surface. The sulcus appears to facilitate size increase during hydration. 'Advanced' gymnosperms (e.g. *Gnetum*) have ellipsoidal, striate or polyplicate pollen, or spherical grains with short spines.

The range of morphological variation in angiosperm pollen is considerable and more detailed studies of pollen morphology are available in Erdtman (1986), Traverse (1988), Faegri & Iversen (1989) and Jarzen & Nichols (in Jansonius & McGregor 1996, vol. 3, pp. 2261–293). Angiosperm pollen can be shed singly (**monads**), in pairs (**dyads**), in groups of four (**tetrads**) or in multiples of four (**polyads**). Individual grains can be **inaperturate**, or have one or more pores (monoporate, diporate, triporate, etc.), or slit-like apertures or **colpi** (monocolpate, tricolpate, etc., Fig. 13.10), or these features can be equatorial (**stephanoporate** or **colpate**, Fig. 13.13f) or distributed over the whole grain (**peri**-). There are numerous variations and combinations of apertureal arrangement. **Triprojectate** pollen (e.g. the extinct *Aquilapollenites* Fig. 13.11e) have apertures on three projecting arms. **Occulate** grains ('Occulata'), typified by the Late Cretaceous to Palaeogene genus *Woodhousia* (Fig. 13.11f), and have an elongate disc-shaped central body surrounded by a spinose flange.

The wall of the pollen grain comprises two layers, the outer, highly resistant **exine** and the inner **intine** that surrounds the cytoplasm (Fig. 13.6c). The exine is divided into two sublayers, the inner **endexine** and the outer **ektexine**. The ektexine consists of a basal layer with projecting **columellae**, these may be free distally (**intectate**), partially connected by a **tectum** (**semitectate**) or completely covered (**tectate**). In pollen grains the clavate condition, by expansion of the tops, may

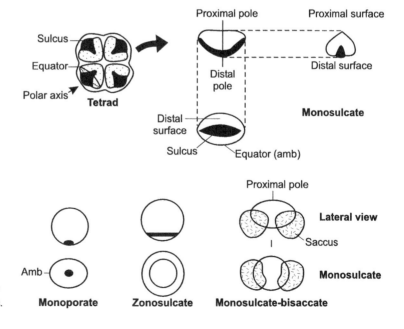

Fig. 13.9 Morphology and terminology of monosulcate and related pollen grains.

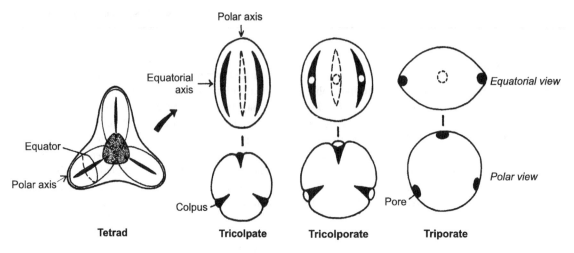

Fig. 13.10 Morphology and terminology of tricolpate and related pollen grains.

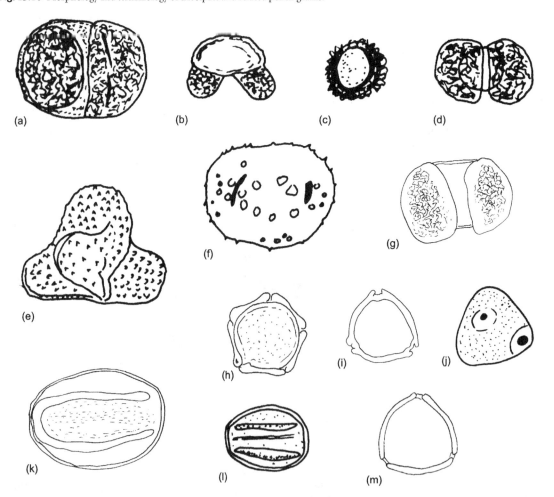

Fig. 13.11 Diagrammatic representation of pollen grains mentioned in the text. (a) *Abies* (Pleistocene), ×250. (b) *Pinus* (Recent), ×350. (c) *Tsuga* (Recent), ×1200. (d) *Podocarpus* (Cretaceous), ×500. (e) *Aquilapollenites* (Cretaceous), ×1400. (f) *Woodhousia* (Cretaceous), ×1140. (g) *Picea*, ×325. (h) *Alnus*, ×1400. (i) *Betula*, ×1600. (j) *Carpinus*, ×2000. (k) *Acer*, ×880. (l) *Quercus*, ×1000. (m) *Corylus*, ×1800. ((a), (d) After Tschundy & Scott 1969; (b), (c), (e), (f) after Traverse 1988; (g)–(m) after Moore et al. 1991.)

give rise to a perforate tectum supported by columellae (i.e. tectate). The tectum surface may be smooth or sculptured in much the same way as outlined above.

Spore and pollen taxonomy

The names of fossil spores (*sporae dispersae*) follow the rules of the International Code of Botanical Nomenclature (ICBN, Greuter & Hawksworth *et al.* 2001). This code formally recognizes whole-plant taxa (**eutaxa**), and form-genera and species (**parataxa**, i.e. dispersed spores, pollen grains, dissociated leaves, roots, fruits, seeds and other parts of plants). Morphology provides the only means of classifying dispersed spores and these are defined on the basis of the nature of the germinal opening, equatorial outline, wall stratification and sculpture and any structural modification or thickening of the spore wall. Generic names often reflect the morphology of the forms or perceived affinities (which can be misleading). Hughes (1989) advocated the abandonment of the Linnean System of taxonomy and nomenclature, suggesting a system based on biorecords. This system though more flexible has not been adopted. Jansonius & Hills (1976, and supplements) provide a catalogue and descriptions of fossil spore and pollen genera.

The most widely used classification scheme for spores is that proposed by Potonié (Potonié & Kremp 1954) with subsequent modifications. The outline classification for common and selected fossil spores and pollen is presented herein (Boxes 13.1–13.4) and follows that proposed by Playford & Dettmann (in Jansonius & McGregor 1996, vol. 1, pp. 227–261). When a morphological continuum occurs between species previously considered taxonomically distinct, the concept of the **morphon** can be used. Some morphons may reflect plant evolution.

Distribution and ecology

In a general way spores and pollen reflect the ecology of their parent plants. Because of size sorting in sediments, however, the leaves, wood, seeds and spores of a plant are rarely preserved together. The habitat and ecology of the spore- and pollen-producing plants can, none the less, be inferred, but an understanding of dispersal and sedimentation must precede this.

Dispersal and sedimentation

The distance travelled by air-borne pollen and spores depends greatly on their size, weight, sculpture and on atmospheric conditions. They are most frequently found about 350–650 m above the land surface during the day, but many sink to the surface at night or are brought down by rainfall. Under favourable conditions pollen grains have been known to drift for at least 1750 km, but about 99% tend to settle within 1 km of the source. Only a very small proportion ever reaches the oceans by aerial dispersal.

Once the pollen grains or spores have settled, they stand a chance of entering the fossil record, either by falling directly into bogs, swamps or lakes, or by being washed into them and into rivers, estuaries and seas. By this stage the pollen record has already been filtered by differential dispersal in the air and may now undergo a similar filtering in water. For example, size sorting across the continental shelf can occur; large miospores, pollen grains and megaspores will tend to settle out in rivers, estuaries, deltas or shallow shelf areas, whereas small miospores and pollen grains may settle out in outer shelf and oceanic conditions. Those which are not buried in reducing sediments will tend to become oxidized and may ultimately be destroyed.

Spores and pollen may suffer several cycles of reworking and redeposition, leading to some confusion in the fossil record. Experienced palynologists detect these reworked forms by differences in preservation (e.g. colour, corrosion, abrasion and fragmentation), ecological or stratigraphical inconsistencies and associated evidence for reworking.

Geological history

Sediments from deltaic and lacustrine deposits of Mid-Ordovician to Early Silurian age yield cryptospore monads, dyads, triads and tetrads. *Nodospora* has thickenings of sporoderm along the contacts between members of the tetrad. Some dyads and tetrads have

Box 13.1 Higher taxonomic categories and diagrams of representative genera found within the Turma Triletes, Suprasubturma Acavatitriletes

Subturma	Infraturma	Infraturma	Infraturma	Infraturma
AZONOTRILETES Wall of more or less uniform thickness	**LAEVIGATI** Wall more or less laevigate *Cyathidites*	**RETUSOTRILETI** Proximo-equatorial surface curvaturate *Retusotiletes*	**APICULATI** Wall sculptured with elongate to more or less isodiametric, non-muronate, positive elements	**MURORNATI** Wall more or less reticulated rugulate *Appendicisporites*
			Subinfra. GRANULATI: wall granulate *Granulatisporites*	
			Subinfra. VERRUCATI: wall verrucate *Verrucosisporites*	
			Subinfra. NODATI: wall echinate (spinose, conate) *Dibolisporites*	
			Subinfra. BACULATI: wall baculate or pilate *Raistrickia*	
ZONOTRILETES Wall structurally differentiated equatorially and/or distally (e.g. cingulum, zona or patina present)	**AURICULATI** With radial, equatorial modifications of wall (valvae, auriculae or radial appendages) *Tripartites*	**TRICRASSATI** With interradial equatorial extensions (coronae) or thickenings (interradial crassitides) *Diatomozonotriletes*	**CINGULATI** With continuous equatorial thickening (cingulum), more or less membranous extension (zona), or combination of these (cinguli-zona) *Contignisporites*	**APPENDICIFERI** Spores with appendages *Elaterites*

Subturma
LAGEOTRILETES Wall with proximal beak- or cone-like apical prominence (gula) or extension associated with laesurae *Lagenicula*

Box 13.2 Higher taxonomic categories and diagrams of representative genera found within the Turma Triletes, Suprasubturma Laminatitriletes (I), Suprasubturma Pseudosaccititriletes and Suprasubturma Perinotriletes

Suprasubturma	Subturma		Infraturma	
LAMINATITRILETES Wall cavate, but intexine in fairly close proximity to exoexine	AZONOLAMINATITRILETES Wall layers not differentially thickened or extended		TUBERCULORNATI Exoexine sculptured with such elements as grana, verrucae, coni, spinae, bacula, etc.	HYSTRICOSPORITES *Hystricosporites*
	ZONOLAMINATITRILETES Wall layers not widely separated; sporoderm equatorially thickened and/or extended	CRASSITI Wall equatorially crassitudinous, but not distinctly cingulate; e.g. *Crassispora*	CINGULICAVATI Exoexine equatorially thickened (cingulate) or extended (zonate); e.g. *Densosporites*	PATINATI Distal hemisphere distinctly thicker than proximal; equatorial sporoderm may also be thickened; e.g. *Tholisporites*

Suprasubturma		Infraturma	
PSEUDOSACCITITRILETES Conspicuously cavate (pseudosaccate) spores, with intexine constituting more or less distinct inner body ('mesospore') within exoexine	MONOPSEUDOSACCITI Exoexine appears as a single comprehensive bladder-like inflation about intexinal body and separated from latter equatorially; **cavum** may extend over most or part of proximal and distal hemispheres; e.g. *Endosporites*	POLYPSEUDOSACCITI Separation and inflation of exoexine from intexine variable equatorially to produce three or more pseudosacci; e.g. *Dulhuntyspora*	

Suprasubturma			
		PERINOTRILITES Exospore enveloped by a perimous or episporous layer; e.g. *Crybelosporites*	

Box 13.3 Higher taxonomic categories and diagrams of representative genera found within the Turma Monoletes, Subturmas Azonomonoletes, Zonomonoletes and Cavatomonoletes, and the Turmas Hilates, Aletes and Cystites

Turma: MONOLETES			
Subturma	Infraturma		Subturma
AZONOMONOLETES Wall of more or less uniform thickness	LAEVIGATOMONOLETI Wall laevigate *Laevigatisporites*	SCULPTATOMONOLETI Wall sculptured *Polypodiidite*	ZONOMONOLETES Wall with equatorial thickening or extension. These are very uncommon. *Speciosporites* is the spore of *Pecopteris*
Subturma	Turma: HILATES	Turma: ALETES	Turma: CYSTITES
CAVATOMONOLETES Wall cavate *Aratrisporites*	Spores hilate; i.e. with proximal or distal hilum *Aequitriradites*	Subturma: AZONOALETES *Fabosporites*	Includes large megaspores produced by arborescent lycopods *Cystosporites*

a membrane that encloses the whole unit. Rocks of Llandovery age yield the first spores with conspicuous trilete marks, typified by *Ambitisporites* spp. (Fig. 13.12b). Palynological preparations of this age can also contain tubes and sheets of cuticle that may represent debris from the first subaerial plants. The first macroplant remains of *Cooksonia* are found in deposits of Late Silurian age. From this time onwards the number of macroplant fossils and spore types found increases dramatically, reflecting a major diversification in primitive plants. By the Ludlow approximately 10 spore genera are present. The parent plants of these early spores appear to have had cosmopolitan distributions.

The Devonian probably marks the acme of pteridophytic plants with the appearance of primitive members of the lycophytes (e.g. *Zosterophyllum* and *Baragwathania*), the trimerophytes (e.g. *Psilophyton*) and possible sphenopsids (e.g. *Protohyenia*). These were joined in the Emsian by the progymnosperms which produced true seeds and pollen grains by the Late Famenian. Initially these primitive pollen grains were indistinguishable from trilete miospores and as a result have been called **pre-pollen**. Increasing provinciality during the Devonian led to distinct equatorial-low latitude (North American-Eurasian), Australian and southern Gondwana floras. This increase in provincialism may have been a response to the greater latitudinal spread of the Devonian continents or global cooling associated with the onset of glacial conditions. By the Siegenian microspores had increased in size to 100 µm (e.g. *Ancyrospora*, Fig. 13.12c) and by the Emsian to 200 µm. *Cystosporites* (Fig. 13.12d) can be over 1 cm in maximum dimension. It comprises one large and three aborted spores and may have functioned as a 'seed megaspore'. The importance of true megaspores seems to have declined after the Carboniferous, until the Jurassic and especially the Cretaceous when they may be common again in non-marine deposits.

Chapter 13: Spores and pollen 117

Box 13.4 Higher taxonomic categories (mainly subturma level) and diagrams of representative genera of saccate spores and pollen

Subturma	Infraturma		
MONOSACCITES	TRILETESACCITI: includes pollen of cycads, seed ferns and cordaitean pre-pollen *Schulzosporas*	ALETESACCITI: pollen of primitive conifers *Florinites*	VESICULOMONORADITI: cycadofilicalean pollen *Potonieisporites*
DISSACITES	DISACCITRILETI: coniferalean pollen *Illinites*	DISACCIATRILETI: medullosan pollen *Pityosporites*	
Subturma	Subturma	Subturma	Subturma
STRIATITES: glossopterid and early conifer pollen *Luekisporites*	PRAECOLPATES: medullosan seed-fern pollen *Monoletes*	POLYPLICITES: gnetalean pollen *Vittatina*	MONOCOLPITES: pollen of various members of the ginkgos and cycads *Moncolpopollenites*

Carboniferous floras are extremely well known due largely to extensive coal deposits. They included a wealth of arborescent, heterosporous lycopsids, no doubt liberating clouds of *Lycospora* (Fig. 13.13b), *Lagenicula* and other spined spore species into the air. The horsetails (with *Calamospora* Fig. 13.12e, *Laevigatosporites* Fig. 13.12f, *Reticulatisporites* Fig. 13.13e), seed ferns (with spores and bisaccate pollen) and cordaitaleans (with *Florinites* pollen, Fig. 13.12g) were also important elements. Carboniferous coal swamps were characterized by lycopsids including the well-known *Lepidodendron* and *Sigillaria*, seed-fern trees and shrubs including *Medullosa*, sphenopsid trees and shrubs including *Calamites*, and shrub cordaitaleans such as cordaites which comprised primitive conifers. Tropical deltas have been used to provide analogues for Carboniferous coal swamps (Scheihning & Pfefferkorn 1984). Many spore–plant associations are known for the Carboniferous. Some plants produced more than one spore type in the same microsporangium. For example, *Densosporites* (Figs 13.12h, 13.13a), commonly found in coal seams, is associated with several Carboniferous lycopsids such as *Porostrobus* and *Sporangiostrobus* and the Devonian lycopsid (?) *Barrandeina*.

By the Permian the seed and pollen habit of the gymnosperms had become the dominant life cycles and pollen grains increasingly replace spores in Mesozoic palynological assemblages, particularly from mid-Cretaceous onwards, following the early evolution of the angiosperms.

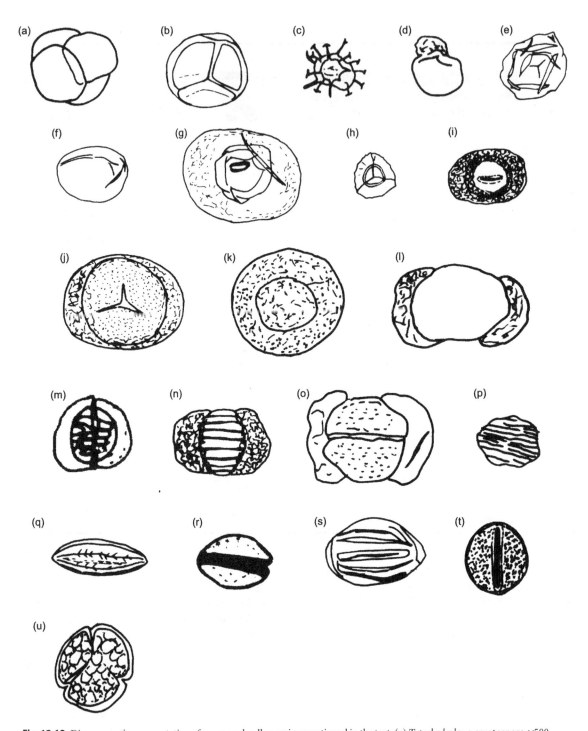

Fig. 13.12 Diagrammatic representation of spores and pollen grains mentioned in the text. (a) *Tetrahedrales*, a cryptospore, ×500. (b) *Ambitisporites*, ×1000. (c) *Ancyrospora*, ×50. (d) *Cystosporites*, ×30. (e) *Calamospora*, ×1000. (f) *Laevigatosporites*, ×350. (g) *Florinites*, ×350. (h) *Densosporites*, ×380. (i) *Potonieisporites*, ×220. (j) *Schulzospora*, ×475. (k) *Wilsonites*, ×670. (l) *Pityosporites*, ×915. (m) *Illinites*, ×420. (n) *Protohaploxypinus*, ×500. (o) *Lueckisporites*, ×560. (p) *Vittatina*, ×320. (q) *Ephedra*, ×1150. (r) *Corollina*, ×1600. (s) *Clavatipollenites*, ×1000. (t) *Eucommiidites*, ×1200. (u) *Tricolpites*, ×500. ((a), (b) After Richardson in Jansonius & McGregor 1996, pp. 555–575); (f)–(j) after Clayton in Jansonius & McGregor 1996, vol. 2, pp. 589–597; (k), (u) after Tschudy & Scott 1969; (l)–(s) after Traverse 1988.)

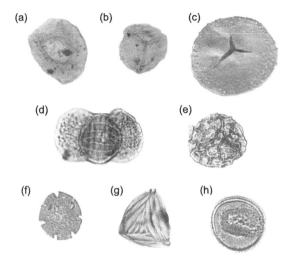

Fig. 13.13 Photomicrographs of selected spores and pollen. (a) *Densosporites annulatus*, a *Lepidodendron* spore found within *Sporangiostrobus* and *Porostrobus* cones, Westphalian B, distal view, ×500. (b) *Lycospora pusilla*, a *Lepidodendron* spore found within the *Lepidostrobus* cone, Westphalian A, proximal view, ×530. (c) SEM photomicrograph of *Tuberculatisporites triangulates*, Westphalian B, proximal view, ×16. (d) *Striatopodocarpites* sp., Permian, ×415. (e) *Reticulatisporites cancellatus*, Visean, ×247. (f) *Nothofagidites brassi*-type, pollen of the Southern Beech, Santonian, ×600. (g) *Appendicisporites* cf *A. potomacensis*, Cenomanian, ×287. (h) *Clavatipollenites hughesii*, Cretaceous, ×695. ((c)–(f) From Traverse 1988 (with the permission of the AASP Foundation); (g) from Playford & Dettmann in Jansonius & McGregor 1996, plate 1, figure 12 (with the permission of Kluwer Academic Publishers).)

The pteridosperms or seed ferns were the first plants to produce pollen. They evolved from the pteridophytes, although the exact nature of this event is unclear; the heterosporous pteridophytes were probably an intermediate stage in their emergence. The oldest known pollen, termed **pre-pollen**, dates from the Late Devonian (Famenian). Chaloner (1970) provided a summary of the morphological differences between spores, pre-pollen and pollen. Gymnosperm pollen with distal germination is first found in Upper Carboniferous deposits. A large number of gymnosperm pollen types evolved in the later Palaeozoic. Saccate pollen grains are the most easily recognized of these and are common among many groups, including the extinct pteridosperms and conifers and cordaitaleans. Monosaccate grains were more common than bisaccates during the Carboniferous, Early Permian and Late Triassic. Carboniferous and Permian genera include *Florinites*, *Potonieisporites* (coniferalean pollen, Fig. 13.12i), *Schulzospora* (pteridosperm pre-pollen, Fig. 13.12j) and *Wilsonites* (cycad pollen, Fig. 13.12k). Upper Palaeozoic bisaccate conifer pollen grains include *Pityosporites* (Fig. 13.12l) and *Illinites* (Fig. 13.12m).

A number of Carboniferous to Triassic gymnosperms produced striate bisaccate pollen grains. Permo-Triassic examples include *Protohaploxypinus* (Fig. 13.12n), *Lueckisporites* (Fig. 13.12o) and *Vittatina* (Fig. 13.12p). Most modern gnetalean gymnosperms produced striate, but non-saccate (**polyplicate**) grains. A modern example is *Ephedra* (Fig. 13.12q). The fossil record of *Ephedra*-like pollen extends from the Mesozoic to the present day.

Circumpolles pollen is unique to certain extinct gymnosperms. These grains have a circumpolar subequatorial groove that divides the grain into two unequal halves, which bear a distal pseudopore and a proximal triangular area. *Corollina* (=*Classopollis*, Fig. 13.12r) is the most well known example. It was produced by a now extinct coniferalean group, the Cheirolepidiaceae. Pollen of this type is common from the mid-Triassic to the mid-Cretaceous. Monosulcate grains are found in cycads and related groups, and are most common in Jurassic samples. Simple monosulcate pollen grains (e.g. some species of *Eucommiidites*, Fig. 13.12t), though resembling primitive angiosperm, pollen were not produced by this group of plants.

Angiosperms evolved from a group of advanced gymnosperms, though the precise relationships are controversial. Angiosperm pollen characteristics include a non-laminate endoexine and a fully differentiated ektexine and many angiosperm pollen grains are triaperturate. The palynological record suggests the angiosperms arose during the Early Cretaceous (Hughes 1976; Hughes & McDougall 1987). Several Late Triassic genera (*Crinopolles* group) have exines of similar structure, but there is no megafossil evidence to support a pre-Cretaceous age for the angiosperms. *Clavatipollenites hughesii* (Barremian, Lower Cretaceous, Figs 13.12s, 13.13h) is one of the earliest angioperm pollen grains; it is monosulcate and has a columellate, tectate exine. *Tricolpites* first appeared in the Albian (Fig. 13.12u) and probably evolved from a *Clavatipollenites*-type ancestor (Chaloner 1970); other tricolpate pollen arose in equatorial latitudes in the Aptian and spread to mid-latitudes by the Albian and polar regions by the Cenomanian (Hickey & Doyle 1977). Either changes in palaeoclimate and palaeogeography may have controlled this geographical spread, or plants evolved rapidly and migrated into cooler latitudes.

The appearance of tricolpate pollen was a major evolutionary innovation and this, plus a seed protected by carpels, was among the reasons for the success of the earliest angiosperms. All the structural features found in modern pollen grains had evolved by the end of the Cenomanian. As angiosperms diversified during the Late Cretaceous they became more provincial in their distribution (Batten 1984).

The modern flora emerged gradually from the Neogene onwards mainly by extinction of relict Cretaceous and Palaeogene species. Two new modern groups that became widespread in the mid-Tertiary are the Asteraceae (the composites) and the Poaceae (the grasses). They arose as a consequence of climate deterioration and have become the most successful of the modern groups, with a vast number of living species. The morphology of their pollen is very different because the grasses are anemophilous and the composites entomophilous. The pollen of the grasses is simple spheroidal and monoporate and is the major cause of hayfever.

The structure of modern plant communities has developed since the last ice age and due to the influence of man some communities have only became established in the last 200 years.

Applications of fossil spores and pollen

Spores and pollen provide a continuous record of the evolutionary history of the vascular plants. Spores were first utilized economically in coal-seam correlation and biostratigraphy (Smith & Butterworth 1967, and references therein) and now have wide-ranging uses in source rock provenance and palaeoenvironmental, palaeoecological and phytogeographical studies.

Silurian to Carboniferous palynozonations are based on spores; pollen grains are more important for dating and correlating younger rocks. Palynozonations can be found for the Silurian in Richardson (in Jansonius & McGregor 1996, vol. 2, pp. 555–575), the Devonian in Streel & Loboziak (in Jansonius & McGregor 1996, vol. 2, pp. 575–589), the Lower Carboniferous in Clayton (in Jansonius & McGregor 1996, vol. 2, pp. 589–597), the Upper Carboniferous in Owens (in Jansonius & McGregor 1996, vol. 2, pp. 597–607), the Permian in Warrington (in Jansonius & McGregor 1996, vol. 2, pp. 607–621) and the Mesozoic and Cenozoic in Batten & Koppelhus (in Jansonius & McGregor 1996, vol. 2, pp. 795–807), Batten (in Jansonius & McGregor 1996, vol. 2, pp. 807–831, 1011–1065) and Frederiksen (in Jansonius & McGregor 1996, vol. 2, pp. 831–843).

Spores and pollen grains are widely utilized in hydrocarbon exploration through thermal maturity studies (Thermal Alteration Index (TAI) and equivalents) and palynofacies analysis (Batten 1996 in Jansonius & McGregor 1996, vol. 3, pp. 1011–1085).

Quantitative spore studies in the 1950s and 1960s demonstrated a clear relationship between spore content and rock type in Carboniferous cyclothems, reflecting changes in vegetation and palaeoenvironments (Smith, 1962, 1968; Chaloner 1968; Eble in Jansonius & McGregor 1996, vol. 3, pp. 1143–1156). Spores and pollen in association with other palynomorphs have an application in delimiting palaeoshorelines (e.g. Frakes *et al.* 1987) and provenance through recycling (Collinson *et al.* 1985).

Pollen analysis

Pollen analysis involves the quantitative examination of spores and pollen at successive horizons through a core, particularly in bog, marsh, lake or delta sediments. This method yields remarkable information on regional changes in vegetation through time, especially in Quaternary sediments where the parent plants are well known, though similar techniques have been used with success in older deposits such as Carboniferous coals. More complete reviews of Quaternary palynology can be found in MacDonald (in Jansonius & McGregor 1996, vol. 2, pp. 879–910). Specific methods

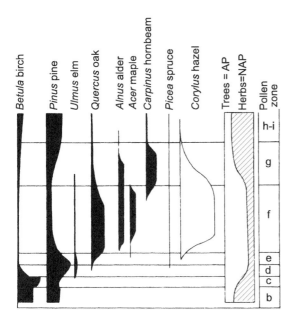

Fig. 13.14 Generalized pollen diagram from the Ipswichian (or Eemian) interglacial deposits. Pollen types are illustrated in Fig. 13.15. AP, arboreal pollen; NAP, non-arboreal pollen. (After West & Pearson, in Tschudy & Scott 1969, modified from figures 17–19.)

and areas of research can be found in Birks & Birks (1980), Faegri & Iversen (1989) and Moore *et al.* (1991).

The relative frequencies of different spore and pollen types are calculated for each of a number of closely spaced sample horizons through the core. Tree pollen (e.g. pine, oak, elm, beech, Fig. 13.11h–m) is often summed together, whilst the non-tree or non-arboreal pollen (NAP, e.g. herbs, grasses) may be documented separately, although expressed as a percentage in relation to the tree pollen. Spores and pollen (e.g. those of sedges, grasses and heather) from bog, heath and lake-vegetation may also be expressed independently, but again in relation to tree pollen. The **pollen spectra** of each species are then arranged alongside to give a **pollen diagram** of palynological changes through the core (Fig. 13.14).

Such diagrams invariably give a biased impression of the flora. Apart from the adverse effects of dispersal, the frequency of flowering or dehiscence, the number of sporangia, cones or flowers, their position relative to the dispersal agencies and the preservation potential of various spores and pollen all have some influence on pollen counts. A large number of statistical techniques are available for analysing pollen data in order to quantify changes in vegetation, rates of migration and vegetation reconstruction through time.

Pollen has been most widely applied in the correlation and palaeoecology of Quaternary deposits. For example, the familiar divisions of the British Recent from Pre-Boreal with birch woodland (about 10,000 years BP) to Sub-Atlantic alder-oak woodland (modern) were based on changing pollen spectra (see West 1968, pp. 279–283, 292–325). Most Quaternary interglacial deposits in temperate latitudes record a change from glacial to cool birch forest with abundant small herbs and shrubs in the late glacial, through pine forest, to a climatic optimum with elm, oak, lime, alder and hazel, followed by a climatic deterioration with pine, birch and then renewed glacial conditions. In England the Flandrian pollen diagram is rather atypical of the Atlantic region because it shows a birch decline at 8500 years BP. At more remote periods the causes of microfloral changes are less certain but ecological successions and biofacies can be recognized (Traverse 1988). A typical Devensian pollen assemblage is shown in Fig. 13.15 with the arboreal component dominated by pine and beech pollen. In North America the continent is too large and the vegetation too variable to produce the same sort of pollen diagrams as for European sections. Classic examples of the influence of Quaternary and Recent climate change on North American vegetation can be found in Davis *et al.* (1980) and Watts (1979), and the effect these changes had on animal populations in Whitehead *et al.* (1982). Comparison of climate models and pollen-derived estimates of climate can be found in Webb *et al.* (1998).

Pollen analysis is also of great assistance to archaeologists, not only because it provides a stratigraphical framework for the Late Quaternary but because of the view it gives of man's early environment and his effect upon it. There was, for example, a curious sudden decline in the tree pollen at the horizon of the late-middle Acheulian (Palaeolithic) hand axe culture in the Hoxnian interglacial (West, in Tschudy & Scott 1969, p. 421) that might have been due to forest clearance. The appearance of human-introduced weeds

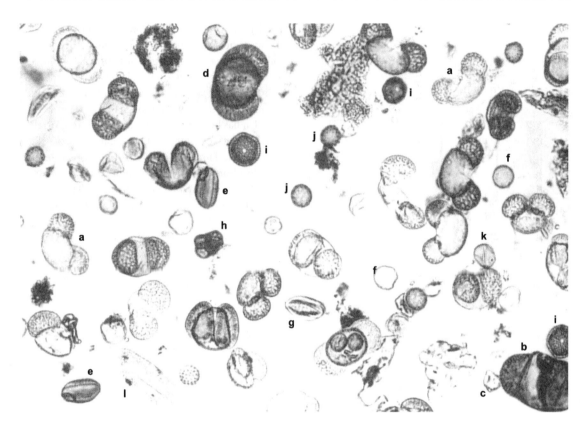

Fig. 13.15 A typical pollen assemblage of the Würmian cold stage (equivalent to the Devensian in Britain and the Wisconsian in central North America) deposits, St Front, France. The arboreal component includes (a) *Pinus*, (b) *Picea*, (c) *Betula* and (d) *Cedrus*. The non-arboreal component includes (e) *Helianthemum* (long axis 45 μm), (f) *Plantago*, (g) *Ephedra*, (h) *Calluna* and pollen from the families (i) Caryophyllaceae, (j) Chenopodiaceae, (k) Poaceae/Gramineae and (l) Liliaceae. (Photomontage from Lowe & Walker 1997, figure 4.1, originally composed by M. Reille and V. Andrieu, reproduced with the permission of Longman, London.)

marks the onset of agriculture, and the spread of heath in Scotland indicates the clearing of forest for grazing (Traverse 1988). Godwin (1967) even outlined the remarkable evidence for cultivation of *Cannabis* in England by Saxons, Normans and Tudors. Leroi-Gourham (1975) showed that 50,000 years BP Neanderthals buried their dead on a blanket of flowers. Palynologists have also studied gut contents and coprolites of various animals to reveal diets and changing climatic conditions at the time. An excellent review of archaeological palynology can be found in Dimbleby (1985).

Pollen and spores can also help sedimentologists to discover the provenance of fine-grained sediments. Sediment samples from the Mississippi delta contain both local and reworked pollen and spores from Devonian upwards. Carboniferous spores are abundant in Recent sediments of the northeast coast of England. Collinson *et al.* (1985) reported reworking of Palaeozoic and Mesozoic megaspores into the Paleocene deposits of southern England. Needham *et al.* (1969) used reworked Carboniferous palynomorphs as tracers of sedimentation patterns in the northwest Atlantic. As with other fossils, pollen and spores can be used to estimate the rate of sedimentation (see Davis 1968).

Although reworking may be a natural hazard for palynologists, Stanley (1967) showed how horizons rich in reworked miospores can be used as correlation markers in deep sea sediments, in this case corresponding with glacial maxima and periods of greatly lowered sea level. Traverse (1974) also noted that reworked spores and pollen were most abundant in Black Sea surface sediments that were deposited during the last glacial maximum. He suggested this was due to rejuvenation and increased erosion as sea levels fell.

Further reading

Invaluable introductions are available in Traverse (1988) and Jansonius & McGregor (1996, 3 volumes). A review of megaspores can be found in Scott & Hemsley (in Jansonius & McGregor 1996, vol. 2, pp. 629–641). Quaternary palynology is reviewed in Lowe & Walker (1997) and (Bradley 1999) and pollen analysis by Moore *et al.* (1991). Further information on palynology can be found at the International Federation of Palynological Studies website http://geo.arizona.edu/palynology/ifps.html and by following the links to other learned societies.

Hints for collection and study

To understand the morphology of fossil spores and pollen, it is particularly worthwhile looking at living material. A collection of common spore and pollen types from trees, shrubs and ferns can readily be made by removing the flowers, cones or sporangia when just on the point of opening. If not examined directly they should be stored in alcoRec. Strew slides can be made by removing the anthers, pollen sacs or sporangia with a scalpel and placing these on a glass slide with a drop of distilled water. While looking down a microscope, bruise the anthers, etc. with a seeker or the blunt edge of a scalpel and spread the released grains over part of the slide. To make the structures more distinct, allow the strew to dry and then add a drop of Gray's spore stain (0.5% malachite green and 0.05% basic fuschin in distilled water; the slide should then be warmed for 1 minute), or basic fuschin stain (0.5% basic fuschin in distilled water) or safranin stain (1 g of safranin 'O' in 50 ml of 95% alcohol plus 50 ml of distilled water). After 10 minutes rinse the slide with a little distilled water and dry at a low temperature. Mount the cover slip with water or glycerine (30% aqueous solution) for temporary preparations, or in Canada Balsam for permanent ones. View with well-condensed transmitted light at 400× magnification or higher. Berglund (1986) contains useful sections on field and laboratory techniques.

Fossil miospores are most readily prepared from plant-bearing muds or shales and from peats, lignites and coals. They can also be very abundant in dark marine shales and mudstones. Palynological laboratories invariably remove the siliceous material with hydrofluoric acid, the calcareous material with hydrochloric acid and the vegetative plant tissues with a variety of strong acids, alkalis and oxidants. Spores and pollen grains can be prepared for study without these sophisticated techniques but the results are inevitably diluted with mineral and vegetable matter. Disaggregation should follow methods A to F (see Appendix), wash as in method G and concentrate as in method H or K. If the organic material is dark and opaque, treat with method E. Temporary mounts in water or glycerine and permanent mounts in glycerine jelly or Canada Balsam can be prepared on glass slides.

Examine the strewn slide with well-condensed transmitted light, using oil immersion objectives for the higher magnifications if possible. Microspores can usually be distinguished from other vegetable matter by their shape, their sharper outlines and often by their amber colour. Much information on palynological techniques can be found in Gray (in Kummel & Raup 1965, pp. 470–706) and in Jones & Rowe (1999).

REFERENCES

Batten, D.J. 1984. Palynology, climate and development of Late Cretaceous floral provinces in the Northern Hemisphere: a review. In: Brenchley, P.J. (ed.) *Fossils and Climate*. John Wiley, New York, pp. 127–164.

Berglund, B.E. 1986. *Handbook of Holocene Palaeoecology and Palaeohydrology*. Wiley, Chichester.

Birks, H.J.B. & Birks, H.H. 1980. *Quaternary Palaeoecology*. Edward Arnold, London.

Bradley, R.S. 1999. *Paleoclimatology: reconstructing climates of the Quaternary*. Academic Press, San Diego.

Chaloner, W.G. 1968. The palaeoecology of fossil spores. In: Drake, E.T. (ed.) *Evolution and Environment*. Yale University Press, New Haven, Conneticut, pp. 125–138.

Chaloner, W.G. 1970. The rise of the first land plants. *Biological Reviews* 45, 353–377.

Collinson, M.E., Batten, D.J., Scott, A.C. & Ayonghe, S.N. 1985. Palaeozoic, Mesozoic and contemporaneous megaspores from the Tertiary of southern England: indicators of sedimentary provenance and ancient vegetation. *Journal of the Geological Society, London* 142, 375–395.

Davis, M.B. 1968. Pollen grains in lake sediments: redeposition caused by seasonal water circulation. *Science* 162, 796–799.

Davis, M.B., Spear, R.W. & Shane, L.C.K. 1980. Holocene climate of New England. *Quaternary Research* 14, 240–250.

Dimbleby, G. 1985. *The Palynology of Archaeological Sites*. Academic Press, London.

Erdtman, G. 1986. *Pollen Morphology and Plant Taxonomy: angiosperms – an introduction to palynology*. E.J. Brill, Leiden.

Faergi, K. & Iversen, J. 1989. *Textbook of Pollen Analysis*, 4th edn. John Wiley, New York.

Frakes, L.A. with 21 other authors 1987. Australian Cretaceous shorelines, stage by stage. *Palaeogeography, Palaeoclimatology, Palaeoecology* 59, 31–48.

Godwin, H. 1967. Pollen analytic evidence for the cultivation of *Cannabis* in England. *Review of Palaeobotany and Palynology* 4, 71–80.

Gray, J. 1985. The microfossil record of early land plants: advances in understanding of early terrestrialization, 1970–1984. *Philosophical Transactions of the Royal Society of London* B309, 167–195.

Greuter, W. & Hawksworth, D.L. (eds) 2001. *International Code of Botanical Nomenclature (Tokyo Code)*. Also online at http://www.bgbm.fu-berlin.de/iapt/nomenclature/code/SaintLouis/0001ICSLContents.htm).

Hickey, L.J. & Doyle, J.A. 1977. Early Cretaceous fossil evidence for angiosperm evolution. *Botanical Review* 43, 3–104.

Hughes, N.F. 1976. The challenge of abundance in palynomorphs. *Geoscience and Man* 11, 141–144.

Hughes, N.F. 1989. *Fossils as Information: new recording and stratal correlation techniques*. Cambridge University Press, Cambridge.

Hughes, N.F. & McDougall, A.B. 1987. Records of angiosperm pollen entry into the English Cretaceous succession. *Review of Palaeobotany and Palynology* 50, 255–272.

Jansonius, J. & Hills, L.V. 1976 et seq. *Genera File of Fossil Spores*. Department of Geology and Geophysics, University of Calgary; Alberta. special publication, with 11 supplements.

Jansonius, J. & McGregor, D.C. (eds) 1996. *Palynology: principles and applications*, vols 1–3. American Association of Stratigraphic Palynologists, Dallas.

Jones, T.P. & Rowe, N.P. (eds) 1999. *Fossil Plants and Spores: modern techniques*. Geological Society, London.

Kummel, B. & Raup, D. (eds) 1965. *Handbook of Paleontological Techniques*. W.H. Freeman, San Francisco.

Leroi-Gourham, A. 1975. The flowers found with Shanidar IV, a Neanderthal burial in Iraq. *Science* 190, 562–564.

Lowe, J.J. & Walker, M.J.C. 1997. *Reconstructing Quaternary Environments*. Longman, London.

Moore, P.D., Webb, J.A. & Collinson, M.E. 1991. *Pollen Analysis*, 2nd edn. Blackwell Scientific Publications, Oxford.

Needham, H.D., Habib, D. & Heezen, B.C. 1969. Upper Carboniferous palynomorphs as a tracer of red sediment dispersal patterns in the northwest Atlantic. *Journal of Geology* 77, 113–120.

Potonié, R. & Kremp, G. 1954. Die Gattungen der paläozoischen *Sporae dispersae* und ihre Stratigraphie. *Geologisches Jahrbuch* 69, 111–194.

Richardson, J.B. 1992. Origin and evolution of the earliest land plants. In: Scopf, J.W. (ed.) *Major Events in the History of Life*. Jones and Bartlett, Boston, pp. 95–118.

Scheihning, M.H. & Pfefferkorn, H.F. 1984. The taphonomy of landplants in the Orinoco Delta: a model for the incorporation of plant parts in clastic sediments of late Carboniferous age Euramerica. *Review of Palaeobotany and Palynology* 41, 205–240.

Schraudolf, H. 1984. Ultrastructural events during sporogenesis of *Anemia phyllitidis* (L.) Sw. *Beiträge zur Biologie der Pflanzen* 59, 237–260.

Smith, A.V.H. 1962. The palaeoecology of Carboniferous peats based on the miospores and petrography of bitumenous coals. *Proceedings. Yorkshire Geological Society* 33, 423–474.

Smith, A.V.H. 1968. Seam profiles and characters. In: Murchison, D.G. & Westoll, T.S. (eds) *Coal and Coal-bearing Strata*. Oliver and Boyd, Edinburgh, pp. 31–40.

Smith, A.V.H. & Butterworth, M.A. 1967. Miospores in the coal seams of the Carboniferous of Great Britain. *Special Papers in Palaeontology* 1, 324pp.

Stanley, E.A. 1967. Palynology of six ocean-bottom cores from the south-western Atlantic Ocean. *Review of Palaeobotany and Palynology* 2, 195–203.

Traverse, A. 1974. Paleopalynology 1947–1972. *Annals of the Missouri Botanical Garden* 61, 203–226.

Traverse, A. 1988. *Paleopalynology*. Unwin Hyman, Boston.

Tschudy, R.H. & Scott, R.A. (eds) 1969. *Aspects of Palynology*. Wiley-Interscience, New York.

Uehara, K., Kurita, S., Sahashi, N. & Ohmoto, T. 1991. Ultrastructural study on microspore wall morphogenesis in *Isoetes japonica* (Isoetaceae). *American Journal of Botany* 78, 1182–1190.

Watts, W.A. 1979. Late Quaternary vegetation of central Appalachia and the New Jersey coastal plain. *Ecological Monographs* 49, 427–469.

Webb, T., Anderson, K.H., Bartlein, P.J. & Webb, R.S. 1998. Late Quaternary climate change in eastern North America: a comparison of pollen-derived estimates with climate model results. *Quaternary Science Reviews* 17, 587–606.

West, R.G. 1968. *Pleistocene Geology and Biology, with Special Reference to the British Isles*. Longman, London.

Whitehead, D.R., Jackson, S.T., Sheehan, M.C. & Leyden, B.W. 1982. Late-glacial vegetation associated with caribou and mastodon in central Indiana. *Quaternary Research* 17, 241–257.

PART 4
Inorganic-walled microfossils

PART A

Inorganic-walled microfossils

CHAPTER 14

Calcareous nannoplankton: coccolithophores and discoasters

Calcareous nannoplankton are a heterogeneous group of calcareous forms, including coccoliths, discoasters and nannoconids, ranging in size from 0.25 to 30 µm. In the fossil record they are found in fine-grained pelagic sediments and can be sufficiently abundant to become rock-forming, for example the Upper Cretaceous chalk. Coccolithophores are unicellular planktonic protozoa with chrysophyte-like photosynthetic pigments, but they differ from most other Chrysophyta in having two flagella of equal length and a third whip-like organ called a **haptonema**. The group is an important constituent of the oceanic phytoplankton, providing a major source of food for herbivorous plankton. Tiny calcareous scales called **coccoliths** (3–15 µm in diameter) form around these cells as a protective armour that eventually falls to the ocean floor to build deep sea ooze and fossil chalks. Being both abundant and relatively easy to recover from marine sediments, coccoliths are used for biostratigraphical correlation of post-Triassic rocks and in palaeoceanographic studies.

The stellate calcareous nannofossils, the **discoasters**, are an extinct group that are exceedingly useful in the biostratigraphy of the Tertiary. Their taxonomy is based on the number of rays and ornamentation in plan view.

Nannoconids are minute, cone-shaped microfossils (5–30 µm) constructed of closely packed, calcite wedges that form a spiral. A canal penetrates the axis of the cone. Up to 12 specimens make up the skeleton of a single organism arranged in a petal-like structure (Trejo 1960). Nannoconids are useful in Cretaceous biostratigraphy in the absence of other groups.

The living coccolithophore

A coccolithophore is generally a spherical or oval unicell, less than 20 µm in diameter, equipped with two golden-brown pigment spots with a prominent nucleus between, two flagella of equal length and a **haptonema**. The small, calcite coccoliths are formed in vesicles within the cell under the stimulus of light. These eventually move to the outside of the cell where old coccoliths are shed. Reproduction is mostly asexual, by simple division of the mother cell into two or more daughter cells. In some living genera there is also an alternation between a motile and a non-motile planktonic or benthic stage. The motile stage has a flexible skeleton with coccoliths embedded in a pliable cell membrane, but in the non-motile cysts, calcification of the membrane can take place, thereby forming a rigid shell called a coccosphere (Fig. 14.1).

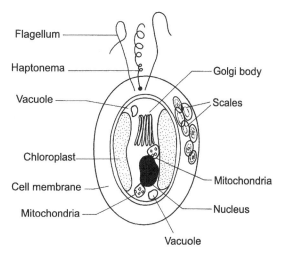

Fig. 14.1 Sketch of a living coccolithophore cell. During the non-motile stage the flagella are absent and the cell is covered by coccoliths. (After Siesser in Lipps 1993, figure 11.14 (with permission).)

Coccoliths

Coccolith morphology is the basis for classification of both living and fossil members of the group. Two basic modes of construction are known from electron microscope studies: **holococcoliths** are built entirely of submicroscopic calcite crystals, mostly rhombohedra, arranged in regular order; **heterococcoliths** are usually larger and built of different submicroscopic elements such as plates, rods and grains, combined together into a relatively rigid structure. As holococcoliths invariably disintegrate after they are shed, it is the heterococcoliths that provide the bulk of the microfossil record. Heterococcoliths vary considerably in form and

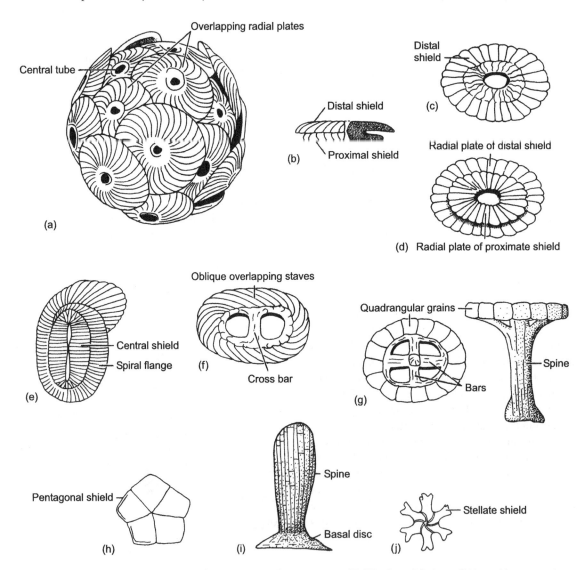

Fig. 14.2 Coccoliths (a) Recent coccolithophore in non-motile stage, ×2780. (b) Side view of *Cyclococcolithina*, with cross-section. (c) *Pseudoemiliania* distal view, ×3600. (d) Same as (c) from proximal side. (e) *Helicopontosphaera*, ×2930. (f) *Zygodiscus*, ×5340. (g) *Prediscosphaera* proximal and side views, ×4000. (h) *Braarudosphaera*, ×2140. (i) *Rhabdosphaera* side view, ×4000. (j) *Discoaster*, ×1000.

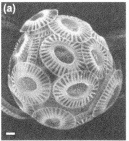 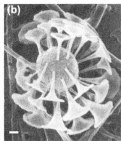 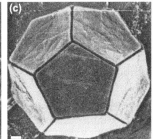 

Fig. 14.3 Electron photomicrographs of living coccolithophores. Scale bar = 1 μm. (a) *Emiliania huxleyi* var. *huxleyi* (Pleist.-Rec.). (b) *Discosphaera tubifera* (Pleist.-Rec.). (c) *Braarudosphaera bigelowii* (Jur.-Rec.). (d) *Scyphosphaera apsteinii* f. *apsteinii* (Eoc.-Rec.). ((a)–(d) From Winter & Siesser 1994 (with permission).)

construction. The majority comprise discs of elliptical or circular outline (shields) constructed of radially arranged plates, enclosing a central area which may be empty, crossed by bars, filled with a lattice or produced into a long spine. The outward-facing (distal) side of the shield is often more convex with a prominent sculpture and may be provided with a spine, whilst the other proximal face is flat or concave and may have a separate architecture (Fig. 14.2).

Coccolithophores have provided a major source for carbonate ooze since the Early Mesozoic and thus the biomineralization of coccoliths is a globally significant rock-forming process, yet little is known about the mechanism of formation of coccoliths (for a review see Piennar, in Winter & Siesser 1994, pp. 13–39). Coccolithophores grown in laboratory cultures produce coccoliths of calcite with small amounts of aragonite and vaterite, however fossil coccoliths are exclusively composed of low magnesium calcite. The Golgi body, reticular body and nucleus are all instrumental for the formation of the coccoliths and it appears that not all groups produce coccoliths in the same way. The simplest method seems to be the secretion of scales and coccoliths in the Golgi body followed by extrusion to the cell surface. In *Coccolithus pelagicus* scales are first produced in the Golgi body, are extruded and then form the nucleation sites for the later development of the coccolith between the cell membrane and an organic pedicle that develops around the cell. *Emiliania huxleyi* (Fig. 14.3a) produces coccoliths in a vesicle adjacent to the nucleus and reticular body by the precipitation of calcite, controlled by an organic matrix. The base of the coccolith is precipitated first followed by the upward and lateral development of the shields. On completion the coccoliths are extruded to form the interlocking external skeleton (Westbroek *et al.* 1984).

It is thought that coccoliths are formed for a number of reasons including protection from intense sunlight, to concentrate light, to provide a site for the disposal of toxic calcium ions or as supporting armour which stabilizes and acts as ballast for the cell.

Some species of coccolithophores are known to be **dimorphic**, for example *Scyphosphaera apsteinii* (Fig. 14.3d) and *Pontosphaerea japonica* are known to occur on the same coccosphere as do *Helicosphaera carteri* and *H. wallichi*. Some living coccolithophores (e.g. *Scyphosphaera*, Fig. 14.3d) produce two layers of morphologically distinct coccoliths (**dithecism**). **Pleomorphism** can also occur with hetero- and holococcolith bearing coccospheres being produced in different phases of the life cycle of a single species. All these phenomena have led to different fossil coccoliths being placed in separate form species when they should have been described as a single species, and as a consequence estimates of coccolith diversity though time may have been grossly overestimated.

Ecology of coccolithophores

Coccolithophores are predominantly autotrophic nannoplankton (i.e. 5–60 μm in size), utilizing the energy from sunlight for photosynthesis. Living cells are therefore largely restricted to the photic zone of the water column (0–200 m depth) with the lighter, smaller cells living near the surface and heavier cells living lower down. As such the distribution of coccolith species is under the direct control of climate. They thrive in zones of oceanic upwelling or of pronounced vertical mixing, as it is here that vital trace minerals are most readily available.

Although a few species are adapted either to fresh or brackish waters, the majority of species are marine. Nannofloras do not typically show nearshore–offshore differentiation though members of the Braarudosphaeraceae (Box 14.1) are found exclusively in inshore waters. Marked seasonal variation occurs in the abundance of some species including *E. huxleyi*, though in most cases the rhythmic millimetre scale laminations in deep sea sediments accumulate over thousands of years and do not reflect annual cycles. The relative abundance of complete coccoliths to

---

Box 14.1 Family level classification of coccoliths with diagrams of typical terms (sketches from photomicrographs in Siesser in Lipps 1990 and Perch-Nielsen in Bolli *et al.* 1985)

Kingdom CHROMISTA Infrakingdom CHROMOBIOTA Phylum HAPTOPHYTA Class PATELLIFERA Order COCCOSPHAERALES		**Ahmuellerellaceae** (Reinhardt 1965). Elliptical coccoliths with a wall of inclined crystal elements and a central area spanned by a cross, aligned with the axis of the ellipse. Trias./E. Jur.-L. Cret/Palaeog.	*Ahmuellerella*
Arkhangelskiellaceae (Bukry 1969). Elliptical coccoliths with a complex rim consisting of three to five elements. L. Jur.-L. Cret.	*Arkhangelskiella* 	**Biscutaceae** (Black 1971). Circular to elliptical coccoliths consisting of two closely appressed shields composed of petal-shaped elements. E. Jur.-Palaeog.	*Biscutum*
Braarudosphaeraceae (Deflandre 1947). Pentagon-shaped coccoliths. E. Cret.-Rec.	*Braarudosphaera* 	**Calciosoleniaceae** (Kamptner 1927). Rhomboidal coccoliths with calcite laths extending inwards from the walls. E. Cret.-Rec.	*Anaplosolenia*
Calyculaceae (Noël 1973). Elliptical to subcircular coccoliths with a central area covered in a grid; cup-like in side view. E.-L. Jur.	*Calyculus* 	**Calyptrosphaeraceae** (Boudreaux & Hay 1969). Holococcoliths with a highly variable morphology. L. Jur.-Rec.	*Zygrhablithus*
Ceratolithaceae (Norris 1965). Horseshoe-shaped coccoliths. Neog.-Rec.	*Ceratolithus* 	**Chiastozygaceae** (Rood *et al.* 1973). Elliptical coccoliths with an X- or H-shaped central structure. Trias./Jur.-Palaeog.	*Chiastozygus*

Box 14.1 (cont'd)

Coccolithaceae (Poche 1913). Elliptical coccoliths with a distal shield of radiating, petal-shaped elements. Proximal shield usually birefringent between cross polars, distal shield is larger and not birefringent. L. Cret.-Rec.

Coronocyclus

Crepidolithaceae (Black 1971). Elliptical coccoliths consisting of a ring of elements lacking imbrication. A large distal process may be present. Palaeog.-Neog.

Conusphaera

Discoasteraceae (Tan 1927). Star- or rose-shaped nannofossils. Palaeog.-Neog.

Discoaster

Eiffellithaceae (Reinhardt 1965). Elliptical coccoliths, distal shield with slightly overlapping elements, proximal shield with radially arranged elements. E. Jur.

Eiffelithus

Fasciculithaceae (Hay & Mohler 1967). Cylindrical nannoliths with a promial column and a distal disc or cone. Palaeog.

Fasciculithus

Goniolithaceae (Deflandre 1957). Pentagon-shaped coccoliths with a wall composed of vertical elements enclosing a granular centre. L. Cret.-Palaeog.

Goniolithus

Helicosphaeraceae (Black 1971). Spiral-walled coccoliths, usually with a flange. Central area open, spanned by a bridge or rarely closed. Palaeog.-Rec.

Helicosphaera

Heliolithaceae (Hay & Mohler 1967). Cylindrical nannofossils with a short proximal column and one or two distal cycles of elements. Palaeog.-Rec.

Heliolithus

Lithostromationaceae (Deflandre 1959). Triangular, hexagonal or nearly circular nannofossils covered in symmetrical arranged depressions. Palaeog.-Neog.

Lithostromation

Microrhabdulaceae (Deflandre 1963). Cylindrical, rod- or spindle-shaped nannofossils. L. Jur.-L. Cret.

Lithoraphidites

Nannoconaceae (Deflandre 1959). Conical nannofossils with a thick wall of wedge-shaped elements perpendicular to and spirally surrounding an axial canal. L. Jur.-L. Cret.

Nannoconus

Podorhabdaceae (Noel 1965). Elliptical coccoliths with a rim consisting of two to three cycles of elements. The wide central area spanned by a variety of structures. E. Jur.-L. Cret.

Cretarhabdus

Polycyclolithaceae (Forchheimer 1972). Cylinder-, block-, star- or rosette-shaped nannofossils. E.-L. Cret.-Palaeog.

Eprolithus

Pontosphaeraceae (Lemmermann 1908). Coccoliths with a raised wall, of varying height, consisting of two cycles of elements and a large central area. Palaeog.-Rec.

Pontosphaera

Box 14.1 (cont'd)

Prediscosphaeraceae (Rood et al. 1971). Circular or elliptical coccoliths, almost always with 16 elements in each of two shields. E.-L. Cret.	*Prediscosphaera*	**Prinsiaceae** (Hay & Mohler 1967). Circular to elliptical coccoliths, distal shield is birefringent between crossed polars. L. Cret.-Rec.	*Gphyrocapsa*
Rhabdosphaeraceae (Lemmermann 1908). Nannofossils with a base consisting of a varying number of cycles of elements. A central process rises from the base. Palaeog.-Rec.	*Rhabdosphaeara*	**Rhagodiscaceae** (Hay 1977). Elliptical coccoliths with a wall composed of inclined elements with a granular central area. L. Jur.-L. Cret.	*Rhagodiscus*
Schizosphaerellaceae (Deflandre 1959). Nannofossils consisting of two overlapping hemispheres. Trias.-L. Jur.	*Schizopharella*	**Sollasitaceae** (Black 1971). Elliptical coccoliths with two shields and a large central opening occupied by a grid or bars, lacking a central process. E. Jur.-Palaeog.	*Sollasites*
Sphenolithaceae (Deflandre 1952). Nannoliths with a proximal shield or column above which are disposed tiers of radiating lateral elements. Palaeog.-Neog.	*Sphenolithus* 	**Stephanolithiaceae** (Black 1968). Circular, elliptical or polygonal coccoliths. The outer wall has vertically arranged elements and may bear lateral spines. E. Jur.-L. Cret.	*Stephanolithus*
Syracospaeraceae (Lemmermann 1908). Coccoliths with a complex wall and a central area partially closed by laths. Neog.-Rec.	*Syracosphaera*	**Thoracosphaeraceae** (Schiller 1930). Spherical or ovoid nannofossils composed of interlocking polygonal elements. L. Jur.-Rec.	*Tharacosphaera*
Triquetrorhabdulaceae (Lipps 1969). Spindle-shaped rods constructed of three blades. Triradial cross-section. Palaeog.-Neog.	*Triquetrorhabdulus*	**Zygodiscaceae** (Hay & Mohler). Coccoliths with one or two cycles of inclined elements in the wall and a bridge aligned with the short axis of the ellipse. E. Jur.-Palaeog.	*Glaucolithus*

broken coccoliths and coccolith flour changes with depth (Fig. 14.4).

In the Atlantic Ocean nannofloral provinces are delimited by temperature (Fig. 14.5) with different assemblages indicating subglacial, temperate, transitional, subtropical and tropical latitudes. It is in tropical areas where they are most abundant and their numbers may reach as many as 100,000 cells per litre of sea water. A similar latitudinal differentiation occurs in the Pacific Ocean but the greatest diversity occurs at

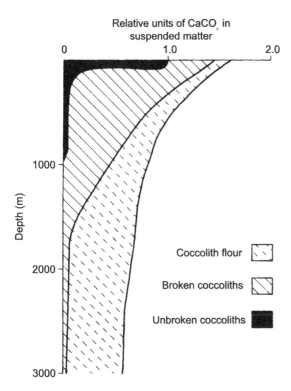

Fig. 14.4 Vertical distribution of coccoliths and coccolith-derived carbonates in the Pacific Ocean. (After Lisitzin in Funnell & Riedel 1971, figure 11.4.)

50°N. Depth stratification also occurs in the Pacific Ocean (see Honjo & Okada 1974; Honjo in Ramsay 1977, pp. 951–972). Of the 10 species cultured by McIntyre et al. (1970) E. huxleyi had the broadest temperature tolerance (1–31°C) and tropical species (e.g. Discosphaera, Fig. 14.3b) the narrowest range (20–30°C). There also appears to be a narrowing of temperature tolerance in species living offshore.

Production of coccoliths is strongly but not completely controlled by light. Whilst E. huxleyi increases abundance with increasing nutrients (in culture and in the oceans), most subtropical, oceanic species do not (Brand, in Winter & Siesser 1994, pp. 39–51).

Coccoliths and sedimentology

After death coccolithophores sink through the water column at about 0.15 m per day and the coccoliths fall away. With increasing depth these scales tend to dissolve or disaggregate into finely dispersed carbonate matter (Fig. 14.4), this process operating first on holococcoliths or delicate heterococcoliths. Therefore coccolith assemblages from sediments deeper than 1000 m are not truly representative of the original nannoflora. At depths of over 3000–4000 m, few coccoliths remain as most of the $CaCO_3$ has gone into solution, at these depths coccolith oozes are replaced by the less-soluble diatom or radiolarian oozes, or by red clays. Many factors may cause this dissolution, including high hydrostatic pressures, high CO_2, low O_2, low pH, low temperatures, low $CaCO_3$ precipitation by organisms, or sluggish recycling of $CaCO_3$ from the land. Honjo (1976) and Philskaln & Honjo (1987) showed, however, that coccoliths (and even whole coccospheres) can reach ocean depths intact by settling rapidly within the faecal pellets of copepod crustaceans. The proportion of coccolithic material in Recent oceanic carbonates is greatest in subtropical and tropical regions underlying waters with high organic productivity. Here they may average 26% by weight of the sediment (Fig. 14.5). Coccoliths are likewise an important constituent of Cretaceous and Tertiary chalks. They are fewest in sediments from subglacial waters (about 1%) where both productivity and preservation conditions are unfavourable.

Unfortunately, there is a tendency for calcite overgrowths or recrystallization to occur in coccoliths, obscuring their morphology. Solution of elements critical to the identification of fossil coccoliths may also present problems. Yet another disadvantage to the biostratigrapher is the ease with which coccoliths are reworked into younger sediments without showing outward signs of wear. The role of coccolithophores in sedimentation is reviewed by Honjo (1976) and Steinmetz (in Winter & Siesser 1994, pp. 179–199).

Classification

Kingdom CHROMISTA
Infrakingdom CHROMOBIOTA
Phylum HAPTOPHYTA
Class PATELLIFERA

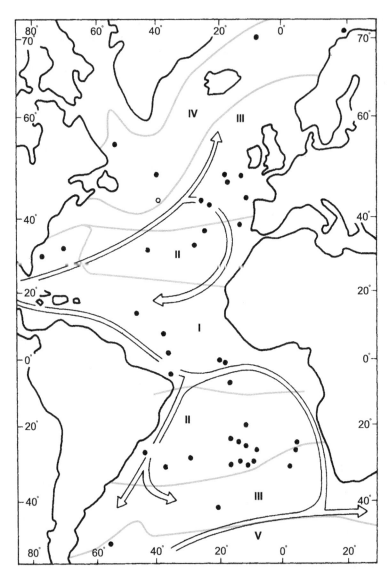

Fig. 14.5 Coccolith concentrations in near-surface sediments of the Atlantic Ocean plotted as percentages. Superimposed are major surface currents and calcareous nannoplankton provinces. Black dots are Deep Sea Drilling Project locations. Roman numerals in the figure correlate with the assemblages that follow. I – Tropical: *Umbellosphaera irregularis, Calcidiscus annulus, Oolithotus fragilis, Umbellosphaera tenuis, Discosphaera tubifer, Rhabdosphaera stylifer, Helicosphaera carteri, Gephyrocapsa oceanica, Emiliania huxleyi, Calcidiscus leptoporus*. II – Subtropical: *Umbellosphaera tenuis, Rhabdosphaera stylifer, Discosphaera tubifer, Calcidiscus annulus, Gephyrocapsa oceanica, Umbilicosphaera sibogae, Helicosphaera carteri, Calcidiscus leptoporus, Oolithotus fragilis*. III – Transitional: *Emiliania huxleyi, Calcidiscus leptoporus, Gephyrocapsa ericsonii, Rhabdosphaera stylifer, Gephyrocapsa oceanica, Umbellosphaera tenuis, Coccolithus pelagicus*. IV – Subarctic: *Coccolithus pelagicus, Emiliania huxleyi, Calcidiscus leptoporus*. V – Subantarctic: *Emiliania huxleyi, Calcidiscus leptoporus*. (After McIntyre & McIntyre in Funnell and Reidel 1971.)

Neither botanists nor palaeontologists have agreed on how to classify the coccolithophores and their relatives. Cavalier-Smith (1993) proposed they be placed in the kingdom Chromista; based upon the nature and location of the chloroplast and 18sRNA phylogenetic studies. He regarded them as belonging to the phylum Haptophyta because they are unicellular, golden-brown algae with two equal flagella and a coat of scales. Traditional micropalaeontological classification schemes retain the coccolith-bearers in the division Chrysophyta, class Coccolithophyceae. Beyond this recent schemes are based on the ultrastructure of coccoliths and their arrangement about the cell, little of which can be seen without the aid of an electron microscope.

Box 14.1 outlines the familial level classification and shows illustrations of eponymous taxa. The following genera exemplify some of the main types of heterococcolith. *Cyclococcolithina* (Olig.-Rec., Fig. 14.2b) has a disc comprising two circular or elliptical rings (termed proximal and distal shields) built of overlapping radial plates arranged around a central, tubular pillar. Such

Chapter 14: Calcareous nannoplankton: coccolithophores and discoasters

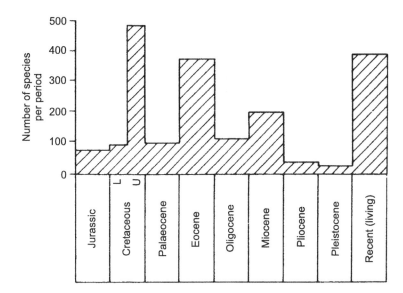

Fig. 14.6 Species diversity of described coccoliths through time. (Based on Tappan & Loeblich 1973.)

an arrangement, with two shields connected by a central tube, is called a **placolith**. In *Pseudoemiliania* (U. Plioc.-L. Pleist., Fig. 14.2c), the radial plates of the two shields do not overlap and are arranged around a central space. The radial plates of *Helicopontosphaera* (Eoc.-Rec., Fig. 14.2e) are distinctively arranged into a single elliptical central shield surrounded by a spiral flange, also of radial elements. The coccolith of *Zygodiscus* (U. Cret.-Eoc., Fig. 14.2f) comprises an elliptical ring built of steeply inclined and overlapping staves spanned by a cross bar. An open ring built of 16 quadrangular grains spanned by cross bars is characteristic of *Prediscosphaera* (M.-U. Cret., Fig. 14.2g). This genus contributed greatly to the deposition of the Cretaceous chalk. *Braarudosphaera* (Cret.-Rec., Figs 14.2h, 14.3c) has five plates arranged with pentaradial symmetry. The solid spine of *Rhabdosphaera* (Plioc.-Rec., Fig. 14.2i) arises from a basal disc of fine and complex construction. Such **rhabdoliths** probably serve to reduce sinking of the cell below the photic zone. Simpler in plan are the **stellate** coccoliths of the discoasters. *Discoaster* (U. Mioc.-Plioc., Fig. 14.2j) had a star-like disc up to 35 μm in diameter, built from 4–30 radiating arms of variable shape. The upper and lower surfaces also differ slightly in appearance. Discoasters are mostly found in fossil deep sea carbonates, especially from warmer latitudes, and play an important role in Cenozoic biostratigraphy.

General history of coccolithophores

Being both a primary source of food in the oceans and a significant producer of atmospheric oxygen, the history of coccolithophores has a bearing on the overall history of life (see Tappan & Loeblich 1973; Tappan 1980). Palaeozoic records are few and dubious. The first generally accepted fossil coccoliths are rare and reported from upper Triassic rocks. Their diversification in the Early Jurassic was a remarkable event that parallels the radiation of the peridinialean dinoflagellate cysts and both may be related to oceanographic changes connected with the opening of the Atlantic Ocean at this time. Their numbers and taxonomic diversity increased steadily until the Late Cretaceous period when there was a major marine transgression and a further, explosive radiation of many planktonic groups (Fig. 14.6). These conditions led to the deposition of chalk over vast areas of the continental platforms. The vast majority of coccolithophores became extinct at the K-T boundary, many of their habitats being filled by the diatoms during the Early Cenozoic. Coccolithophores have since regained their dominance in tropical and temperate waters but are significantly less diverse than in the Mesozoic.

There was another resurgence of forms in the Eocene, including the discoasters, many of them rosette-shaped with numerous rays. The latter died out at the end of the Eocene after which time there was a general dwindling in the diversity of coccoliths and discoasters, leading to the extinction of the discoasters at the end of the Pliocene. This may have been due to climatic cooling and regression. Certain of the placolith-bearing coccolithophores, however, thrived in the cooler waters of the Quaternary Era.

Applications of coccoliths

The biostratigraphical value of coccoliths and discoasters is unrivalled in the Mesozoic and Cenozoic and they have become the standard biostratigraphical index fossils for the Cenozoic. Mesozoic and Cenozoic biostratigraphical zonations are summarized in Bown (1998) and Perch-Nielsen (in Bolli et al. 1985, pp. 329–554). Examples of coccolith and discoaster evolution are given by Prins (in Brönnimann & Renz 1969, vol. 2, pp. 547–559), Gartner (1970), Bukry (1971) and Siesser (in Lipps 1993, pp. 169–203).

The increasingly large database relating coccolith assemblages to modern day water masses and latitudinal provinciality means coccoliths are extremely important in oceanographical studies. The distribution of coccolithophores has changed significantly over time. In the Cretaceous they were cosmopolitan (Tappan 1980) and abundant in both coastal and oceanic waters and from the poles to the tropics. Now the highest diversity is found in the subtropical gyres or in areas of nutrient-rich upwelling. Most species live in stratified water and the degree of stratification affects abundance (Winter 1985; Verbeek 1989; Brand 1994, in Winter & Siesser 1994, pp. 39–51; Roth 1994, in Winter & Siesser 1994, pp. 199–219).

During the last glacial maximum (c. 18,000 BP) North Atlantic water masses and their constituent nannofloras shifted 15 degrees southwards of their present location. Vertical changes in nannofloras in sediment cores from cool- to warm-water assemblages reflect the glacial–interglacial cycling of the Pleistocene climate (Fig. 14.7). Similar whole-scale shifts in nannofloras have also been documented from the Miocene though the direct climatic implications are poorly understood

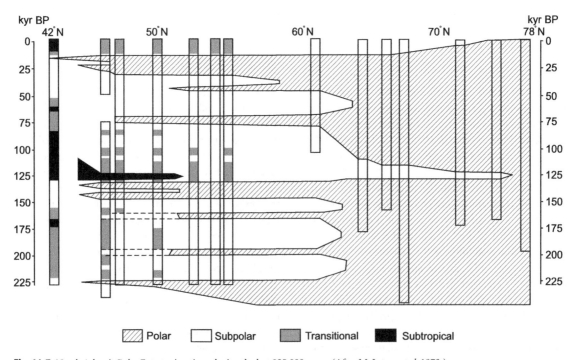

Fig. 14.7 North Atlantic Polar Front migrations during the last 225,000 years. (After McIntyre et al. 1972.)

(Haq 1980). Haq & Lohmann (1977) have plotted the apparent migrations of 'warm' and 'cold' coccolith assemblages through the Cenozoic and estimated from this the changes in palaeotemperature.

Coccolith morphology is also known to vary with temperature. The cold-water variety *of E. huxleyi* has a solid proximal shield whereas in warm water this shield is open and the rim is composed of many more elements. The ratio between coccoliths of warm and cool water type (e.g. *Discoaster, Chiasmolithus*) is a useful tool for indicating the changing palaeotemperature through Late Cenozoic time (see Bukry 1973, 1975) but becomes decreasingly reliable for more remote periods. Worsley (1973) discussed similar palaeoclimatic aspects and the determination of depositional depth in coccolith-bearing sediments.

The analysis of stable isotopes from calcareous nannoplankton is hampered by their small size and problems caused by diagenetic overgrowths; typically bulk sediment samples are analysed. Anderson & Arthur (1983) and Steinmetz (in Winter & Siesser 1994, pp. 219–231) have reviewed the difficulties and provide case examples. In general stable oxygen isotope values in the $CaCO_3$-living coccolithophores reflects the influence of temperature and vital effects. Experiments in culture have shown that many species do not grow in chemical equilibrium with the sea water. Despite these problems there is a strong correlation between the $\delta^{18}O$ values from planktonic foraminifera and coccolithophores through the Pleistocene (Fig. 14.8). The progressive enrichment in $\delta^{18}O$ values from benthic to planktonic forams to coccolithophores probably reflects their depth of growth. Margolis *et al.* (1975) noted the $\delta^{13}C$ profile from coccolithophores paralleled curves derived from benthic and planktonic forams. Data from Cretaceous and Cenozoic DSDP cores indicate $\delta^{13}C$ values from coccolithophores are a better indicator of surface water chemistry and reflect surface productivity (Kroopnick *et al.* 1977).

Further reading

Good general introductions to all aspects of calcareous nannoplankton can be found in Siesser (in Lipps 1993,

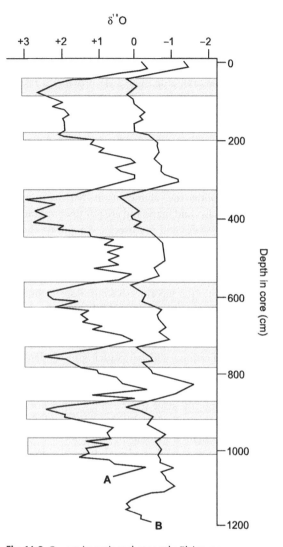

Fig. 14.8 Oxygen isotopic analyses on the Pleistocene Caribbean core P6304-4. A, *Globigerinoides sacculifer*; B, coccolith size fraction (3–25 μm; data from Steinmetz & Anderson 1984). Shaded areas are glacials. (Based on Steinmetz in Winter & Siesser 1994, figure 3.)

pp. 169–203) and Haq (1983) and coccolithophores in Winter & Siesser (1994). Further information on collection, examination and identification to generic level can be found in Hay (in Ramsay 1977, pp. 1055–1200). Identification of genera and species may also be assisted by reference to Farinacci (1969 to date). Some aspects of their classification, ecology, distribution and evolution are brought together in a chapter by Haq in Haq & Boersma (1998). A comprehensive biostratigraphical treatment of the Mesozoic and Cenozoic in Britain can be found in Bown (1998). Perch-Nielsen (in Bolli et al. 1985, pp. 329–554) provides a taxonomic and biostratigraphical synthesis of Cenozoic nannofossils and can be used for identification.

Hints for collection and study

Fossil coccoliths are abundant in Mesozoic and Cenozoic chalks and marls and are not uncommon in fossiliferous shales and mudstones. To extract them for study is relatively simple. Pulverize about 5–50 g of fresh sample (as in method A, see Appendix) and pour the liquid into a glass container to a depth of about 20 mm. After vigorous shaking allow the liquid to separate for about 2 minutes and then pipette some of the supernatant liquid on to a glass slide. For a temporary mount, add a cover slip and examine the slide at 800× magnification (or higher) with highly condensed transmitted light under a petrographic microscope. The light should be polarized with crossed nicols so that rotation of the stage (or the slide) brings out the position of the small wheel-like coccoliths with black cross optical figures. Permanent mounts can be prepared from strews dried on glass slides: add a drop of Caedax or Canada Balsam to the cover slip and place this over the strew mount.

REFERENCES

Anderson, T.F. & Arthur, M.A. 1983. Stable isotopes of oxygen and carbon and their application to sedimentologic and paleoenvironmental problems. In: Arthur, M.A., Anderson, T.F., Veizer, J. & Land, L.S. (eds) *Stable Isotopes in Sedimentary Geology*, SEPM Short Course No. 10, pp. 1–151.

Bolli, H.M., Saunders, J.B. & Perch-Nielsen, K. 1985. *Plankton Stratigraphy*. Cambridge University Press, Cambridge.

Bown, P.R. (ed.) 1998. *Calcareous Nannofossil Biostratigraphy*. British Micropalaeontological Society, Kluwer Academic Publishers, Dordecht.

Brönnimann, P. & Renz, H.H. (eds) 1969. *Proceedings of the First International Conference on Planktonic Micro-fossils, Geneva 1967*, vols 1, vol. 2. E.J. Brill, Leiden.

Bukry, D. 1971. Discoaster evolutionary trends. *Micropalaeontology* 17, 43–52.

Bukry, D. 1973. Coccolith and silicoflagellate. stratigraphy, Tasman Sea and southwestern Pacific Ocean. 21, 885–891.

Bukry, D. 1975. Coccolith and silicoflagellate stratigraphy, northwestern Pacific Ocean, DSDP Leg 32. 32, 677–701.

Cavalier-Smith, T. 1993. Kingdom Protoza and its 18 phyla. *Microbiological Review* 57, 953–994.

Farinacci, A. 1969 to date. *Catalogue of Calcareous Nannofossils*. Edizioni Tecnoscienza, Rome.

Funnel, B.M. & Riedel, W.R. (eds) 1971. *The Micropalaeontology of Oceans*. Cambridge University Press, Cambridge.

Gartner Jr, S. 1970. Phylogenetic lineages in the lower Tertiary coccolith genus *Chiasmolithus*. *Proceedings. National American Paleontological Convention 1969, Part G*, 930–957.

Haq, B.U. 1980. Biogeographic history of Miocene calcareous nannoplankton and paleoceanography of the Atlantic Ocean. *Micropaleontology* 26, 414–443.

Haq, B.U. (ed.) 1983. Calcareous nannoplankton. *Benchmark Papers in Geology* 78, 338.

Haq, B.U. & Boersma, A. (eds) 1998. *Introduction to Marine Micropaleontology*. Elsevier, Amsterdam.

Haq, B.U. & Lohmann, G.P. 1977. Calcareous nannoplankton biogeography and its paleoclimatic implications. Cenozoic of the Falkland Plateau (DSDP Leg 36) and Miocene of the Atlantic Ocean. 36, 745–759.

Honjo, S. 1976. Coccoliths: production, transportation and sedimentation. *Marine Micropalaeontology* 1, 65–79.

Honjo, S. & Okada, H. 1974. Community structure of coccolithophores in the photic layer of the Mid Pacific. *Micropaleontology* 20, 209–230.

Kroopnick, P.M., Margolis, S.V. & Wong, C.S. 1977. ^{13}C variations in marine carbonate sediments as indicators of the CO_2 balance between the atmosphere and the oceans. In: Andersen, N.R. & Malahouf, A. (eds) *The Fate of Fossil Fuel CO_2 in the Ocean*. Plenum Press, New York, pp. 295–321.

Lipps, J. (ed.) 1993. *Fossil Prokaryotes and Protists*. Blackwell Scientific Publications, Oxford.

McIntyre, A. & Bé, A.W.H. & Roche, M.B. 1970. Modern Pacific coccolithophorida: a paleontological thermometer. *Transactions of the New York Academy of Science* 32, 720–731.

McIntyre, A., Ruddiman, W.F. & Jantzen, R. 1972. Southward penetrations of the North Atlantic Polar Front: faunal and floral evidence for large-scale surface water mass movements over the last 225,000 years. *Deep sea Research* 19, 61–77.

Margolis, S.V., Kroopnick, P.M., Goodney, D.E., Dudley, W.C. & Mahoney, M.E. 1975. Oxygen and carbon isotopes from calcareous nannofossils as paleoceanographic indicators. *Science* **189**, 555–557.

Philskaln, C.H. & Honjo, S. 1987. The fecal pellet fraction of biogeochemical particle fluxes to the deep sea. *Global Biogeochemical Cycles* **1**, 31–48.

Ramsay, A.T.S. (ed.) 1977. *Oceanic Micropalaeontology*, 2 vols. Academic Press, London.

Steinmetz, J.C. & Anderson, T.F. 1984. The significance of isotopic and palaeontologic results on Quaternary calcareous nannofossil assemblages from Caribbean core P6304-4. *Marine Micropalaeontology* **8**, 403–424.

Tappan, H. 1980. *The Paleobiology of Plant Protists*. W.H. Freeman, New York.

Tappan, H. & Loeblich Jr, A.R. 1973. Evolution of the ocean plankton. *Earth Science Reviews* **9**, 207–240.

Trejo, M.H. 1960. La Familia Nannoconidae y su alcance estratigrafico en America (Protozoa, Incertae saedis). *Boletin. Asociatión Mexicana de Géologos Petroleros* **XII**, 259–314.

Verbeek, J.W. 1989. Recent calcareous nannoplankton in the southernmost Atlantic. *Polarforschung* **59**, 45–60.

Westbroek, P., De Jong, E.W., Van Der Wal, P., Borman, A.H., De Vrind, J.P.M., Kok, D., De Bruijn, W.C. & Parker, S.B. 1984. Mechanism of calcification in the marine alga *Emiliania huxleyi*. In: Miller, A., Phillips, D. & Williams, R.J.P. (eds) *Mineral Phase in Biology*. Royal Society, London, pp. 25–34.

Winter, A. 1985. Distribution of living coccolithophores in the California Current System, southern California Borderland. *Marine Micropalaeontology* **9**, 385–393.

Winter, A. & Siesser, W.G. (eds) 1994. *Coccolithophores*. Cambridge University Press, Cambridge.

Worsley, T.R. 1973. Calcareous nannofossils: Leg 19 of Deep Sea Drilling Project. **19**, 741–750.

CHAPTER 15

Foraminifera

The Foraminiferida are an important order of single-celled protozoa that live either on the sea floor or amongst the marine plankton. The soft tissue (**cytoplasm**) of the foraminiferid cell is largely enclosed within a shell or test (Fig. 15.1a) variously composed of secreted organic matter (**tectin**), secreted minerals (calcite, aragonite or silica) or of **agglutinated** particles. This test consists of a single (**unilocular**) chamber or multiple (**multilocular**) chambers mostly less than 1 mm across and each interconnected by an opening, the **foramen**, or several openings (foramina). The group, which takes its name from these foramina, is known from Early Cambrian times through to recent times, and has reached its acme during the Cenozoic.

Foraminiferid tests can be very abundant; in the modern ocean they comprise over 55% of Arctic biomass and over 90% of deep sea biomass. In marine sediments, foraminiferid tests typically vary from a few individuals per kilogram to rock-forming *Globigerina* ooze and Nummulitic limestone.

Foraminifera (as they are informally referred to) are important as biostratigraphical indicators in marine rocks of Late Palaeozoic, Mesozoic and Cenozoic age because they are abundant, diverse and easy to study. **Planktonic foraminifera** are widespread and have had rapidly evolving lineages, factors which greatly aid the inter-regional correlation of strata in the Cretaceous (28 zones), Palaeogene (22 zones) and Neogene (20 zones). **Smaller benthic foraminifera** are the most common and are widely used for regional stratigraphy. **Larger benthic foraminifera** are typically larger than 2 mm in diameter and 3 mm^3 in volume and have complex internal structures which, when studied in thin section, are useful for the biostratigraphy of Tethyan and other tropical limestone. These include the largest single-celled organisms known, reaching up to 180 mm across. Because the developmental stages and foraminiferid life history are preserved in the test, they are well suited to evolutionary studies.

Foraminifera have a wide environmental range, from terrestrial to deep sea and from polar to tropical. Ecological sensitivity renders the group particularly useful in studies of recent and ancient environmental conditions. Changes in the composition of foraminiferal assemblages may be used to track changes in the circulation of water masses and in sea-water depth. They are especially important in studies of Mesozoic to Quaternary climate history because isotopes within their $CaCO_3$ tests record changes in temperature and ocean chemistry.

Living foraminifera

The cell

The cytoplasm of a foraminiferid comprises a single cell differentiated into an outer layer of clear **ectoplasm** and an inner layer of darker **endoplasm** (Fig. 15.1a). The ectoplasm forms a thin and extremely mobile film around the test which gives rise to fans of numerous, finely branching **granular** and **reticulose pseudopodia** whose form is ever changing. Foraminifera feed by trapping and engulfing small organisms and organic particles with these sticky pseudopodia, which are used to draw in the food material towards the test and later to expel it. Food requirements vary between species but include bacteria, diatoms and other protozoa, small crustaceans, molluscs,

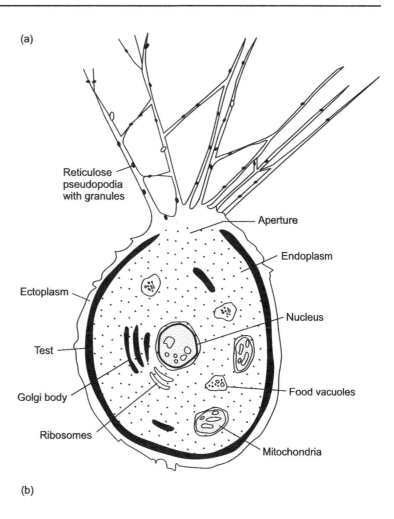

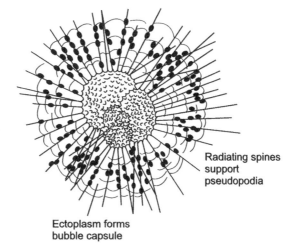

Fig. 15.1 (a) A living, single-chambered benthic foraminiferid, as seen in cross-section with transmitted light. (b) A living multichambered planktonic foraminiferid surrounded by radiating spines and pseudopodia (not all drawn) which support photosymbionts and frothy ectoplasm of the bubble capsule, as seen in transmitted light.

nematodes and invertebrate larvae. A few foraminifera are thought to be parasitic. Pseudopodia are further employed, in benthic forms, as a means of pulling the test along and for anchorage. The ectoplasm is connected with the inside of the test by means of an **aperture**, which acts as the 'front door' for the passage of cytoplasm, food, excretory products and reproductive cells.

The endoplasm is the storehouse and factory of the cell and is always protected by the test. It contains either a single nucleus (**uninucleate**) or several nuclei (**multinucleate**) which house the chromosomes and control protein synthesis. **Food vacuoles** contain engulfed food items which are subjected to enzymatic action that releases small molecules which are then absorbed by the cell. The endoplasm also contains numerous small organelles such as mitochondria, Golgi apparatus and ribosomes (Fig. 15.1a).

Naked **photosymbionts**, especially diatoms and dinoflagellates, are also found in the endoplasm of many larger benthic and planktonic foraminifera (Fig. 15.1b). Symbionts release photosynthates and O_2 to the host and benefit themselves from P, N and respiratory CO_2 released by the host. Planktonic foraminifera typically bear long, stiff, radiating pseudopodia borne on skeletal spines (Fig. 15.1b). Symbionts move to the ends of these spines during the day and retire to the protection of the test at night. Planktonic forms may also have their ectoplasm frothed into a **bubble capsule**, to aid buoyancy (Fig. 15.1b).

Life cycle

The life cycle of foraminifera is characterized by an alternation between two generations: a **gamont** generation which reproduces sexually, and an **agamont** generation which reproduces asexually (Fig. 15.2). While the life cycle may be completed within a year in tropical latitudes, it can take two or more years at higher latitudes. This alternation, however, is not always strictly followed and there are many variations.

Asexual reproduction in the agamont begins with the withdrawal of cytoplasm into the test. The cytoplasm then splits, by **multiple fission**, into numerous, tiny haploid daughter cells, each with a nucleus or

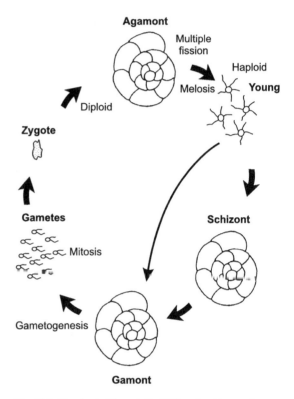

Fig. 15.2 The classical foraminiferid life cycle with a regular alternation of generations between gamont and agamont. (Diagrammatic after Goldstein, in Sen Gupta, 1999, figure 3.14.)

several nuclei containing only half the chromosomal compliment found in the parent nucleus. Chamber formation then begins and this new gamont generation is released into the water to disperse. When mature the cytoplasm is again withdrawn and it divides mitotically to form gametes (**gametogenesis**) retaining the same, haploid chromosome number as the parent. In most cases the gametes bear two whip-like flagella. When released from the parent test, two gametes may fuse (sexual reproduction) to form the next, agamont generation with a full, diploid chromosome number. The parent test is typically left empty after the dispersal of the juveniles.

In smaller benthic foraminifera, tests of these two generations are slightly different in appearance (**dimorphic**). Those of the gamont are more common, and may have a large initial chamber called a **proloculus**, which is therefore described as **megalospheric**. Those of the agamont, which originate from tiny gametes, may have a relatively small (**microspheric**) proloculus but a larger test (e.g. Fig. 15.2).

In larger benthic foraminifera the life cycle is thought to be **trimorphic**. Here, a **schizont generation** is added to the classical life cycle. These are diploid, megalospheric and multinucleate and are produced from the agamont by multiple fission without meiosis. The test of the microspheric agamont (or B) generation is typically much larger than those of the megalospheric gamont (A1) and schizont (A2) generations and commonly reveals more developmental stages. In these forms, the life cycle is thought to take from one to several years.

Planktonic foraminifera are thought to reproduce sexually every 28 days in relation to the lunar cycle. It is widely assumed that they do not reproduce asexually but this requires further study. According to theory, sexual reproduction (in all foraminifera) should be favoured in physically variable environments. This is because the greater genetic variety that arises from the sexual recombination of genes provides for a wider adaptive range.

The test

The test is thought to reduce biological, physical and chemical stress. Biological pressures include, for example, the risk of accidental ingestion by worms, crustaceans, gastropods, echinoderms and fish that deposit feed or browse on detritus on the sea floor. Others, such as scaphopods and certain gastropods, actually prey upon benthic foraminifera, while the tests may also risk being infested by parasitic nematode worms. Physical stresses include harmful radiation (including ultraviolet light) from the Sun, water turbulence and abrasion. Test strength is therefore likely to be important. Chemical stresses encompass fluxes in salinity, pH, CO_2, O_2 and toxins in the water. In all these cases, the cytoplasm can withdraw into the inner chambers leaving the outer ones as protective 'lobbies', or a detrital plug may close the aperture. $CaCO_3$ shells may also help to buffer the acidity of organic-rich, oxygen-deficient environments or digestive tracts.

Additional advantages of the test include the negative buoyancy it gives to a group of organisms especially adapted to a benthic way of life. Surface sculpture may variously assist positive buoyancy in planktonic forms (e.g. spines and keels), improve adherence, strengthen the test against crushing and help to channel ectoplasmic flow to and from the apertures, pores and umbilicus. Without a shell, the build up of biomass to a greater extent than seen in any other protozoa would also prove difficult. General overviews of form and function can be found in Brasier (1986), Murray (1991), Lee & Anderson (1991) and in various papers in Sen Gupta (1999).

Wall structure and composition

The structure and composition of the test wall is important to the classification of the group.

Organic-walled forms belong to the suborder Allogromiina. These have a thin, non-rigid test of proteinaceous or pseudochitinous matter generally termed **tectin** (Fig. 15.3). Similar material is also present as a thin lining to the chambers of most hard-tested foraminifera, where it may act as a template for mineralization.

The suborder Textulariina encompasses forms with **agglutinated** tests. In these, organic and mineral matter from the sea floor is bound together by an organic, calcareous or ferric oxide cement (Fig. 15.3). The grains are commonly selected for size, texture or composition (e.g. coccoliths, sponge spicules and heavy minerals).

Tests of the suborder Fusulinina are **microgranular** and appear dark in thin section when viewed with transmitted light and opaque (usually brown or grey) when viewed in reflected light. The microgranules may be packed randomly or aligned normal to the surface of the test and interspersed with mural pores, thereby giving the wall a fibrous appearance (Fig. 15.3), especially in more advanced forms. These granular and fibrous layers of microgranular calcite are often combined in the structure of a single, multilayered wall.

Calcareous tests are by far the most abundant and occur in all the remaining suborders. There are three main types of calcareous wall: **porcelaneous**

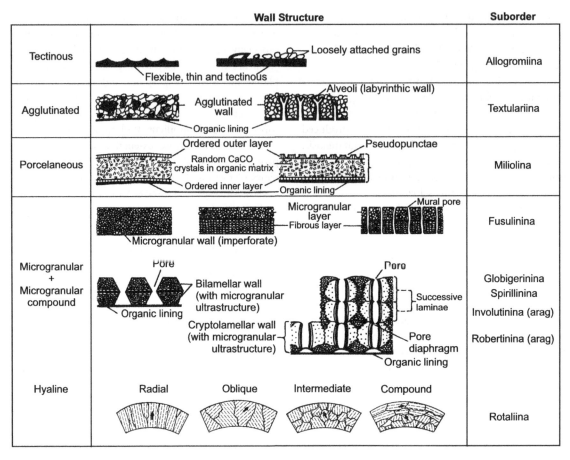

Fig. 15.3 Examples of wall structures in the foraminifera (diagrammatic, mainly based on studies using scanning electron microscopy).

imperforate, **microgranular** and **hyaline perforate**. Porcelaneous imperforate tests are characteristic of the suborder Miliolina. These lack mural pores and are a distinctive milky white in reflected light and an amber colour in transmitted light. They are constructed of tiny needles of high magnesium calcite randomly arranged for the most part, but the outer and inner surfaces are coated with a layer of horizontally arranged needles (Fig. 15.3). A single taxon of the Miliolina, *Miliammellus*, has a comparable wall constucted of needles of opaline silica. In the Miliolina, the test is built by the secretion of biomineral needles within tiny vesicles in the cytoplasm, which are then exported to the outer margin of the cell. In the remaining suborders, a tectinous template is laid down first, upon which carbonate is then precipitated.

Recent **hyaline perforate** tests are generally glassy when viewed with reflected light and grey to clear in transmitted light. However, thick walls, fine dense perforations, granules, spines, pigments and diagenesis may all obscure this clarity. Such hyaline perforate tests are found within six suborders that comprise the majority of Mesozoic to Cenzoic foraminifera, and may either be of low to high magnesium calcite (suborders Spirillinina, Globigerinina, Rotaliina) or aragonite (Involutinina), Robertinina). In the Spirillinina, the walls consist of a single crystal of calcite (**monocrystalline**). In the other groups, the polycrystalline hyaline perforate walls are commonly constructed of a mosaic of rhomboidal calcite or aragonite crystals, each about 1 μm in diameter, whose c-axes are perpendicular to the test surface. This optically radial ultrastructure may give a black-cross polarization figure with coloured rings when viewed with crossed nicols. In some hyaline tests, it is the a-axes that are radial so that the c-axes are oblique, giving minute flecks of colour but no polarization figure when viewed under crossed nicols. These optically granular walls are now known to be of only limited use for classification because they vary widely between related species.

The walls of many foraminifera are traversed by small straight **mural pores** or branched alveoli through which fluids and gases may pass by osmosis, linking ectoplasm and endoplasm (Fig. 15.3). Finely perforate organic diaphragms across these pores appear to act as semi-permeable membranes. Such **radial pores** are characteristic of the hyaline perforate foraminifera (discussed below) but are also found in certain of the more complex Textulariina and Fusulinina. They give to the wall a **pseudo-radial** or **pseudo-fibrous** appearance in thin section.

Test growth

Feeding adds continually to the bulk of the cytoplasm. In related **testate amoebae**, which live for only 2–4 days, the test is not enlarged and consists of a single chamber (**unilocular**) which is vacated on reproduction. In foraminifera, which generally live for between 1 month and several years, several strategies for test enlargement have arisen.

The tests of many **primitive** foraminifera are unilocular, although test form varies greatly (Fig. 15.4a,c). Such tests may be said to show **contained growth** because there is little or no capacity for enlargement. The foraminiferid must therefore expend energy in rebuilding the wall, vacate the test and grow a new one, or reproduce. These limitations of contained growth have been overcome in some primitive lineages by the addition of a second tubular chamber, which shows **continuous growth** (Fig. 15.4d–j). Such simple forms predominated in the lower Palaeozoic but can still be found today, especially in marginal marine and abyssal habitats. Unilocular tests are also found in modern parasitic forms.

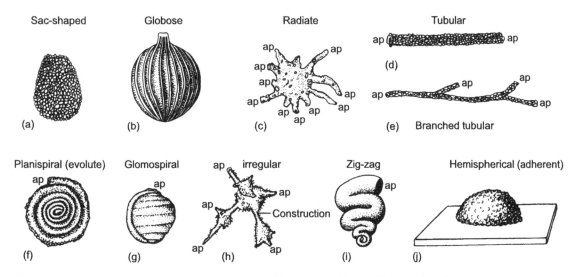

Fig. 15.4 Unilocular and bilocular tests. (a) *Pleurophrys* ×200. (b) *Lagena* ×53. (c) *Astrorhiza* ×49. (d) *Bathysiphon* ×7. (e) *Rhizammina* ×12. (f) *Ammodiscus* ×17. (g) *Usbekistania* ×66. (h) *Aschemonella* ×3. (i) *Ammovertella* about ×2. (j) *Hemisphaerammina* about ×16; ap. = aperture. ((c) (h) (i), (j) After Loeblich & Tappan 1964; (a) modified from Loeblich & Tappan 1964 after Saedeleer; (b) after Loeblich & Tappan 1964 from H.B. Brady; (g) after Loeblich & Tappan 1964 from Suleymanov (all from the Treatise on Invertebrate Paleontology, courtesy of and © 1964, Part C, The Geological Society of America and The University of Kansas).)

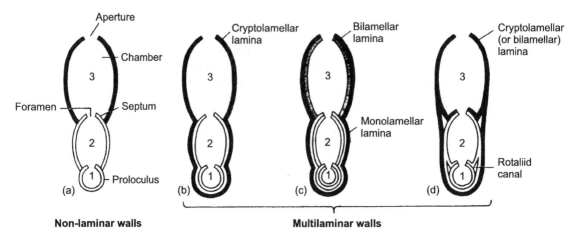

Fig. 15.5 Diagrammatic axial sections illustrating different modes of chamber addition. (a) Non-laminar. (b) Multilaminar cryptolamellar. (c) Multilaminar, monolamellar bilamellar. (d) Multilaminar, cryptolamellar, with septal flaps and canals. The uniserial growth is shown here for simplicity.

In **multilocular** forms (Fig. 15.5) protoplasmic growth is gradual but test growth is periodic, with a new and larger chamber being added at regular intervals. Each chamber is provided with a distinct apertural face (**septum**) that confines the aperture and improves protection of the endoplasm. Chamber addition begins with the construction of a loosely bound growth cyst, composed largely of food debris. The pseudopodia are then withdrawn to occupy the space of the new chamber, building first a thin organic wall and then an agglutinated or calcareous one on the outer side, or on both sides. This **simple septate growth** condition is predominant today, being found in planktonic foraminifera and many smaller benthic foraminifera.

In **complex septate growth**, the chambers' shape is greatly modified and the chambers may be subdivided by partitions into **chamberlets** which have multiple apertures. This condition is typically found in larger benthic foraminifera which have photosymbionts.

Wall ultrastructure

Foraminifera can have either **lamellar** (Fig. 15.5a) or **multilamellar** ultrastructure (Fig. 15.5b–d). In hyaline perforate forms, septate growth can bring about changes in the fine structure of the test when seen in thin section. Where there is no overlap of previous chamber walls by the new wall, the arrangement is termed **non-laminar** (Fig. 15.5a). This is the typical arrangement in non-hyaline, imperforate foraminifera. **Monolamellar** structure occurs where each chamber is composed of a single layer which also overlaps previous chambers (Fig. 15.5b), as in the finely perforate suborder Lagenina. In the majority of hyaline forms, multilamellar ultrastucture (Fig. 15.5b–d) is seen. Each chamber wall is here composed of two distinct lamellae of calcite (i.e. **bilamellar**, Fig. 15.5c) on either side of a tectin membrane, of which only the outer lamella coats previous chambers. Multilamellar ultrastructure has the advantage of increasing the strength of the test with growth. It also allows the development of complex architecture not seen in non-hyaline groups, such as the spines of planktonic foraminifera. In the suborder Rotaliina, the inner lamella also coats the previous apertural face and forms a **septal flap** together with a space called a **rotaliid canal** (Fig. 15.5d). Such canal systems provide for the rapid extrusion of cytoplasm during chamber construction and reproduction.

Chamber architecture

Foraminiferid tests may appear to represent a bewildering array of modes of growth. Although the variation is remarkable it is possible to impose a degree of order by recognizing that most multilocular test types arise as the result of interaction between three variables during growth: the **rate of translation** (i.e. the net rate of movement along the growth axis to the net movement away from the growth axis), the **rate of chamber expansion** and the **chamber shape** (Fig. 15.6).

Different rates of translation produce the four common growth plans of the foraminiferid test: **planispiral**, **trochospiral**, **biserial** and **uniserial**. In **planispiral** tests the rate of translation is zero, the chamber or chambers being arranged more or less symmetrically in a plane coil about the growth axis. This growth plan may be further modified by different rates of chamber overlap towards the coiling axis (e.g. **evolute** to **involute** form) and in involute forms by an extension of growth along the coiling axis (e.g. **discoidal** to **fusiform**; Fig. 15.6).

Where material is added in a helical coil the test is called **trochospiral**. Such tests have a **spiral** side and an **umbilical** side, which are more evolute and more involute respectively (Fig. 15.8a). In multilocular tests, a successive decrease in the spiral angle may ultimately bring about a reduction in the number of chambers per whorl to three (**triserial**), although triserial and biserial forms with wide spiral angles are known. Further reduction may obscure or eliminate the spiral component resulting generally in biserial and **uniserial** growth plans (i.e. two and one chamber per whorl respectively, Fig. 15.6). Not infrequently, some of these arrangements may be found together in one test, with developmental changes from planispiral or triserial to uniserial.

Chamber shape

The rate of chamber expansion may be defined as the rate of increase in volume (or of width, length or depth) from one chamber to the next. In most foraminifera this remains a fairly constant logarithmic trend, at least through early ontogeny. However, the number of chambers per whorl in a species can change through life or between localities and is therefore an unreliable taxonomic character.

Chamber shape varies widely. Unilocular tests may be flask-shaped globose, tubular, branched, radiate or irregular (Fig. 15.4). Although the chambers of multilocular forms generally remain of constant shape through ontogeny, their arrangement and ornament can vary. Common shapes include globular, tubular, compressed lunate and wedge-shaped. Expansion rate and chamber shape are closely linked. Figure 15.7 demonstrates tests with identical rates of volumetric expansion but differing chamber proportions and growth plans.

Apertures and foramina

The aperture is found in the wall of the final chamber and serves to connect the external pseudopodia with the internal endoplasm, allowing passage of food and contractile vacuoles, nuclei and the release of the daughter cells. Its position remains more or less constant through ontogeny so that each chamber is linked to the next by a foramen or several foramina (Fig. 15.5). In forms that lack apertures, foramina may be secondarily developed by resorbtion of the chamber wall.

The **primary aperture(s)** may be **single** or **multiple** in number and terminal, areal, basal, extraumbilical or umbilical or in position (Figs 15.8, 15.12). Their shape varies widely, for example rounded, bottle-necked (phialine), radiate, dendritic, sieve-like (cribrate), cruciform, slit- or loop-shaped. Apertures can be further modified by the presence of an apertural lip or flap (termed a **labiate** aperture, Fig. 15.8c), teeth (**dentate** aperture, Fig. 15.8e), a cover plate (**bullate** aperture, Fig. 15.8f) or an **umbilical boss** (Fig. 15.8g). **Secondary apertures** may also be added, for example along the sutures or the periphery of the test (Fig. 15.8d). Such apertural and foraminal structures are used for classification, especially below the subordinal level.

Sculpture

The external surface of the test may bear spines (termed **spinose**), keels (**carinate**), rugae (**rugose**), fine striae (**striate**), coarser costae (**costate**), granules (**granulate**) or a **reticulate** sculpture. These features should be used with caution in distinguishing certain genera and species for they vary through ontogeny and with environment.

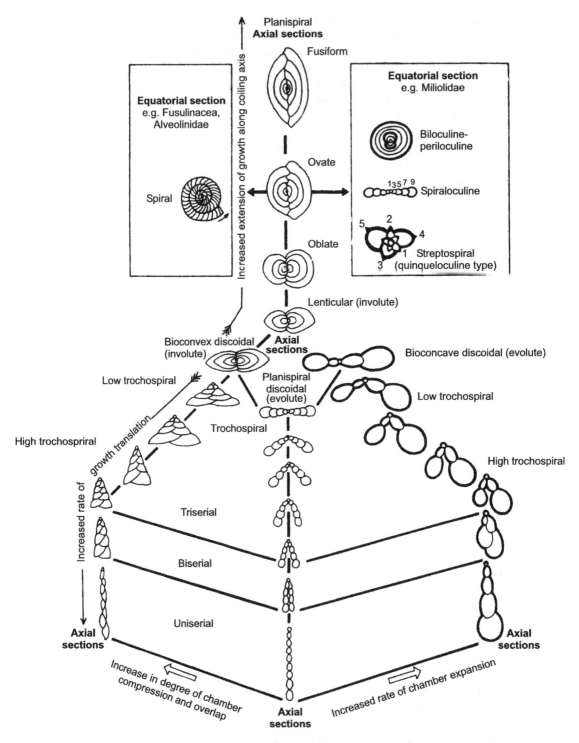

Fig. 15.6 The main growth forms in multilocular tests of foraminifera. Axial sections are those cut parallel to and including the main axis of symmetry and growth. Equatorial sections (*sensu lato*) are cut at right angles to this axis, at the widest point on the test.

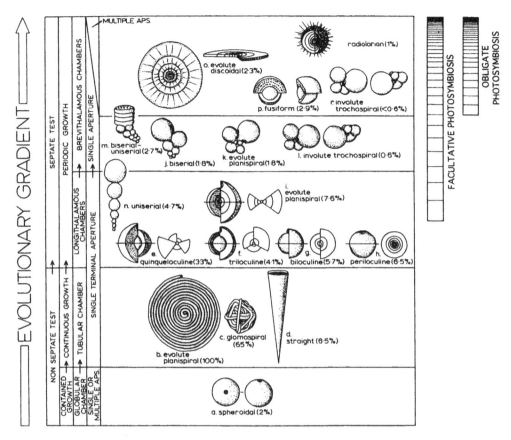

Fig. 15.7 The evolutionary gradient in foraminiferid test morphospace, from primitive (below) to advanced (above). Models of unit volume are used to compute the minimum line of communication (MinLOC), here given as a standardized percentage in brackets, relative to the evolute planispral form in 'b'. Forms known to have photosymbionts show relatively short lines of communication within the test. (Adapted from the models of Brasier 1982a, 1982b, 1984, 1986, 1995.)

Architectural evolution

The main pattern of evolution seen in Foraminiferid test architecture is shown in Fig. 15.7. In the Lower Palaeozoic, the tests were mainly agglutinated and had contained growth (Fig. 15.7a, e.g. *Saccammina*) or were enlarged by continuous growth of tubular chambers (Fig. 15.7b, e.g. *Ammodiscus*; 15.7c, e.g. *Glomospira*). By the late Devonian, **septate periodic** growth had evolved. At this stage, larger body size was enabled by more flaring chambers (Fig. 15.7d, e.g. *Hyperammina*) whose openings to the outside were protected by the formation of a **septum** around the single aperture. In primitive foraminifera (Fig. 15.7e, *Quinqueloculina*), banana-shaped chambers reflect the ancestral condition in being longer than they are wide (**longithalamous**). A progressive shortening of the internal minimum line of communication (**MinLOC**) is seen in more advanced stocks, through the formation of increasingly tight coiling (e.g. from uniserial – biserial – triserial – trochospiral) combined with **brevithalamous** chambers that are wider than long. In later stocks, the aperture is often positioned to maintain the shortest possible MinLOC for that growth plan, being **basal** or **umbilical** (Figs 15.7j–k, e.g. *Bolivina*, *Elphidium*; Fig. 15.8). In advanced foraminifera, **multiple apertures** also help to maintain short lines of communication between each chamber. The shortest possible lines of communication with septate growth are found in those foraminifera that

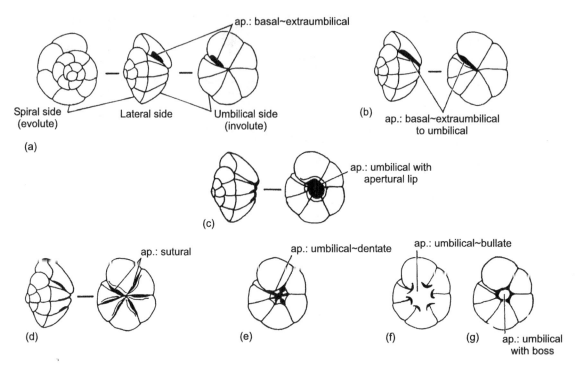

Fig. 15.8 (a)–(g) Trochospiral tests with different kinds of aperture (ap).

culture photosymbiotic protists. In planktonic foraminifera, this is combined with globular chambers that reduce the mass of the test to retard sinking (Fig. 15.7r, *Globigerinoides*). In larger benthic foraminifera, short lines of communication are combined with attempts to maximize the relative surface area for photosymbiosis (Fig. 15.7p, e.g. fusulinids, alveolinids; Fig. 15.7o, e.g. *Orbitolites*). It is these advanced forms with minimum lines of communication, and inferred symbiosis, that appear to have been most vulnerable during mass extinctions (Brasier 1988, 1995). Primitive forms, like those at the bottom of Fig. 15.7, seem to have survived by migration into the deep ocean.

Foraminiferal ecology

Smaller benthics

About 5000 species of living smaller benthic foraminifera are known. They are especially important as environmental indicators because they have colonized marine habitats from the most extreme tidal marshes to the deepest trenches of the oceans (see Murray 1991). Exploitation of resources across this wide range of habitats is reflected in adaptations of test morphology (Murray 1991).

Light The zone of light penetration in the oceans (the **photic zone**) is affected by water clarity and the incident angle of the Sun's rays. Hence the photic zone is deeper in tropical waters (<200 m) and decreases in depth towards the poles where it also varies with marked seasonality. Primary production by planktonic and benthic protozoa, and the protection and substrates provided by algae and sea grasses, render this zone attractive to foraminifera, especially the Miliolina. The porcelaneous wall of miliolines such as *Quinqueloculina* (Fig. 15.20e) is thought to protect the cytoplasm from damage in shallow equatorial waters by scattering the short wavelength, ultraviolet light.

Food Foraminifera play a prominent role in marine ecosystems as micro-omnivores, i.e. they feed on small bacteria, protozoa and invertebrates. Epifaunal forms living in the photic zone feed especially upon diatoms so that their numbers may fluctuate in relation to the seasonal cycle. These often have tests that are flattened on one or both sides (e.g. *Discorbis*, Fig. 15.25a). Some smaller benthic forms are known to culture photosymbionts (e.g. *Elphidium*, Fig. 15.27b, 15.31l). Others live infaunally within the sediment or below the photic zone and feed on dead organic particles or graze upon bacteria. The tests of active forms tend to be lenticular (e.g. *Lenticulina*, Fig. 15.22d) or elongate. Those living on the abyssal plains, such as *Bathysiphon* (Fig. 15.4d), may extend their pseudopodia into the water column to capture the seasonal rain of phytodetritus. Such forms tend to have erect, tubular, often branched, tests that are fixed to the substrate. Some hyaline foraminifera have degenerate unilocular tests (e.g. *Lagena*, Fig. 15.4b, 15.22e) and may lead a parasitic mode of life.

Substrate Those foraminifera which prefer hard substrates (i.e. rock, shell, sea grasses and algae) are normally attached, either temporarily or permanently, by a flat or concave lower surface. Typical growth forms are hydrodynamically stable and include discoidal, plano-convex, concavo-convex, dendritic and irregular shapes. *Cibicides* (Fig. 15.25e) and its relatives are typical of this life habit and many other examples occur throughout the order. Adherent forms often develop a relatively thin test and will tend to exhibit greater morphological variability than seen in sediment-dwelling and planktonic forms.

Although foraminifera have been found living up to 200 mm below the sediment surface, the majority are found within the top 10 mm (e.g. infaunal *Cassidulina*, Fig. 15.29c) or live at the surface (e.g. epifaunal *Elphidium*, Figs 15.27b, 15.31l). The larger pore spaces of higher-energy sands and gravel of the inner shelf may only support sparse populations. Foraminifera from these coarser substrates tend to be either adherent forms or free-living and thick-shelled, heavily ornamented forms of lenticular or globular shape. Low-energy habitats with silty and muddy substrates typical of lagoons, and the mid-shelf to bathyal slope, are often rich in organic debris and the small pore spaces tend to encourage bacterial blooms. Such substrates are therefore attractive to free-living foraminifera and can support large but patchy populations. Many of the infaunal species are thin-shelled, delicate and elongate (e.g. *Bolivina*, Fig. 15.24c, 15.31h; *Nodosaria*, Fig. 15.22b), and their activities can produce minute burrow systems.

Salinity The majority of foraminifera are adapted to normal marine salinities (about 35‰) and the highest diversity assemblages are found here. The low salinity of brackish lagoons and marshes favours low-diversity assemblages of agglutinated foraminifera (mostly with non-labyrinthic, imperforate walls and organic cements that may become secondarily siliceous or ferruginous; e.g. *Reophax*, Fig. 15.13a) and certain hyaline forms (e.g. *Ammonia*, Fig. 15.27a, 15.31i; *Elphidium*, Fig. 15.27b, 15.31l). The tectinous imperforate Allogromiina are also found in fresh and brackish waters, but their delicate tests are rarely encountered as fossils. The high carbonate ion concentrations of hypersaline waters, where salinities are in excess of 40‰, appear to favour the porcelaneous Miliolina (especially the Nubecularidae and Miliolidae, e.g. *Quinqueloculina*, Fig. 15.20e) but deter most other groups.

It seems that the imperforate tests of Textulariina and Miliolina are better at protecting the endoplasm from the stressful osmotic gradients of extreme salinity. Triangular plots of the relative proportions of Textulariina, Miliolina and hyaline forms have therefore proved useful as indices for palaeosalinity. Samples from certain habitats usually fall within the proscribed fields (Fig. 15.9; see Murray 1991). This method can give misleading results, however, where there has been selective post-mortem reworking, solution or fragmentation of tests. Nor can the method be used much before the Tertiary because it is only from that time that hyaline forms have occupied brackish water environments.

Nutrients and oxygen The biolimiting nutrients of phosphate and nitrate exert considerable control over the rates of primary productivity in seas and oceans. Where the rates of food supply are low, as in the deep sea, foraminiferal densities tend to be low (<10/10 cm^2) but diversity can be high. In upwelling zones where

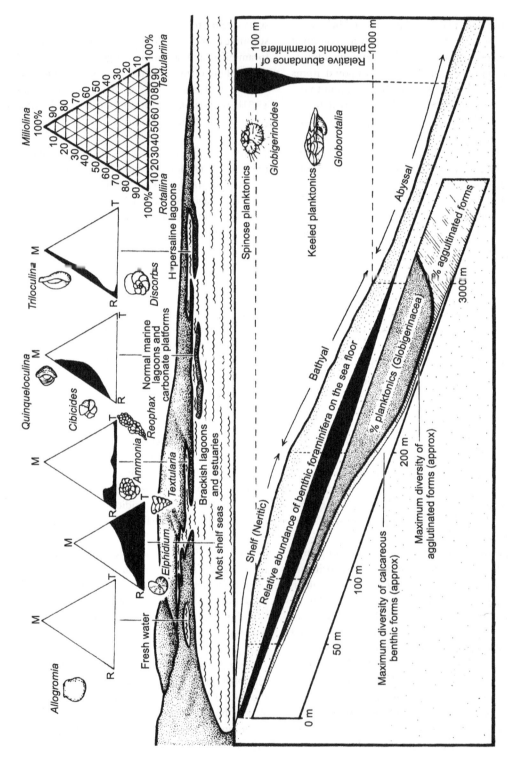

Fig. 15.9 How benthic and planktonic foraminiferid assemblages (and some typical taxa) change with depth and latitude in the Pacific Ocean, especially in relation to temperature (based partly on Saidova 1967).

rates of nutrient supply to the surface are high, foraminiferal diversities tend to be reduced for several reasons. High rates of nutrient flux tend to discourage photosymbiosis, so that planktonic and larger benthic foraminifera which culture symbionts and other **oligotrophic** species are discouraged. High rates of primary production at the surface also lead to anaerobic bacterial blooms in the oxygen minimum zone of mid-waters and on the sea floor beneath. In anaerobic conditions, foraminifera may be scarce but in dysaerobic conditions eutrophic benthic foraminifera may dominate the biota, with densities over 1000/10 cm^2. Such assemblages are typified by small, thin-shelled, unornamented calcareous buliminaceans (e.g. *Bulimina*, Fig. 15.24b; *Bolivina*, Figs 15.24c, 15.31h; *Uvigerina*, Fig. 15.10) or primitive agglutinated forms (e.g. *Ammodiscus*, Figs 15.4f, 15.31b). Oxygen deficiency does not entirely eliminate microscopic organisms such as foraminifera, presumably because of their low oxygen demand and the high diffusion rates associated with a high surface area–volume ratio. Brasier (1995a, 1995b) has reviewed the use of microfossils as nutrient indicators.

Temperature Each species is adapted to a certain range of temperature conditions, the most critical being that range over which successful reproduction can take place. Generally, this range is narrowest for low-latitude faunas adapted to stable, tropical climates. However, stratification of the oceans results in the lower layers of water being progressively cooler, as for example in tropical waters where the surface may average 28°C but the bottom waters of the abyssal plains may average less than 4°C. These cooler, deeper waters are characterized by cool-water benthic assemblages that otherwise are found at shallower depths nearer the Poles (Fig. 15.10).

Water mass history Until the 1970s it was widely thought that certain smaller, hyaline, benthic foraminiferal species were adapted to specific water depths, largely controlled by temperature, and could therefore be used to estimate ancient water depth (**palaeobathymetry**). Research has since shown that these species are closely tied to specific water masses. For example, *Epistominella* is typical of North East Atlantic Deep Water, *Fontbotia* of North Atlantic Deep Water and *Nutallides* of Antarctic Bottom Water. This means that the ancient distribution of such benthic species can be used to reconstruct the history of a specific water mass in relation to changes in global climate or in basin geometry.

Diversity This refers to the number of taxa in an assemblage. To measure diversity it is important to use a technique which is not dependent on sample size, such as the **alpha index** (see Murray 1991). In living assemblages one species is normally found to be more abundant than any other and is said to be **dominant**. Species dominance is commonly expressed as a percentage of the population, and lower dominance tends to be found with higher diversity.

The diversity of modern benthic foraminiferal assemblages from marginal marine habitats is less than that of normal marine and deep sea habitats. Higher diversity of the latter may be taken to suggest greater partitioning of resources among species. This is typical of stable habitats, especially where food is scarce and assemblages are likely to include large **K-strategists** with relatively large tests and long life spans, such as some deep sea foraminifera. Conversely, oscillations in environmental stability, such as found in marshes and lagoons, result in foraminiferid blooms of great abundance but lower diversity. These opportunistic species are **r-strategists** that must reach maturity quickly and therefore tend to be of relatively small size.

Larger benthics

Larger benthic foraminifera are K-strategists that live largely in oligotrophic reef and carbonate shoal environments where terrestrial and seasonal influences are slight. They culture endosymbiotic diatoms, dinoflagellates, rhodophytes or chlorophytes, in much the same way as do the hermatypic corals (e.g. living *Archaias*, Fig. 15.21b). These endosymbionts release photosynthates to their hosts and also take up respiratory CO_2 during photosynthesis, which allows for high rates of $CaCO_3$ precipitation during test growth. It follows that larger foraminifera are very sensitive to light levels. Many have their chambers partitioned

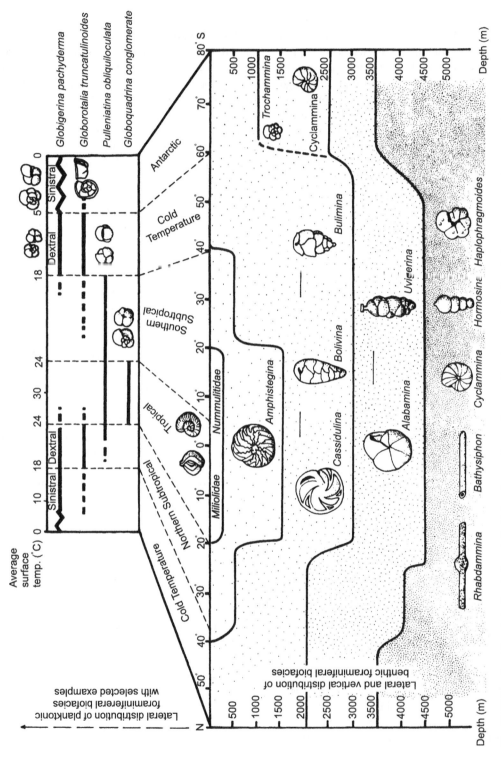

Fig. 15.10 How benthic and planktonic foraminiferid abundance and general composition change with depth and salinity. Some typical genera are shown.

into small chamberlets with translucent outer walls, which allow for more efficient culturing of the symbionts. Some, such as *Amphistegina* (Fig. 15.25g), are known to increase their surface area–volume ratio (i.e. become flatter) and thin their outer walls with increasing water depth and decreasing light intensity. Pillars of calcite that radiate through the test may even have functioned as fibre-optic lenses in fossil *Nummulites* (Fig. 15.28a,b,d). The depth distribution of living larger benthic foraminiferal taxa is also closely related to the light wave lengths required by their symbionts, from shallowest to deepest: *Archaias* (0–20 m, chlorophytes, red light); *Peneropolis* (0–70 m, rhodophytes, yellow light), *Amphistegina* (0–130 m, diatoms, blue light). It therefore appears that fossil larger benthic foraminifera, which have evolved repeatedly since the Carboniferous, have achieved their great size (up to 180 mm in Oligocene *Lepidocyclina*, Fig. 15.26b,c) and skeletal complexity through co-evolution with endosymbionts (see Hallock 1985).

Many larger foraminifera have adapted to a life in mobile carbonate sands and their tests are therefore robust and fusiform (e.g. fusulinids, Fig. 15.17; alveolinids, Fig. 15.19c), conical (e.g. orbitolinids, Fig. 15.15e) or biconvex (e.g. *Amphistegina*, Fig. 15.25g; nummulitids, Fig. 15.28). Those reclining on sediments in the deeper part of the photic zone tend to be large and discoidal in shape (e.g. *Spiroclypeus*, Fig. 15.28c). Forms adapted for adherence to seagrass or algal fronds tend to be small and flat (e.g. *Peneropolis*, Fig. 15.21a) or have robust spines for anchorage (e.g. *Calcarina*, Fig. 15.27c).

Large test size and rapid rates of growth mean that larger benthic foraminifera are major contributors to modern carbonate sedimentation, producing as much as 2800 g $CaCO_3/m^2$ every year in modern tropical oligotrophic settings (Murray 1991). Vast areas of carbonate ramp environments have also been colonized, and at times built up, by larger fossil foraminifera, especially during the Carboniferous and Permian (e.g. fusulinids, Fig. 15.17) and the Tertiary (e.g. *Nummulites*, Fig. 15.28). Nummulitic sands, in particular, are important as hydrocarbon reservoirs in the Middle East, where they may host as much as 60% of the petroleum reserves of the planet.

Planktonic foram ecology

The environmental controls on planktonic foraminifera are much better understood than those for benthics, because the only major ecological factors here are temperature and salinity. Species are distributed in large latitudinal provinces showing some bipolar distribution (e.g. Oberhänsli 1992), with temperature as the dominant control. This characteristic has been of great value in estimating Quaternary sea-surface temperatures, from the fossil record of extant species (e.g. Arnold & Parker in Sen Gupta 1999, pp. 103–123).

Depth and food

There are about 100 species of living planktonic foraminifera. They tend to be small (mostly <100 μm) and short lived (about 1 month) with tests that are adapted to retard sinking. Most modern species reproduce in the surface layers of the ocean. Towards the end of adult life, they sink slowly through the water column. Each species tends to end up in an oceanic layer of a particular temperature and density range. Shallow species live mainly in the upper 50 m of the photic zone. Those forms that live in oligotophic, central oceanic water masses feed on zooplankton, especially copepods. They supplement their diet by culturing dinoflagellate or chrysophyte photosymbionts. Long spines and globular chambers with high porosity (and hence low relative mass) may help to improve buoyancy, while secondary apertures may allow increased mobility of the symbionts. Intermediate species live mainly at 50–100 m (except as juveniles) and include spinose forms with symbionts adapted to oligotrophic waters (e.g. *Orbulina universa*, Fig. 15.23f) and non-spinose forms without symbionts that are adapted to more eutrophic waters (e.g. *Globigerina bulloides*, Fig. 15.23e). Deeper species living mainly below 100 m (except as juveniles) include forms with club-shaped (**clavate**) chambers (e.g. *Hastigerinella adamsi*, Fig. 15.23g) or lack spines but bear keels that may help to retard the settling velocity (e.g. *Globorotalia menardii*, Fig. 15.23d,h). These species are adapted to cooler, denser, more eutrophic water masses and hence have fewer buoyancy problems and consequently a lower test porosity than those from warmer or shallower waters. Deep-water planktonic forms have to cope, however, with the effects of $CaCO_3$ solution (due to higher pressure, lower pH and other factors) which may account for the extra **crust** of radial, hyaline calcite seen in some forms (e.g. *Globorotalia*, Fig. 15.23d,h). Species that live below the photic zone are thought to scavenge the sinking phytodetritus.

Temperature and latitude

Modern assemblages can be arranged into biogeographic provinces: Arctic; Subarctic; Transitional; Tropical; Subtropical; Transitional; Subantarctic; Antarctic (Fig. 15.11). A number of trends should be noted here. The distributions are **bipolar**, so that *Globorotalia truncatulinoides*, for example, is characteristic of both northern and southern subtropical waters. The number of endemic forms, and hence diversity, increases towards the tropics. Keeled forms (*Globorotalia* spp.), for example, are not found at higher latitudes in waters cooler than 5°C. Test porosity of shallow and intermediate species (e.g. *Orbulina universa*) also increases towards the equator, presumably in relation to the lower density of warmer water. In *Globigerina pachyderma*, subpolar and polar populations can be distinguished by a predominance of left- (sinistral) or right-handed (dextral) coiling (Fig. 15.10). Sinistral coiling tests have the aperture on the left when the spire is uppermost. The distribution of these assemblages shows a strong correlation with surface circulation pattern. The history of Quaternary oceanic and temperature fluctuations can therefore be determined from the distribution of planktonic foraminifera preserved in deep sea cores.

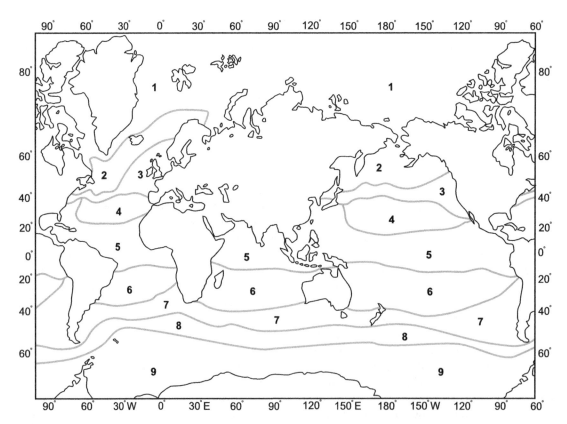

Fig. 15.11 Modern planktic foraminiferal provinces. 1, Arctic; 2, subarctic; 3, transitional; 4, subtropical; 5, tropical; 6, subtropical; 7, transitional; 8, subantarctic; 9, antarctic. (Based on data in Belyaeva 1963.)

Planktonic foraminiferal densities can be very high around the margins of oceanic gyres, where upwelling and mixing take place and nutrient levels are high. Where seasonal perturbations take place at lower latitudes (e.g. with monsoonal upwelling), then an ecological succession of species is found.

In a transect across the inner shelf to bathyal slope (e.g. Fig. 15.9) there is typically an increase in the ratio of planktonic to benthic tests within the total foraminiferal assemblage. This takes place in part because an increase in water depth increases the biomass of plankton above a given area of sea floor and in part because the food supply reaching the sea floor tends to diminish as water depth increases. The ratio is only a crude index of **palaeobathymetry**, however, because local conditions can vary the test production rate of either planktonic or benthic foraminifera. For further information on modern planktonic foram ecology, see Hemleben *et al.* (1989).

Globigerina ooze

Planktonic foraminifera are important contributors to deep sea sedimentation and, with coccoliths, account for more than 80% of modern carbonate deposition in seas and oceans. At present the foraminifera contribute more than the coccolithophores, although this was not the case with earlier chalks and oozes. Three factors are important in controlling the deposition of *Globigerina* ooze (i.e. ooze in which over 30% of sediment is globigerinacean): climate, depth of the lysocline and terrigenous sediment supply. The position and strength of currents, especially diverging and upwelling currents, are greatly affected by climate and hence affect the plankton productivity. Berger (1971) estimated than from 6 to 10% of the living population of planktonic foraminifera leave empty tests every day, mostly as a result of reproduction. These tests settle quite rapidly and are less susceptible to dissolution than coccoliths (which lack organic outer layers), except when they approach the lysocline which usually lies between 3000 and 5000 m depth. Fluctuations in the depth of the calcite compensation depth (see below) during the Mesozoic and Cenozoic are now known to have caused cycles of deposition and dissolution, selectively removing some of the smaller or more delicate forms and rendering the fossil record of the deep sea incomplete. Even where the conditions are otherwise favourable, *Globigerina* oozes cannot accumulate where there is an influx of terrigenous clastics, hence they are rarely found on continental shelves. At present such oozes are mainly accumulating between 50°N and 50°S at depths between about 200 and 5000 m, especially along the mid-oceanic ridges. In many cases, though, they are diluted with the siliceous remains of diatoms and radiolarians.

Calcite compensation depth (CCD)

The solubility of $CaCO_3$ is less in warm than in cool waters. This in part favours the thicker tests and the occurrence of foraminiferid limestones and oozes at low latitudes. More important, however, is the vertical change in $CaCO_3$ solubility, which also increases with greater pressure, and hence with greater depth in the ocean. The partial pressure of CO_2 also increases with depth because there is no photosynthesis below the photic zone, although animals and bacteria continue to respire. These factors led to a decrease in pH with depth, from about 8.2 to as low as 7.0. The level in the water column at which $CaCO_3$ solution equals $CaCO_3$ supply is called the **calcium carbonate compensation depth** (or CCD). As this is impractical to locate in the geological record, the concept of the **lysocline** (i.e. the level of maximum change in the rate of solution of foraminiferal test calcite) is widely used. The net result, of course, is a drop in the number of calcareous organisms with depth, there being few below 3000 m. For this reason, benthic agglutinated foraminifera (e.g. ammodiscaceans as in Fig. 15.12c–f) dominate populations from abyssal depths.

Classification

Kingdom PROTOZA
Phylum SARCODINA
Class RHIZOPODA
Order FORAMINIFERIDA

Foraminifera are included in the phylum Rhizopoda (Corliss 1994) or Reticulosa (Cavalier-Smith 1993) or

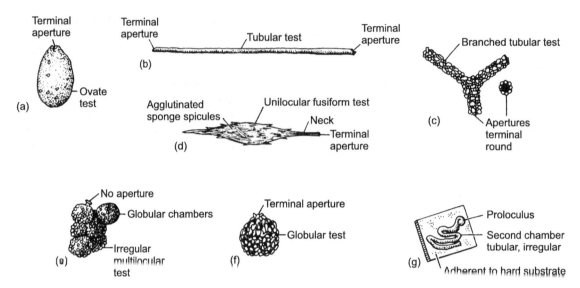

Fig. 15.12 Suborder Allogromiina. (a) *Allogromia* ×23. (b) *Shepheardella* ×8. Suborder Textulariina, superfamily Ammodiscacea. (c) *Rhabdammina* ×10. (d) *Technitella* ×17. (e) *Sorosphaera* ×7.5. (f) *Saccammina* ×10.5. (g) *Tolypammina* ×12.5; ((e) After Loeblich & Tappan 1964 (from the *Treatise on Invertebrate Paleontology*, courtesy of and © 1964, Part C, The Geological Society of America and The University of Kansas).)

considered a separate phylum (Cavalier-Smith 1998). Reconciling the classification scheme proposed by Cavalier-Smith (1993) with that widely used for fossil groups and preferred herein (see Hart & Williams in Benton 1993, pp. 43–66) is problematic. In the Cavalier-Smith scheme the traditional suborders Allogromiina etc. are elevated to subclasses.

The distinction for the major groups of foraminifera is based on the composition and structure of the test wall (Loeblich & Tappan 1988) and takes the following features into account, in order of importance: wall structure and composition, chamber shape and arrangement, aperture and ornament. This reflects the long history of study and utilization mainly by micropalaeontologists. The salient features of the currently recognized suborders and superfamilies are noted in Table 15.1; notes below highlight features of common fossil forms. The extent to which wall structure indicates evolutionary relationships is highly questionable and recent advances in the molecular systematics and the cladistical analysis of the Foraminifera are challenging many traditional hypotheses of relationships. Evidence from scanning electron microscopy has also led to a better understanding of wall structures. The classification followed here emphasizes features visible with an optical microscope.

Suborder Allogromiina

These foraminifera have an entirely organic test with only one chamber. They are rarely encountered as fossils, being found largely in Recent, fresh or brackish water sediments. They are known in marine sediments since Late Cambrian times. *Allogromia* (Rec., Fig. 15.12a) has an ovate test with a rounded terminal aperture. *Pleurophrys* (Rec., Fig. 15.4a) is similar but smaller. *Shepheardella* (Rec. Fig. 15.12b) has a long tubular test with an aperture at each end. Both larger and planktonic types are unknown in this suborder.

Suborder Textulariina

The Textulariina are characterized by non-laminar, agglutinated tests. The Ammodiscacea range from Early Cambrian to Recent times and all would be considered smaller benthic foraminifera and are mostly

Table 15.1 A guide to the morphological character of foraminiferid suborders and superfamilies. (Based in part on Culver, in Lipps 1993, table 12.1.)

Suborder	Wall structure	Septation	Chamber architecture	Range
Allogromiina	Organic, may have iron encrustations of some agglutionated particles	Unilocular	Irregular, sac-, flask- or tube-shaped	Upper Cambrian to Recent
Examples	*Allogromina* *Shepeardella*			
Textulariina	Agglutinated, with organic or mineral cement	Unilocular or multilocular	Wide variety of shapes from uniserial globular, branching or tubular, triserial, planispiral and trochospiral	Lower Cambrian to Recent
Superfamilies	Ammodiscacea Coscinophragmatacea Hormosinacea Rzehakinacea Astrorhizacea Dicyclinidea Lituolacea Textulariacea Ataxophragmiacea Haplophragmiacea Loftusiacea Trochamminacea Biokovinacea Hippocrepinacea Orbitolinidae Verneuilinacea			
Examples	*Ammobaculites Cyclolina Rhabdammina Tolypammina* *Ammovertella Cyclopsinella Rhizammina Trochammina* *Astrorhiza Dicyclina Saccammina Usbekistania* *Aschemonella Hormosina Sorosphaera Verneuilina* *Ammodiscus Loftusia Spirocyclina* *Bathysiphon Milammina Technitella* *Bigenerina Orbitulina Textularia* *Coskinolina Reophax* *Cycloammina*			
Fusulinina	Homogeneous microgranular calcite. Advanced forms may have two or more layers	Unilocular or multilocular, the latter with chamberlets	Predominantly planispiral, mostly fusiform some ovate or discoid. However there are a wide variety of shapes from uniserial globular, branching or tubular, triserial, planispiral and trochospiral	Lower Silurian to Upper Permian
Superfamilies	Archaediscacea Fusulinacea Nodosinellacea Tetrataxacea Earlandiidae Geinitzinacea Parathuramminacea Tournayellacea Endothyracea Moravamminacea Ptychocladiacea			
Examples	*Earlandinita Neoschwagerina Profusulinella Schwagerina* *Endothyra Nodosinella Saccaminopsis Tetrataxis* *Fusulina Palaeotextularia*			
Involutina	Calcareous, perforate, radiate, originally aragonite but commonly recrystallized to homogenous microgranular structure	Proloculus followed by enrolled tubular second chamber		Lower Permian to Upper Cretaceous, Recent
Examples	*Involutina*			

Table 15.1 (cont'd)

Suborder	Wall structure	Septation	Chamber architecture	Range
Spirillinina	Calcite, optically a single or rarely a mosaic of crystals; a-axis along axis of coiling, c-axis parallel to umbilical surface. May have pseudopores filled with organic matter and sieve plates. Wall formed by marginal accretion by pseudopodia not by calcification	Proloculus followed by undivided chamber, or few chambers per whorl, chambers can be secondarily subdivided	Planispiral or high trochospiral	Upper Triassic to Recent
Examples	*Patellina* *Spirillina*			
Carterinina	Test attached. Wall with an organic inner lining and outer layer of rod-like or fusiform secreted spicules, each a single low magnesium calcite crystal. Each spicule is embedded in a mass or small spicules held together by an organic matrix	Early chambers semicircular, later ones crescentic to irregular or spreading	Trochospiral. Early chambers simple, later ones may have secondary septa formed from an infolding of the wall	Eocene, Recent
Example	*Carterina*			
Miliolina	Porcelaneous high magnesium calcite, commonly with organic lining, generally imperforate but pores may occur in proloculus of some	Unilocular or multilocular with chamberlets	Fusiform	Carboniferous to Recent
Superfamilies	Alveolinacea Miliolacea Squamulinacea Cornuspiracea Soritacea			
Examples	*Archaias Fasciolites Peneroplis Articulina Nubeculinella Quinqueloculina Cyclogyra Orbitolites Triloculina*			
Silicoloculinina	Imperforate, or secreted opaline silica	Unilocular or multilocular, simple	Planispiral, trochospiral, biserial or cyclical	Upper Miocene to Recent
Example	*Miliammellus*			
Lagenina	Generally monolamellar, optically and ultrastructurally radiate calcite, c-axes normal to surface: crystal units surrounded by organic membranes. Advanced forms may have second lamella	Unilocula		Upper Silurian to Lower Devonian: Lower Carboniferous to Recent
Superfamilies	Nodosariacea Robuloidacea			
Example	*Frondicularia Lagena Polymorphina Guttulina Lenticulina Nodosaria*			

Robertinina	Hyaline, perforate, ultrastructurally or optically radiate aragonite, hexagonal prisms with c-axis normal to wall surface, prisms in bundles surrounded by organic matrix	Planispiral to trochospirally enrolled	Middle Triassic to Recent
Superfamilies	Ceratobuliminacea Conorboidiacea Duostominacea Robertinacea		
Example	*Ceratobulimina Duostomina Hoeglundina Robertina*		
Globigerinina	Perforate hyaline calcite; optically radiate, c-axes normal to surface. Primarily bilamellar with addition of further lamellae on growth of new chamber	Planispiral, trochospiral, uncoiled biserial or uniserial	Middle Jurassic to Recent
Superfamilies	Globigerinacea Globotruncanacea Heterohelicacea Rotaliporacea Globorotaliacea Hantkeninacea Planomalinacea		
Examples	*Globigerina Hastigerinoides Heterohelix Globorotalia Hastigerinella Orbulina Globotruncana*		
Rotaliina	Perforate hyaline lamellar calcite, formed by calcification on either side of an organic membrane. May be optically radial or granular. Surface may be highly ornamented	Wide variety of shapes. Predominantly planispiral and trochospiral. Uncoiled biserial or uniserial	Triassic to Recent
Superfamilies	Acervulinacea Cibicides Eouvigerinacea Planorbulinacea Annulopatellinacea Discocyclina Fursenkoinacea Rotaliacea Asterigerineacea Discocyclina (Aktinocyclina) Nonionacea Siphoninacea Bolivinitacea Discorbis Nummulitaceadea Stilostomellacea Bolivinacea Elphidium Orbitoidacea Turrilinacea Buliminacea Islandiella Lepidocyclina (Eulepidina)		
Examples	*Ammonia Cibicides Lepidocyclina (Lepidocyclina) Pavonina Amphistegina Discocyclina Linderina Planorbulina Asterigerina Discocyclina (Aktinocyclina) Loxostomum Pleurostomella Bolivina Discorbis Melonis Rectobolivina Bulimina Elphidium Nonion Siphonina Buliminella Islandiella Nummulites Spiroclypeus Calcarina Lepidocyclina (Eulepidina) Osangularia Tretomphalus Cassidulina Virgulinella*		

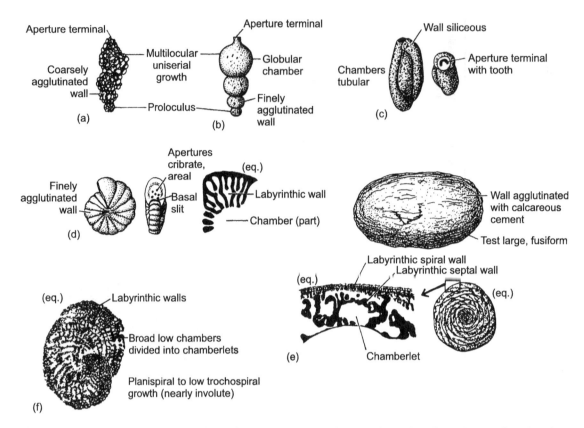

Fig. 15.13 Suborder Textulariina, superfamily Lituolacea. (a) *Reophax* ×18. (b) *Hormosina* ×6. (c) *Miliammina* ×33. (d) *Cyclammina* ×4. (e) *Loftusia* above ×0.7, lower left ×92, upper right ×3.5. (f) *Spirocyclina* ×9.5. The following abbreviations are used on this and the following figures in this chapter: (eq), equatorial section; (ax), axial section. ((a), (b) Modified from Loeblich & Tappan 1964; (e) adapted from Loeblich & Tappan 1964 after Carpenter & Brady; (f) adapted from Loeblich & Tappan 1964 after Maync.) The following abbreviations are used on this and the following figures in this chapter: eq, equatorial section; ax, axial section.

unilocular, however *Astrorhiza* (Fig. 15.4c) can be up to 10 mm in diameter. *Saccammina* (Sil.-Rec., Figs 15.12f, 15.31a) is a simple globular form with a terminal aperture. Irregularly arranged chambers of similar type are found in the multilocular *Sorosphaera* (Sil.-Rec., Fig. 15.12e). In *Technitella* (Olig.-Rec., Fig. 15.12d) the test is fusiform and built of carefully selected sponge spicules. Tubular tests generally have several apertures and may be simple and unbranched as in *Bathysiphon* (?Camb., Ord.-Rec., Fig. 15.4d), branched as in *Rhizammina* (Rec., Fig. 15.4e) or radiating from a central point as in *Astrorhiza* (?M. Ord.-Rec., Fig. 15.4c), *Aschemonella* (U. Dev.-Rec., Fig. 15.4h) and *Rhabdammina* (Ord.-Rec., Fig. 15.12c).

Planispiral coiling is seen in *Ammodiscus* (Sil.-Rec. Figs 15.4f, 15.31b) and **glomospiral coiling** (like a skein of wool) in *Usbekistania* (Jur.-Rec., Fig. 15.4g). Adherent forms are irregularly branched or may meander and zig-zag across the substrate (e.g. *Ammovertella* (L. Carb.-Rec., Fig. 15.4i); *Tolypammina* (U. Ord.-Rec., Fig. 15.12g)).

The tests of the Lituolacea are more complex than those of the Ammodiscacea. The simplest of the smaller benthic forms are commonly straight uniserial (e.g. *Reophax* (U. Dev.-Rec., Fig. 15.13a), *Hormosina* (Jur.-Rec., Fig. 15.13b) or the biserial *Textularia* (U. Carb.-Rec., Fig. 15.14b). Both kinds of growth are combined in different stages of *Bigenerina* (U. Carb.-Rec., Fig. 15.14c). Triserial tests are also common in

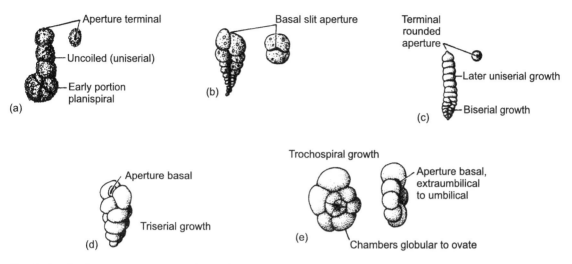

Fig. 15.14 Suborder Textulariina, superfamily Lituolacea. (a) *Ammobaculites* ×20. (b) *Textularia* ×12.5. (c) *Bigenerina* ×11.5. (d) *Verneuilina* ×13.5. (e) *Trochammina* ×29. ((a) After Pokorny 1963 from d'Orbigny; (b) & (d) after Morley Davies 1971 from H.B. Brady with permission from Kluwer Academic Publications; (c), (e) after Loeblich & Tappan 1964, from the *Treatise on Invertebrate Paleontology* (courtesy of and © 1964, Part C, The Geological Society of America and The University of Kansas).)

the group (e.g. *Verneuilina* (Jur.-Rec., Fig. 15.14d) and *Miliammina* (L. Crec.-Rec., Fig. 15.13c) is coiled like a miliolid (see below).

Coiled growth plans are also common as in the planispiral *Cyclammina* (Cret.-Rec., Fig. 15.13d) and the trochospiral *Trochammina* (L. Carb.-Rec., Fig. 15.14e). A combination of planispiral and uniserial growth is seen in the uncoiled test of *Ammobaculites* (L. Carb.-Rec., Fig. 15.14a).

The 'larger' agglutinated foraminifera have tests mostly constructed of calcareous particles with a mineral cement. Examples in the Lituolacea are found in rocks formed in warm shallow facies of Jurassic and Cretaceous age. *Spirocyclina* (U. Cret., Fig. 15.13f) and its relatives had almost planispiral, compressed tests and labyrinthic walls. *Loftusia* (U. Cret., Fig. 15.13e) resembles the more ancient fusulines in having a planispiral fusiform test with a labyrinthic wall, irregular septa and chamberlets. The Dicyclinidea were long ranged (U. Trias.-M. Eoc.), comprise discoidal or low conical forms with cyclical chambers that may be subdivided into chamberlets (e.g. *Cyclolina*, U. Cret., Fig. 15.15b; *Cyclopsinella*, U. Cret., Fig. 15.15c; *Dicyclina* U. Cret., Fig. 15.15d). Conical forms belonging to the Orbitolinidae (L. Cret.-U. Eoc.) are uniserial stacks of saucer-shaped chambers following an early trochospiral stage (e.g. *Coskinolina*, L. Cret.-U. Eoc., Fig. 15.15a; *Orbitolina*, L.-U. Cret., Fig. 15.15e). Radial septulae subdivide these chambers into an outer radial zone of tubular chamberlets. Smaller horizontal and vertical plates may form, within these chamberlets a marginal zone of minute cellules. In the centre of the chambers there is a reticulate zone in which the radial chamberlets are further subdivided by vertical pillars.

Suborder Fusulinina

The Fusulinina contains those foraminifera with calcareous, microgranular walls; advanced forms may have two or more layers. The group was largely Palaeozoic in age, becoming extinct in the Triassic.

The Parathuramminacea were small benthic forms with simple microgranular walls. The architecture was also simple, ranging from unilocular to straight uniserial (e.g. *Saccaminopsis*, Ord.-Carb., Fig. 15.16a; *Earlandinita*, L.-U. Carb., Fig. 15.16b). This group is known with certainty from Ordovician through to Carboniferous times.

The Endothyracea (U. Sil.-Trias,) were small, multilocular foraminifera with walls generally differentiated

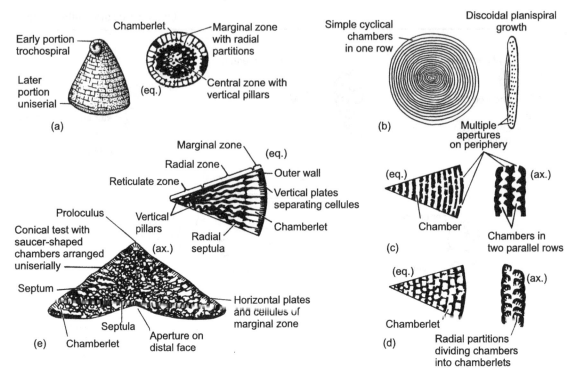

Fig. 15.15 Suborder Textulariina, superfamily Lituolacea. (a) *Coskinolina* ×9.5. (b) *Cyclolina* ×11.5. (c) *Cyclopsinella* ×16. (d) *Dicyclina* ×16. (e) *Orbitolina* left ×13, above right ×9.1. ((a) After Morley Davies 1971 with permission from Kluwer Academic Publications; (e) after Loeblich & Tappan 1964 from Egger (from the *Treatise on Invertebrate Paleontology*, courtesy of and © 1964, Part C, The Geological Society of America and The University of Kansas).)

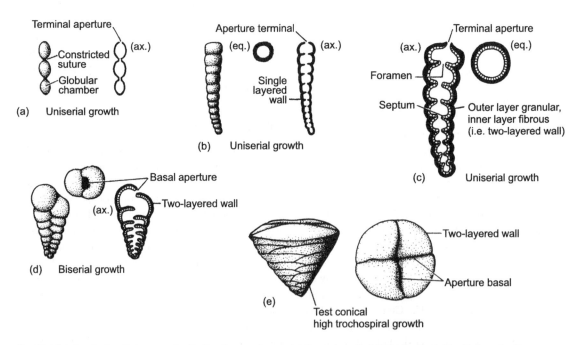

Fig. 15.16 Suborder Fusulinina, superfamily Parathuramminacea. (a) *Saccaminopsis* ×1.5. (b) *Earlandinita* ×40. Superfamily Endothyracea. (c) *Nodosinella* ×16.5. (d) *Palaeotextularia* ×23. (e) *Tetrataxis* ×34. ((a) After Loeblich & Tappan 1964 from H.B. Brady; (b), (c) redrawn after Cummings 1955; (d) after Loeblich & Tappan 1964 from Galloway & Ryniker; (e) after Loeblich & Tappan 1964 ((a), (d), (e) from the *Treatise on Invertebrate Paleontology*, courtesy of and © 1964, Part C, The Geological Society of America and The University of Kansas).)

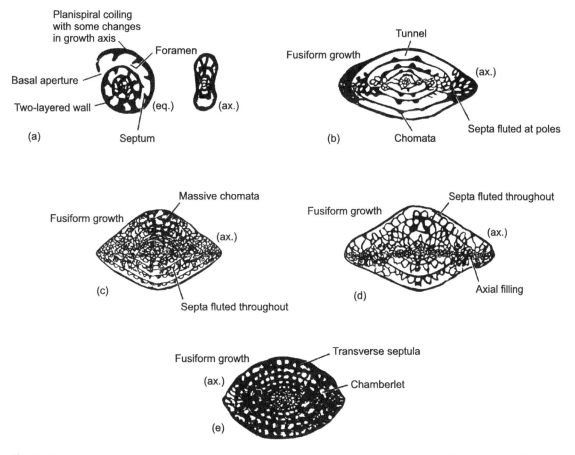

Fig. 15.17 Suborder Fusulinina, superfamily Endothyracea. (a) *Endothyra* ×22. Superfamily Fusulinacea. (b) *Profusulinella* ×90. (c) *Fusulina* ×7. (d) *Schwagerina* ×7. (e) *Neoschwagerina* ×13. (All after Loeblich & Tappan 1964; (a) from Zeller; (b) from Rauzer-Chernousova; (c), (d), (e) after Thompson (from the *Treatise on Invertebrate Paleontology*, courtesy of and © 1964, Part C, The Geological Society of America and The University of Kansas).)

into an outer granular layer and an inner fibrous layer, also microgranular but of fibrous appearance owing to the perforations. The architecture was variable and included uniserial forms (e.g. *Nodosinella*, U. Carb.-Perm., Fig. 15.16c), biserial (e.g. *Palaeotextularia*, L. Carb.-Perm., Fig 15.16d), high trochospiral (e.g. *Tetrataxis*, L. Carb.-Trias., Fig. 15.16e) and planispiral forms (e.g. *Endothyra*, L. Carb.-Perm., Fig. 15.17a).

The Fusulinacea were larger forms which also had microgranular perforate tests but with chambers arranged planispirally in a discoidal to fusiform plan. Two kinds of wall structure are found. The ancestral, fusulinid wall is primarily two-layered with a dark, partly organic outer tectum and an inner, clear **diaphanotheca** (Fig. 15.18b). Secondary deposition of a dark epitheca within the chamber may give the inner walls a four-layered appearance. The schwagerinid wall lacks this secondary thickening and the mural pores are much enlarged to form **alveoli** (Fig. 15.18c). This gives the clearer inner layer a fibrous appearance termed the **keriotheca**. The schwagerinid wall is typical of the larger fusulines of the later Pennsylvanian (U. Carb.) and Permian periods.

The early chambers of microspheric fusulines indicate they had an ancestor like the small planispiral *Endothyra* (Fig. 15.17a). Evolutionary trends included changes in shape, size and wall structure. For example, there was a progressive folding of the septa in some

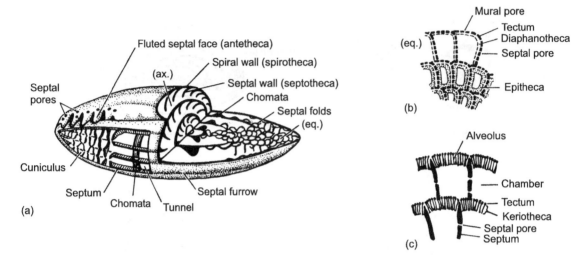

Fig. 15.18 (a) Schematic fusuline, based on *Parafusulina* and *Fusulinella*. (b) 'Fusulinid' wall. (c) 'Schwagerinid' wall.

lineages, the forward folds of one septum generally meeting the backward folds of the next (e.g. *Profusulinella*, U. Carb., Fig. 15.17b; *Fusulina*, U. Carb., Fig. 15.17c). A small passage (**cuniculus**) connected adjacent chamberlets (Fig. 15.18a). In some forms a tunnel was formed by selective resorption of the septa and secretion of two bordering ridges called **chomata** (see Figs 15.17b, 15.18a), thereby connecting the mid-floor of each chamber. In the Permian schwagerinids there was a tendency to fill the central axial chambers with secondary calcite (e.g. *Schwagerina*, Perm., Fig. 15.17d). The Late Permian verbeekinids had flat septa with foramina and spiral walls bearing axial and transverse projections (septulae) into the chambers (e.g. *Neoschwagerina*, U. Perm., Fig. 15.17e). These highly specialized foraminifera were adapted to carbonate and reefal facies in the Late Carboniferous and Permian but became extinct at the end of that period.

Suborder Involutina

These are calcareous foraminifera with perforate, radiate walls that were originally aragonite. In fossils forms this has recrystallized to a homogenous, microgranular structure. The proloculus is followed by an enrolled tubular second chamber (e.g. *Involutina*, Jur., Fig. 15.19a; *Planispirillina*, Jur.-Rec., Fig. 15.31n).

Suborder Spirillinina

Calcitic forms with planispiral to high trochospiral coiling, or with a few chambers per whorl. The proloculus may be followed by an undivided tubular chamber. They are small benthic forms often found adhering to algae or hard substrates. It is possible that they developed independently from the other hyaline superfamilies. *Spirillina* (Jur.-Rec. Figs 15.19b, 15.31f) has a long planispiral second chamber and terminal aperture. The wall is optically a single crystal of calcite, with the a-axis orthogonal to the direction of coiling and the c-axis parallel to the umbilical surface. *Patellina* (L. Cret.-Rec., Fig. 15.19c) has a trochospiral to biserial test in which the chambers are subdivided by a scroll-like median septum and numerous transverse septulae.

Suborder Carterinina

The test is attached with the early chambers semicircular, later ones becoming crescent-shaped and finally irregular. The wall has an organic lining and an outer layer of rod-like, single, spicular crystals of low magnesium calcite, in a matrix of small spicules and organic material. The Carterinina are represented by the single genus *Carterina* (Rec., Figs 15.19d, 15.31e). Unfortunately, the tests disintegrate after death and are not

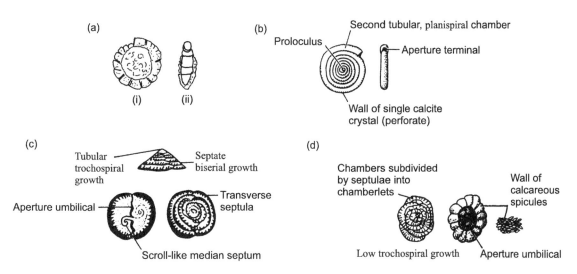

Fig. 15.19 Suborder Involutina. (a) *Involutina*, side (i) and apertural (ii) views ×54.8. Suborder Spirillinina. (b) *Spirillina* ×50. (c) *Patellina* ×33. Suborder Carterinina. (d) *Carterina* ×14. ((a) Redrawn after Cushmann 1948; (b), (c), (d) after Loeblich & Tappan 1964 (from the *Treatise on Invertebrate Paleontology*, courtesy of and © 1964, Part C, The Geological Society of America and The University of Kansas).)

known as fossils. The aperture of *Carterina* is large and umbilical in position and the chambers are thick spines divided into chamberlets by septulae.

Suborder Miliolina

The Miliolina have imperforate calcareous tests of porcelaneous appearance with a planispirally coiled proloculus. Subsequent growth may continue planispirally (e.g. *Cyclogyra*, Carb.-Rec., Fig. 15.20a), uncoil and develop uniserially (e.g. *Nubeculinella*, U. Jur., Fig. 15.20b) or coil streptospirally. **Streptospiral** coiling here involves the addition of tubular chambers (generally half a whorl in length) arranged lengthwise about a growth axis. When added in the same plane (i.e. at 180 degrees to one another) the arrangement is called **spiroloculine** if the chambers are evolute and **biloculine** if they are involute (Fig. 15.6). More commonly, however, chambers are added at angles of 144 degrees leaving five chambers visible from the outside (**quinqueloculine** e.g. *Quinqueloculina*, Jur.-Rec., Fig. 15.20e). In *Triloculina* and *Milonella* (Rec., Fig. 15.31d), the chambers are added at angles of 120 degrees and only three chambers are visible from outside the test (**triloculine**). Such streptospiral growth forms may later unroll to uniserial as in *Articulina* (M. Eoc.-Rec., Fig. 15.20d).

Larger porcelaneous foraminifera fall mainly into two superfamilies: the Soritacea and the Alveolinacea. The Soritacea have thrived in reefal and carbonate habitats since the Late Triassic period. These have a test which is perforate in the earliest stages and may be pseudopunctate throughout (Fig. 15.3), but like all other milioline tests they are properly regarded as imperforate. Coiling is basically discoidal planispiral further modified to cyclical, fan-shaped (**flabelliform**) or straight uniserial in the later stages of growth (e.g. *Peneropolis*, Eoc.-Rec., Fig. 15.21a). Interseptal buttresses or septulae subdivide the chambers into chamberlets in genera such as *Archaias* (M. Eoc.-Rec. Fig. 15.21b). The all-embracing, annular chamber addition in forms like *Orbitolites* (U. Palaeoc.-Eoc., Fig. 15.21c) is called cyclical.

The Alveolinacea also have imperforate tests with a perforate proloculus (e.g. *Fasciolites*, L. Eoc., Fig. 15.20c). Coiling is fusiform to ovate planispiral. The chambers are divided by septulae into numerous tubular chamberlets arranged in one or more rows. This group exhibits remarkable convergence with the Palaeozoic fusulines but is much younger, evolving repeatedly from Early Cretaceous to Recent times.

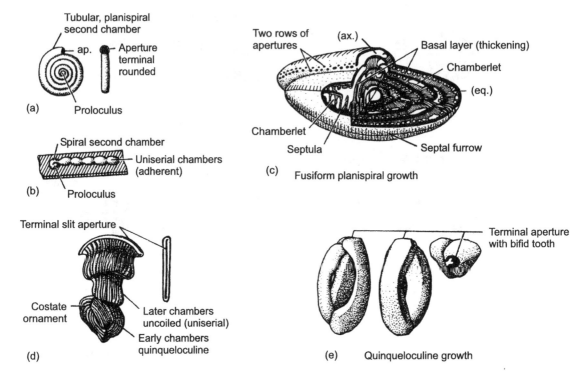

Fig. 15.20 Suborder Miliolina. (a) *Cyclogyra* ×40. (b) *Nubeculinella* ×37. (c) Schematic diagram of the alveolinid *Fasciolites* ×21.5. (d) *Articulina* ×33. (e) *Quinqueloculina* ×23. ((c) Modified after Loeblich & Tappan 1964 after Neumann; (e) after Loeblich & Tappan 1964 (from the *Treatise on Invertebrate Paleontology*, courtesy of and © 1964, Part C, The Geological Society of America and The University of Kansas); (d) redrawn after Pokorny 1963 from H.B. Brady.)

Suborder Silicoloculinina

Includes foraminifera with an imperforate wall or secreted opaline silica (e.g. *Miliammellus*, Mioc.-Rec., Figs 15.22a, 15.31p). These foraminifera can be uni- or multilocular.

Suborder Lagenina

This suborder includes forms which are monolamellar and composed of optically and ultrastructurally radiate calcite. The c-axes are orthogonal to the surface. The Nodosariacea have walls of optically radial calcite known to be of bilamellar ultrastructure under the electron microscope but monolamellar when viewed optically. Such a hidden ultrastructure should be called cryptolamellar. An aperture of radially arranged slits is typical, except in the unilocular genus *Lagena* (Jur.-Rec., Fig. 15.4b, 15.22e). *Nodosaria* has a simple uniserial test (Perm.-Rec., Fig. 15.22b). In *Frondicularia* (Perm.-Rec., Fig. 15.22c) the test is also uniserial but the chambers are compressed and V-shaped. *Dentalina* is also uniserial and had globular chambers (Trias.-Jur., Fig. 15.31g). *Lenticulina* (Trias.-Rec., Fig. 15.22d) is a common involute planispiral form. Biserial growth is seen in *Polymorphina* (Palaeoc.-Rec., Fig. 15.22f) and streptospiral (quinqueloculine) growth in *Guttulina* (Cret.-Rec., Fig. 15.22g).

Suborder Robertinina

The Robertinina have optically radial, bilamellar walls composed of aragonite instead of calcite, although this may revert to the latter mineral with time in the fossil state. The aperture is typically a basal slit extending up

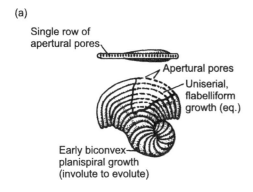
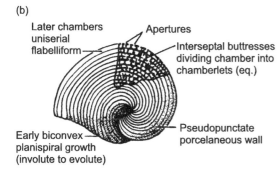
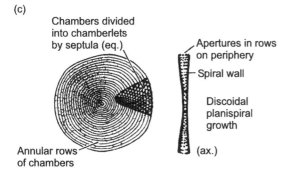

Fig. 15.21 Suborder Miliolina. (a) *Peneropolis* ×20. (b) *Archaias* ×19.5. (c) *Orbitolites* ×7.

the face of the last chamber. In *Robertina* (L. Eoc.-Rec., Fig. 15.22h) the test is high trochospiral, each elongate chamber subdivided by transverse partitions. *Ceratobulimina* (U. Cret.-Rec., Fig. 15.22i) has a moderately low trochospiral test whilst that of *Hoeglundina* (M. Jur.-Rec., Fig. 15.22j) is provided with a keel and peripheral slits marking the primary and relict (supplementary) apertures.

The Duostominacea are an extinct group. The wall structure consists of both optically radial and microgranular calcite. In *Duostomina* (M. Trias., Fig. 15.22k) the low trochospiral test has a basal aperture divided into two by a flap.

Suborder Globigerinina

The planktonic globigerinids typically have trochospirally coiled shells with inflated, coarsely perforate chambers bearing fine spines during life (e.g. *Hastigerinella*, Rec., Fig. 15.23g). These spines support a frothy ectoplasm, the pseudopodia being connected to the endoplasm through the coarse perforations. Inflated chambers, spines and frothy ectoplasm are all adaptations for greater buoyancy. The wall of the Globigerinacea is composed of optically radial and bilamellar, low-magnesium calcite. Although the primary aperture is usually basal it may be modified through evolution to areal or terminal. Secondary sutural or areal apertures are also found. These apertures may be partially covered by one or several flaps called **bullae**. Although inflated chambers are characteristic, some genera have curious club-shaped (**clavate**) chambers (e.g. *Hastigerinoides*, L.-U. Cret., Fig. 15.23b) while others, including the Rotaliporacea, Globotruncanacea and Globorotalinacea have keeled outer margins (e.g. *Globotruncana*, U. Cret., Fig. 15.23c; *Globorotalia*, Palaeoc.-Rec., Fig. 15.23d,h). Ornament is not prominent but the tests often have a rugose or pustulose surface, and rarely, longitudinal costae. Forms which are not trochospiral include the ancestral Heterohelicacea (high trochospiral-biserial-uniserial, e.g. *Heterohelix*, U. Cret., Fig. 15.23a).

Widely used for correlation are species of *Globotruncana*, *Globorotalia*, *Globigerina* (Palaeoc.-Rec., Fig. 15.23e) and *Orbulina* (Mioc.-Rec., Fig. 15.23f). The final, spherical, **orbuline** chamber of *Orbulina* completely envelopes the earlier, globigerine coil. This orbuline trend has occurred in several lineages and represents one of the most efficient adaptations for the maintenance of buoyancy.

Suborder Rotaliina

Rotaliine foraminifera have a calcareous hyaline test

172   Part 4: Inorganic-walled microfossils

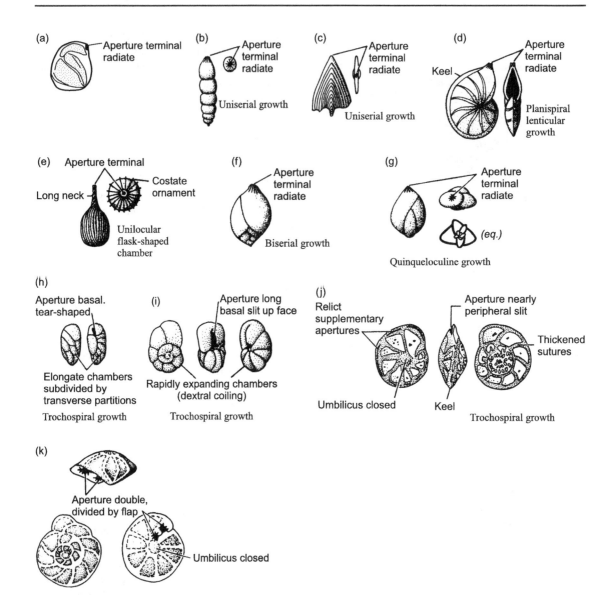

Fig. 15.22 Suborder Silicoloculina. (a) *Miliammellus* ×131. Suborder Lagenina. (b) *Nodosaria*, ×10. (c) *Frondicularia*, ×5. (d) *Lenticulina*, ×8. (e) *Lagena* ×30. (f) *Polymorphina*, ×19.5. (g) *Guttulina*, ×21.5. Suborder Robertinina. (h) *Robertina* ×18.5. (i) *Ceratobulimina* ×30. (j) *Hoeglundina* ×10. (k) *Duostomina* ×60. ((a) After Resig *et al.* 1980; (b), (d–f) after Morley Davies 1971 from H.B. Brady with permission from Kluwer Academic Publications; (d) after Morley Davies 1971 from von Hantken; (g) after Loeblich & Tappan 1964 from d'Orbigny; (h) after Loeblich & Tappan 1964 from Hoglund; (i), (j) after Loeblich & Tappan 1964; (k) after Loeblich & Tappall 1964 from Kristan-Tollmann ((g)–(k) from the *Treatise on Invertebrate Paleontology*, courtesy of and © 1964, Part C, The Geological Society of America and The University of Kansas).)

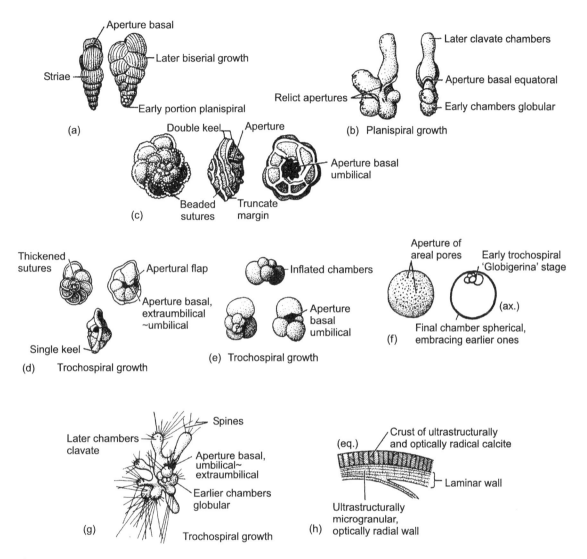

Fig. 15.23 Suborder Globigerinina. (a) *Heterohelix* ×97. (b) *Hastigerinoides* ×65.5. (c) *Globotruncana* ×36.5. (d) *Globorotalia* ×15.5. (e) *Globigerina* ×18.5. (f) *Orbulina* ×20. (g) *Hastigerinella* ×5.7. (h) Diagram of the outer wall structure of deep-water *Globorotalia*. ((a) After Loeblich & Tappan 1964 from Loeblich; (b) after Loeblich & Tappan 1964; (c) after Glaessner 1945; (d) after Loeblich & Tappan 1964 from Bolli, Loeblich & Tappan; (e) after Morley Davies 1971 from H.B. Brady; (g) after Morley Davies 1971 from Rhumbler; (h) redrawn after Pessagno & Miyano 1968 ((a)–(d) from the *Treatise on Invertebrate Paleontology*, courtesy of and © 1964, Part C, The Geological Society of America and The University of Kansas; (e), (g) with permission from Kluwer Academic Publications.)

which is both multilaminar and perforate. Subdivision into superfamilies is based largely on knowledge of wall structure. Larger forms are found mainly in the Rotaliacea and Orbitoidacea. The Buliminacea also have optically radial, cryptolamellar calcite walls, but the aperture is generally a basal, tear-shaped slit. Biserial growth is very common as in *Bolivina* (U. Cret.-Rec., Figs 15.24c, 15.31h) or triserial, as in *Bulimina* (Palaeoc.-Rec., Fig. 15.24b). In *Rectobolivina* (M. Eoc.-Rec., Fig. 15.24d) and *Pavonina* (Mioc.-Rec.,

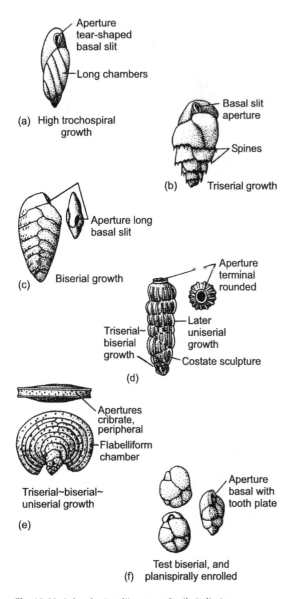

Fig. 15.24 Suborder Rotaliina, superfamily Buliminacea. (a) *Buliminella* ×90. (b) *Bulimina* ×33. (c) *Bolivina* ×20. (d) *Rectobolivina* ×38. (e) *Pavonina* ×55. (f) *Islandiella* ×11. ((a) Redrawn after Pokorny 1963 from d'Orbigny; (b) after Loeblich & Tappan 1964 from Cushman & Parker; (c) redrawn after Pokorny 1963 from Cushman; (d) after Loeblich & Tappan 1964; (e) after Loeblich & Tappan 1964 from Nørvang ((b), (d), (e) from the *Treatise on Invertebrate Paleontology*, courtesy of and © 1964, Part C, The Geological Society of America and The University of Kansas).)

Fig. 15.24e) these plans are modified in later stages to uniserial growth, the former with globular and the latter with C-shaped, flaring (flabelliform) chambers. In *Islandiella* (Palaeoc.-Rec., Fig. 15.24f) the biserial arrangement is even planispirally enrolled. The long chambers of *Buliminella* (U. Cret.-Rec., Fig. 15.24a) are arranged in a high trochospiral coil.

The Discorbacea now contains genera known to have walls of either cryptolamellar or bilamellar optically radial calcite. The tests of discorbaceans are often trochospiral and fresh specimens may be coloured brown. *Discorbis* (Eoc.-Rec., Fig. 15.25a) has a plano-convex profile, as does the juvenile stage of *Tretomphalus* (Rec. Fig. 15.25b), but the final chamber is a globular **float chamber** for planktonic dispersal in the latter genus. *Siphonina* (Eoc.-Rec. Fig. 15.25c) has a biconvex profile and an areal aperture borne on a short neck. The umbilical boss seen in many discorbaceans is covered by a rosette of secondary chambers in *Asterigerina* (Cret.-Rec., Fig. 15.25d). A similar development occurs in *Amphistegina* (Eoc.-Rec., Fig. 15.25g), but the sutures are more angular and the trochospiral growth is hidden by overlap of the chambers. *Cibicides* (Cret.-Rec. Figs 15.25e, 15.31m) is a common genus that deviates from the normal in having a basal aperture that extends from the umbilical side to the spiral side. This spiral side is flat or concave whilst the umbilical side is convex. *Planorbulina* (Eoc.-Rec. Fig. 15.25f) has a *Cibicides*-like early stage followed by more irregular addition of chambers in a planispiral manner. The essentially discoidal, planispiral growth of chambers in *Linderina* (Eoc.-Mioc., Fig. 15.25h) is rendered into a stronger, lenticular test by the lateral secretion of layers of calcite.

The orbitoids (Orbitoidacea) are a Late Cretaceous to Miocene group of larger foraminifera which originated in the tropical Americas. Their tests are radial hyaline and perforate, with a discoidal mode of growth. The chambers are arranged in annular cycles rather than plane spirals. A median (equatorial) layer of chambers is differentiated from the lateral chambers seen most clearly in axial thin sections (e.g. *Discocyclina*, Eoc., Fig. 15.26a,d). Radiating calcite pillars give rise to granules on the outer surface. Equatorial sections are important both for taxonomy and biostratigraphic zoning, note being made of the form of

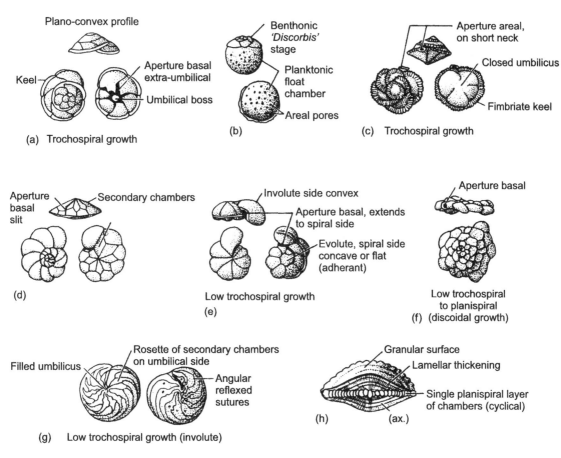

Fig. 15.25 Suborder Rotaliina, superfamily Discorbacea. (a) *Discorbis* ×57. (b) *Tretomphalus* ×34.5. (c) *Siphonina* ×31. (d) *Asterigerina* ×20. (e) *Cibicides* ×22.5. (f) *Planorbulina* ×15. (g) *Amphistegina* ×15. (h) *Linderina* ×47.5. ((a) After Loeblich & Tappan 1964 from Pokorny; (c) after Loeblich & Tappan 1964 from Reuss; (d) redrawn after Pokorny 1963 from d'Orbigny; (e) after Morley Davies 1971 from Macfadyen; (f) after Morley Davies 1971; (g) after Morley Davies 1971 from H.B. Brady; (h) modified from Morley Davies 1971 after Nuttall ((a), (c) from the *Treatise on Invertebrate Paleontology*, courtesy of and © 1964, Part C, The Geological Society of America and The University of Kansas; (e–h) with permission from Kluwer Academic Publications.)

the embryonic chambers and the shape of the median chambers (e.g. *Lepidocyclina*, Eoc.-M. Mioc., Fig. 15.26b,c).

The tests of the Rotaliacea are built of optically radial, bilamellar calcite. They are distinguished by the presence of rotaliid septal flaps and canals (Fig. 15.5d). Although primary apertures may be absent, basal foramina form by secondary resorption of the chamber wall. Generally, growth is planispiral or trochospiral, with a biconvex, lenticular test profile. In the commonly brackish-water genus *Ammonia* (Mioc.-Rec., Figs 15.27a, 15.31i) the umbilicus is partly filled by small calcite pillars. *Elphidium* (L. Eoc.-Rec., Figs 15.27b, 15.31l) is another common genus with an involute planispiral test. A sutural canal system opens at the surface through sutural pores, the latter defined by backward-projecting rods called **retral processes**. *Calcarina* (Rec., Fig. 15.27c) is a tropical genus in which the trochospiral test bears robust spines from a thick outer wall. *Nummulites* are rotaliacean larger foraminifera widely used in correlating Eocene rocks from around the Old World Tethys Ocean but their

176 Part 4: Inorganic-walled microfossils

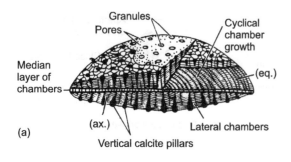

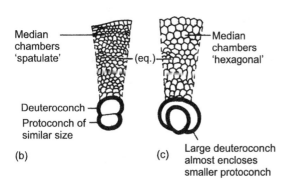

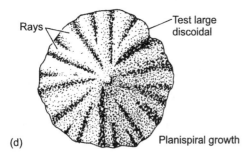

Fig. 15.26 Suborder Rotaliina. Superfamily Orbitoidacea. (a) *Discocyclina* (*Discocyclina*) ×3. (b) *Lepidocyclina* ×13. (c) *Lepidocyclina* (*Eulepidina*) ×27. (d) *Discocyclina* (*Akyinocyclina*) ×7. ((a), (d) After Loeblich & Tappan 1964 from Neumann, from the *Treatise on Invertebrate Paleontology*, courtesy of and © 1964, Part C, The Geological Society of America and The University of Kansas).)

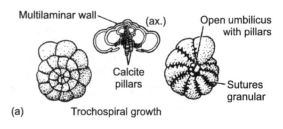

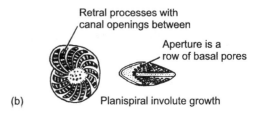

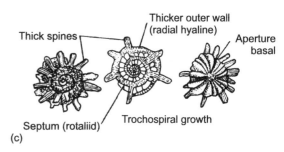

Fig. 15.27 Suborder Rotaliina. Superfamily Rotaliacea. (a) *Ammonia* ×22.5. (b) *Elphidium* ×32. (c) *Calcarina* ×6. ((a) Redrawn after Banner & Williams 1973 and Morley Davies 1971 from Macfadyen; (c) after Loeblich & Tappan 1964 from Cushman Todd & Post (from the *Treatise on Invertebrate Paleontology*, courtesy of and © 1964, Part C, The Geological Society of America and The University of Kansas).)

descendants are still found today in the Indo-Pacific seas. Their tests are radial hyaline and perforate with rotaliid septa. Coiling is biconvex planispiral. Involute forms reveal V-shaped cavities in axial sections, and lateral extensions of these cavities that are called **alar prolongations** (e.g. *Nummulites*, Palaeoc.-Rec., Fig. 15.28a,b,d). Although distinctive, these alar prolongations are no more than an earlier or later chamber extended into the plane of section by the great curvature of the involute planispiral coil. Alar prolongations are not present in the evolute forms. Chambers may be simple or differentiated into **median** (i.e. equatorial) and **lateral** layers. They can also be subdivided into chamberlets (e.g. *Spiroclypeus*, Eoc.-L. Mioc., Fig.

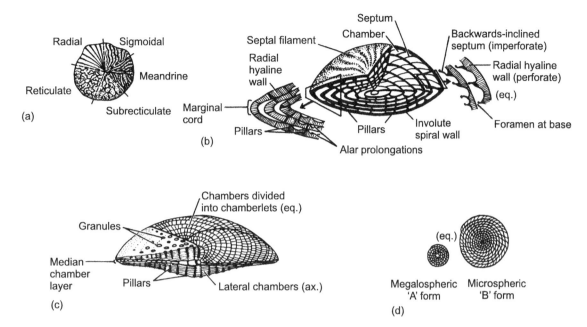

Fig. 15.28 Suborder Rotaliina. (a) Five main types of septal filament on *Nummulites* tests. (b) Centre, *Nummulites* ×3.5 approx.; left, detail of axial section ×7; right, detail of spiral section ×10. (c) *Spiroclypeus* ×7 approx. (d) Megalospheric and microspheric forms of *Nummulites obesus* ×0.67. ((b) Partly after Morley Davies 1971 with permission from Kluwer Academic Publications; (b), (c) after Loeblich & Tappan 1964 from van der Vlerk & Umbgrove (from the *Treatise on Invertebrate Paleontology*, courtesy of and © 1964, Part C, The Geological Society of America and The University of Kansas).)

15.28c). The course of the septa is indicated on the outer surface of the test by markings called septal filaments (Fig. 15.28b). These are the sutures between sinuously curved chambers. In some Late Eocene and Oligocene *Nummulites*, the sinuosity of the chambers is so great that successive chambers and their sutures overlap to give a distinctive reticulate appearance to the filaments. Granules are the surface representation of radiating pillars of calcite (Fig. 15.28c). Microspheric forms were often several times the size of megalospheric forms of the same species (Fig. 15.28d). Unfortunately, this has resulted in many species having two or more names, of which only the first one given remains valid.

The Cassidulinacea comprise small benthic foraminifera with optically granular, cryptolamellar calcite walls and slit-, tear- or loop-shaped apertures, generally areal or terminal. In *Cassidulina* (Eoc.-Rec., Fig. 15.29c) the test is lenticular, consisting of biserially arranged chambers coiled in a plane spiral. Straight biserial followed by uniserial growth is seen in *Loxostomum* (U. Cret.-Palaeoc., Fig. 15.29b) and *Virgulinella* (Mioc.-Plioc., Fig. 15.29d), the latter with supplementary sutural apertures. *Pleurostomella* (L. Cret.-Rec., Fig. 15.29a) is uniserial throughout, with a terminal aperture and two 'teeth'.

The wall structure of nonionacean tests is of optically granular, cryptolamellar or bilamellar calcite. The aperture is generally a basal slit. The involute planispiral tests of the genera *Nonion* (Palaeoc.-Rec., Fig. 15.30a) and *Melonis* (Palaeoc.-Rec., Fig. 15.30b) differ largely in the degree of chamber inflation. The test in *Osangularia* (L. Cret.-Rec., Fig. 15.30c) is trochospiral with a keel and a closed umbilicus.

Molecular phylogeny of Foraminifera

Foraminifera are the most intensively studied group of non-cultured protozoa for which over 900 rDNA

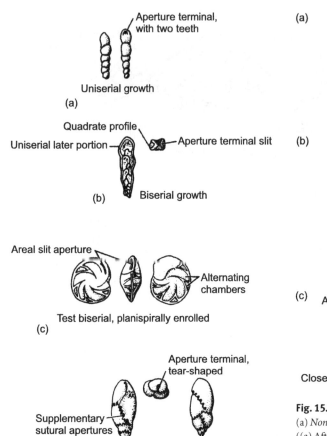

Fig. 15.29 Suborder Rotaliina, superfamily Cassidulinacea. (a) *Pleurostomella* ×16. (b) *Loxostomum* ×34.5. (c) *Cassidulina* ×26. (d) *Virgulinella* ×21.5. ((a), (b), (d) After Loeblich & Tappan 1964; (c) after Loeblich & Tappan 1964 from Montanaro Gallitelli (from the *Treatise on Invertebrate Paleontology*, courtesy of and © 1964, Part C, The Geological Society of America and The University of Kansas).)

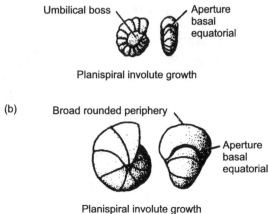

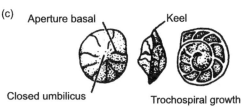

Fig. 15.30 Suborder Rotaliina. Superfamily Nonionacea. (a) *Nonion* ×33. (b) *Melonis* ×37.5. (c) *Osangularia* ×37. ((a) After Loeblich & Tappan 1964 from Voloshinova; (b) after Morley Davies 1971 from H.B. Brady with permission from Kluwer Academic Publications; (c) after Loeblich & Tappan 1964) ((a), (c), from the *Treatise on Invertebrate Paleontology*, courtesy of and © 1964, Part C, The Geological Society of America and The University of Kansas).)

sequences are known. Foram genes are characterized by the presence of several specific insertions and point mutations not found in any other eukaryotes (Pawlowski 2000; Pawlowski & Holzmann 2002). Foraminiferal DNA can also be detected in sediment samples. This wealth of molecular sequence data is revolutionizing our understanding of foraminifera. Recent studies have shown that the traditional view of an essentially marine group, divided by the presence of a membranous, agglutinated or calcareous test, can no longer be sustained, and that this group includes both testate and naked species. Foraminifera are also found in terrestrial, freshwater and marine environments. There also appears to be a high taxonomic diversity within single-chambered forms and cryptic species are common.

Origin and evolution

The origin of the foraminifera is problematic. According to rDNA sequence data, foraminifera diverged amongst

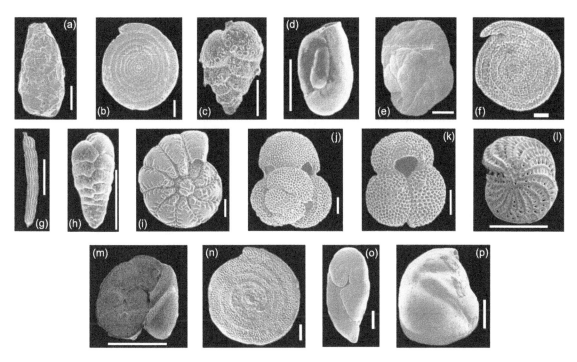

Fig. 15.31 Electron photomicrographs of selected foraminifera. (a) *Saccammina* (Textulariina). (b) *Ammodiscus* (Textulariina). (c) *Siphotextularia* (Textulariina). (d) *Miliolinella* (Miliolina). (e) *Carterina* (Carterinina) dorsal view. (f) *Spirillina* (Spirillinina). (g) *Dentalina* (Lagenida). (h) *Bolivina* (Buliminida). (i) *Ammonia* (Rotaliacea) ventral side. (j), (k) *Globigerinoides* (Globigerinina) spiral and umbilical views. (l) *Elphidium* (Rotaliina). (m) *Cibicides* (Rotaliina). (n) *Planispirillina* (Involutinina). (o) *Robertinoides* (Robertinina). (p) *Milammellus* (Silicoloculinina). Scale bars = 500 μm in (b), (g), (l), (m); = 100 μm in all others. ((a), (h), (i) From Sen Gupta 1999 after Platon; (b), (f), (g), (j), (k) from Sen Gupta 1999; (c) from Sen Gupta 1999 after Jones; (e) from Sen Gupta 1999 after Deutsch & Lipps; (n) from Sen Gupta 1999 after Piller; (o) from Sen Gupta 1999 after Resig (reproduced with the permission of Kluwer Academic Publishers).)

the earliest mitochondriate lineages, contrasting with their relatively late appearance in the Early Cambrian fossil record. Phylogenetic analysis of actin genes shows both the Foraminiferida and Cercozoa (cercomonad flagellates) branching together in the middle of the eukaryote tree, contrasting with the analysis of β-tubulin protein which separates these two groups (Keeling 2001).

Existing hypotheses of relationships imply the progressive transformation of the foram test from the primitive membranous, through agglutinated to secreted calcareous wall (Hansen 1979). The earliest representatives of the group were either single-chambered, organic-walled species, placed in the class Athalamida and resembling the Recent Allogromiina (Tappan & Loeblich 1988) or agglutinated tubular forms such as *Platysolenites* (McIlroy *et al*. 2001). This gave rise to single-chambered agglutinated species from which the multichambered suborders the Textulariina and Rotaliina arose (Grigelis 1978). The suborder Miliolina was considered to have arisen independently from the Allogromiina (Tappan & Loeblich 1988). Molecular studies challenge some of the main axioms of this hypothesis.

Phylogenetic analysis of rDNA sequences from the athalamid *Reticulomyxa filosa*, a giant freshwater amoeba, shows this species branches within the clade of Foraminiferida, among the single-chambered species. The separation of the Athalamida and the foraminifera is therefore artificial and *R. filosa* must have lost its test in adapting to a freshwater habitat (Pawlowski & Holzmann 2002).

Molecular studies show that all examined allogromiids cluster together at the base of the foraminiferid tree and that membranous and agglutinated tests evolved independently in several lineages (Holzmann & Pawlowski 1997). In early studies of rDNA phylogenies the Miliolina appeared to be the earliest group of the foraminifera to diverge (e.g. Pawlowski et al. 1997), before the origin of tests. The latest analysis (Pawlowski & Holzmann 2002) indicates the Miliolina branch within the foraminifera and that no naked forms occur at the base of the miliolinid clade, suggesting that they diverged at a later stage from more evolved agglutinated or calcareous lineages.

Cryptic diversity

Foraminifera are diagnosed on the basis of test characteristics, making species determination difficult (Loeblich & Tappan 1988; Haynes 1990), particularly in separating sibling species and phenotypically variable forms in which ecophenotypes and genotypes are morphologically similar. The first evidence for cryptic speciation in foraminifera was found in *Globigerinella siphonifera*, a planktonic species (Huber et al. 1997). Two genetic types were distinguished on rDNA sequences; Type I also have more negative $\delta^{18}O$ and $\delta^{13}C$ values and larger pores than Type II. High genetic variability is also found in *Orbulina universa*, *Globigerinoides ruber* and *Globigerina bulloides*, each being divided into two types (Darling et al. 1999). Populations from the North Atlantic, North Sea, Mediterranean Sea, Red Sea and Pacific yield 10 distinct genotypes, of which only a few species can be distinguished on morphology (Holzmann & Pawlowski 1997). Similar high genetic variability is displayed by benthic species such as *Ammonia beccarii*.

Geological history of foraminifera

The oldest fossil foraminifera are simple, agglutinated tubes in the earliest Cambrian resembling the modern genus *Bathysiphon* (McIlroy et al. 2001) indicating shelled protozoa appeared at the same time as shelled invertebrates. Agglutinated foraminifera became more abundant in the Ordovician but true multichambered forms did not appear until the Devonian, during which period the Fusulinina began to flourish, culminating in the complexly constructed tests of the Fusulinacea in Late Carboniferous and Permian times. This super-family died out at the end of the Palaeozoic. Miliolina and Lagenina first appeared in the Early Carboniferous.

Important Mesozoic events include the appearance and radiation of the Rotaliina (largely from endothyracean stock), Miliolina and complex Textulariina in the Jurassic, soon followed by the appearance of the first unquestionably planktonic foraminifera (e.g. Oxford et al. 2002). Cretaceous tropical regions witnessed a flowering of larger miliolines and rotaliines while the widespread chalk seas and newly opened Atlantic Ocean favoured a thriving planktonic population. The planktonic Globotruncanidae became extinct at the end of the Cretaceous.

In the low latitude Tethys Ocean about 75% of species disappeared at or near the K-T boundary. Extinction was highly selective and only cological generalists (e.g. heterohelicoids, gueribelitoids, hedbergellids and globigerinellids) survived. This mass extinction pattern coincides with dramatic changes in temperature, salinity, oxygen and nutrients across the boundary, the result of both long-term environmental changes (e.g. climate, sea level, volcanism) and short-term effects such as the proposed bolide impact (Keller et al. 2002).

A relatively rapid radiation followed in the Palaeocene with the appearance of the planktonic Globigerinidae and Globorotalidae and in the Eocene with the development of *Nummulites* and soritids in the Old World and orbitoids in the New World, although they eventually became almost worldwide. Orbitoids died out in the Miocene, since which time larger foraminiferal stocks have progressively dwindled in distribution and diversity, mostly because of climatic deterioration. Planktonics have also diminished in diversity since Late Cretaceous times (Fig. 15.33). Figure 15.32 summarizes the current consensus view of the subordinal phylogeny of the Foraminiferida.

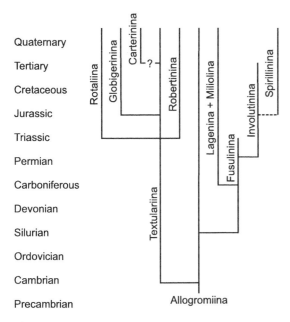

Fig. 15.32 Subordinal phylogeny of the Foraminiferida. (Modified from Tappan & Loeblich 1988, figure 9.)

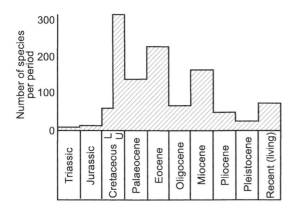

Fig. 15.33 Changes in the specific diversity of planktonic foraminifera through time. Because of the complex evolutionary history, the likely existence of many cryptic taxa and the varied life habits and habitats measures of standing diversity in the foraminifera are probably less meaningful than in other groups. (For further details see Tappan & Loeblich 1988.) (Based on Tappan & Loeblich 1988.)

Applications of foraminifera

Foraminifera are in many respects ideal zonal indices for marine rocks, being small, abundant, widely distributed and often extremely diverse. Many also have an intricate morphology in which evolutionary changes can be readily traced. Planktonic foraminifera provide the basis of important schemes for intercontinental correlation of Mesozoic (especially upper Cretaceous) and Cenozoic rocks (see various papers in Bolli *et al.* 1985, and for British sections Jenkins & Murray 1989). Benthic foraminifera tend to be more restricted in distribution but provide useful schemes for local correlation (e.g. Bolli *et al.* 1994).

Environmental interpretations that use fossil foraminifera are founded mainly on comparisons with the numerous studies of modern ecology, aspects of which are brought together by Murray (1991), Sen Gupta (1999) and Haslett (2002). For example, dramatic changes in depth, salinity and climate can be traced in late glacial and postglacial raised beaches and beach deposits from studies of their foraminifera (e.g. Bates *et al.* 2000; Roe *et al.* 2002).

The value of benthic foraminifera as indicators of the depth of deposition has been based on the known depth distribution of modern foraminifera. Trends in species diversity, planktic–benthic ratios, shell type ratios and morphology have been utilized to plot changes in depth. In general terms, species diversity increases offshore to the continental slope, as does the planktic–benthic ratio. Planktonic life assemblages are depth stratified and so give rise to higher-diversity death assemblages in deeper waters than in shallower waters (Kafescioglu 1971). Benthic depth-related assemblages can also be recognized in Cretaceous sediments (Olsson, in Swain 1977, pp. 205–230). The planktic–benthic ratio can be used for the interpretation of Jurassic and younger rocks (e.g. Stehli & Creath 1964; Hart & Carter 1975). Test types, the agglutinated–porcelaneous–haline proportions, vary with habitat and this appears to hold into the Palaeogene. Modern marginal marine species are strongly influenced by changes in salinity (Sen Gupta, in Sen Gupta 1999, pp. 141–159). In water of normal marine salinity numerous workers have recognized distinctive foraminiferal assemblages in the inner and outer continental shelves, upper slope and deep sea.

Recognition of patterns and distribution of deep sea benthic foraminifera are beginning to emerge from the many studies of the upper parts of DSDP and ODP cores. The biogeography of modern foraminifera is related to the distribution of water masses and ocean currents. The palaeobiogeographical patterns of benthic and planktonic foraminifera are therefore indispensable in inferring palaeoceanography. Hass & Kaminski (1997) provide a case study on the micropalaeontology and palaeooceanography of the North Atlantic from the Paleogene to the Recent. *Cibicides wuellerstorfi* is the dominant benthic species in North Atlantic Deep Water and *Nuttallides umbonifera* in Antarctic Bottom Water (Sen Gupta 1988). There is now increasing evidence that availability of organic matter (from surface productivity) affects the abundances of deep sea foraminifera. *Epistominella exigua* is a species whose population density is dependent on phytodetritus falls (e.g. Gooday 1993). At bathyal depths there is a strong correlation between the oxygen minimum zone and the foraminiferal assemblage (e.g. Hermelin & Shimmield 1990). Benthic foraminifera are also indicators of productivity in areas of upwelling (Schnitker 1994). The relative abundances of *Cibicides wuellerstorfi* and *Bulimina alazanensis* are related to changes in the advection of North Atlantic Deep Water during the Quaternary (Schmiedl & Mackensen 1997). Cretaceous current patterns (e.g. Sliter 1972) and ocean stratification (D'Hondt & Arthur 2002) have also been reconstructed from the distribution and stable isotope chemistry of foraminifera. Price & Hart (2002) used both $\delta^{13}C$ and $\delta^{18}O$ values in benthic and planktonic foraminifera to document changing oceanic temperature gradients in the Early-Middle Albian of the Pacific Ocean. An increase in ocean temperature (and/or decrease in salinity) in the Cenomanian suggests a reduction of the poleward heat flux, promoting the build up of limited polar ice. Stable isotopes in planktic and benthic foraminifera also indicate a 100–500 kyr long period of instability in oceanic bottom-water temperature and sea level prior to the K-T boundary transition at Stevns Klint, Denmark. During the latest Maastrichtian bottom-water temperatures gradually cooled by about 1.5°C as surface water temperatures remained constant, perhaps consistent with the initiation of a thermohaline circulation and the formation of some polar ice (Schmitz *et al.* 1992).

The narrow temperature ranges of living planktonic species have become useful tools in palaeoclimatology especially of Quaternary sediments (see various papers in Haslett 2002 and Bradley 1999). Plots of the changing proportions of warm- to cold-water species, of selected indicator species, or of coiling directions through a cored interval, may allow the construction of palaeotemperature curves. *Neogloboquadrina pachyderma* and *Globigerina bulloides* have been extensively used as a paleotemperature proxy from the Late Miocene through the Quaternary, exploiting the modern polar affinity of the sinistrally coiled forms. However, the relationship between sea surface temperature and coiling direction is not a simple one. Pliocene and Pleistocene sinistral forms of *N. pachyderma* are morphologically and ecologically different and the modern sinistral form only appeared ~1 Myr. This suggests that *N. pachyderma* (sinistral) should not be used for calibrated paleoceanographical reconstructions prior to the Middle Pleistocene. *N. pachyderma* (sinistral) may have evolved in response to the onset of the 100 kyr climate regime in the Middle Pleistocene (Kucera & Kennett 2002). The proportions of sinistral and dextral forms of *N. pachyderma* and *G. bulloides* have also been shown to change in response to the vigour of oceanic upwelling (Naidu & Malmgren 1996).

Studies on the oxygen isotope ratios of calcareous foraminiferid shells have become one of the primary tools in palaeoceanographic and palaeoclimatic studies far too numerous to list (for a review of these techniques refer to Rohling & Cooke, in Sen Gupta 1999, pp. 239–259). A case study on the palaeoceanography of the North Atlantic, utilizing foraminifera, can be found in Hass & Kaminski (1997). Waelbroeck *et al.* (2002) used benthic foraminiferal oxygen isotopic ratios to model the relative sea-level history of the North Atlantic and Equatorial Pacific Ocean over the last climatic cycle. Information on palaeotemperatures, such as the long Cenozoic history of climatic cooling and glaciations shown from Antarctic waters, can be found in Shackleton & Kennett (1975) and subsequent studies. Paired Mg/Ca and $^{18}O/^{16}O$ measurements are being made on both benthic foraminifera to separate global temperature and ice volume changes during the Cenozoic (e.g. Billups & Schrag 2002) and even to challenge those hypotheses that relate to ice advance to orbital forcing (Shackleton 2000). Differences in the $\delta^{18}O$ and $\delta^{13}C$ between shallow- and deep-living planktonic foraminifera are proxies for the stratification of surface waters (e.g. Mulitza et al. 1997).

Foraminifera are particularly useful in palaeoecology and palaeo-oceanography when used in association with other palaeoceanographic proxies, with case studies available from the Jurassic (e.g. Dill & Dultz 2001) and Cretaceous (e.g. Paul et al. 1999; Price & Hart 2002). Many case studies are available from the Cenozoic or document global cooling events at the Eocene–Oligocene boundary, Miocene and during the Quaternary with the most studied events. The onset and orbital forcing of the Messian Salinity Crisis has been documented with the help of foraminifera (Blanc-Valleron et al. 2002) and even changes in solar irradiance have been inferred from the $\delta^{13}C$ profile in *Globigerinoides ruber* over the last 2000 years (Castagnoli et al. 2002).

The trace element composition has also become an important way of elucidating past oceanographic conditions (Lea, in Sen Gupta 1999, pp. 259–281). Four broad areas of research have emerged, nutrient proxies (e.g. Cd, Ba), physical proxies (e.g. Mg, Sr, F and B isotopes), chemical proxies (e.g. Li, U, V, Sr and Nd isotopes) and diagenetic proxies (e.g. Mn).

Less use has been made of the relationship between foraminiferid test morphology, habitat and environment, despite some preliminary studies relating to depth (Bandy 1964), substrate stability (Brasier 1975a) and general environmental factors (Chamney, in Schafer & Pelletier 1976, pp. 585–624). This approach, for example, helped Brasier (1975b) to trace the gradual dispersal of seagrass communities from Cretaceous to Recent times. This approach plus an understanding of functional morphology and the mode of life of foraminifers enabled Geel (2002) to map microfacies changes and shallowing and deepening trends in Palaeogene sedimentary sequences.

Studies on the effects of pollution on foraminifera in coastal waters indicate they can be used in pollution monitoring (Yanko et al., in Sen Gupta 1999, pp. 217–239). Many foraminifera have adapted to extreme habitats including hydrothermal vents, hypersalinity and pack ice (examples in Sen Gupta 1999).

Further reading

Information on the biology, ecology and classification of living foraminifera can be found in the books by Haynes (1981), Lee & Anderson (1991), Murray (1991), Loeblich & Tappan (1988), Hemleben et al. (1989) and Sen Gupta (1999). Specimen identification may be assisted by reference to the Treatise (Loeblich & Tappan 1964, 1988) for genera, and to the Catalogue (Ellis & Messina 1940 to date) for species. Additional papers on foraminiferal palaeoecology can be found in Curtis (1976) and Moguilevsky & Whatley (1996). Applications of foraminifera in the petroleum industry can be found in Jones (1996) and details of biostratigraphy in Bolli et al. (1985) and Jenkins & Murray (1989). Murray (1979) is useful for identifying British nearshore foraminifera.

Hints for collection and study

To collect living specimens of foraminifera, gather samples of relatively fibrous seaweed from marine or estuarine rock pools and tidal flats or scrape up the top 5 mm of mud from intertidal mudflats. The weed samples should then be placed in a bucket of nearby water and shaken vigorously to detach the foraminifera. Remove the seaweed and strain the water and sediment through a 125 µm mesh sieve. The mud samples should likewise be washed through a 125 µm mesh sieve. The sieved residues are then flushed into a container with more sea water for later examination in a petri dish with transmitted light. Living foraminifera can generally be distinguished from dead ones by their dark, cytoplasm-filled chambers and by adherent food debris. Patient observation at high magnifications with condensed light should also reveal pseudopodia, locomotion and feeding habits. Arnold (in Hedley & Adams

1974, pp. 154–206) gives some useful tips for the collection and culture of living foraminifera.

Foraminiferid tests can be very abundant in marine sediments. Recent beach sands and lagoonal and estuarine muds are all excellent sources of material. Fossil foraminifera can be obtained from almost any post-Triassic marine sediment which has not undergone much leaching or become acid. To extract foraminifera from partially indurated argillaceous and marly rocks, methods C to E (especially D) are generally satisfactory (see Appendix). Good assemblages can also be coaxed out of chalks and other limestones by method B, but the hardest limestones will have to be thin sectioned or peeled (see method N). Disaggregated sediments can then be washed, dry sieved, concentrated and mounted by methods G, I, J and O. Most smaller foraminifera are studied in reflected light, sometimes stained with a solution of malachite green or a food dye to bring out the surface structures more clearly. Wall structure and growth plan are, however, better seen if the specimen is wetted and viewed with transmitted light. Larger foraminifera (and some smaller forms within indurated limestones) are generally studied in thin sections, preferably through both the equatorial plane and the growth axis. Thin sections of isolated specimens can also be prepared by embedding them in polyester resin: pour a little resin into the bottom of one cup from a polystyrene egg box or a plastic ice cube cup; scatter a dozen or so specimens over the resin and then cover with a further layer of resin. Bubbles can be discouraged if the cup is then placed in a vacuum. When dry, remove the block and prepare standard thin sections of the foraminiferid-rich portion. Further ideas on foraminiferid techniques are given by Todd *et al.* (pp. 14–20) and Douglass (pp. 20–25) in Kummel & Raup (1965) and Green (2002).

REFERENCES

Bandy, O.L. 1964. General correlation of foraminiferal structure with environment. In: Imbrie, J. & Newell, N.D. (eds) *Approaches to Palaeoecology*. John Wiley, New York, pp. 75–90.

Banner, F.T. & Williams, E. 1973. Test structure, organic skeleton and extrathalmous cytoplasm of *Ammonia* Brünnich. *Journal of Foraminiferal Research* 3, 49–69.

Bates, M.R., Bates, C.R., Gibbard, P.L., Macphail, R.I., Owen, F.J., Parfitt, S.A., Preece, R.C., Roberts, M.B., Robinson, J.E., Whittaker, J.E. & Wilkinson, K.N. 2000. Late Middle Pleistocene deposits at Norton Farm on the West Sussex coastal plain, southern England. *Journal of Quaternary Science* 15, 61–89.

Belyaeva, N.V. 1963. The distribution of planktonic foraminifers over the Indian Ocean bottom. *Voprosi Mikropaleontologii* 7, 209–222 [in Russian].

Benton, M. (ed.) 1993. *The Fossil Record 2*. Chapman & Hall, London.

Berger, W.H. 1971. Planktonic foraminifera: sediment production in an oceanic front. *Journal of Foraminiferal Research* 1, 95–118.

Billups, K. & Schrag, D.P. 2002. Paleotemperatures and ice volume of the past 27 Myr revisited with paired Mg/Ca and O-18/O-16 measurements on benthic foraminifera. *Paleoceanography* 17, article no. 1003.

Blanc-Valleron, M.M., Pierre, C., Caulet, J.P., Caruso, A., Rouchy, J.M., Cespuglio, G., Sprovieri, R., Pestrea, S. & Di Stefano, E. 2002. Sedimentary, stable isotope and micropaleontological records of paleoceanographic change in the Messinian Tripoli Formation (Sicily, Italy). *Palaeogeography, Palaeoclimatology, Palaeoecology* 185, 255–286.

Bolli, H.M., Saunders, J.B. & Perch-Nielsen, K. 1985. *Plankton Stratigraphy*. Cambridge University Press, Cambridge.

Bolli, H.M., Beckmann, J-P. & Saunders, J.B. 1994. *Benthic Foraminiferal Biostratigraphy of the South Caribbean Region*. Cambridge University Press, Cambridge.

Bradley, R.S. 1999. *Paleoclimatology: reconstructing climates of the Quaternary*. International Geophysics Series, vol. 64. Harcourt Academic Press, San Diego.

Brasier, M.D. 1975a. Morphology and habitat of living benthonic foraminiferids from Caribbean carbonate environments. *Revista Espanola de Micropaleontologia* 7, 567–578.

Brasier, M.D. 1975b. An outline history of seagrass communities. *Palaeontology* 18, 681–702.

Brasier, M.D. 1982a. Architecture and evolution of the foraminiferid test – a theoretical approach. In: Banner, F.T. & Lord, A.R. (eds) *Aspects of Micropalaeontology*. George Allen & Unwin, London, pp. 1–41.

Brasier, M.D. 1982b. Foraminiferid architectural history: a review using the MinLOC and PI methods. *Journal of Micropalaeontology* 1, 95–105.

Brasier, M.D. 1984. *Discospirina* and the pattern of evolution in foraminiferid architecture. In: *Second International Symposium on Benthic Foraminifera, Pau, April 1983*. Elf Aquitaine, pp. 87–90.

Brasier, M.D. 1986. Form, function and evolution in benthic and planktic foraminiferid test architecture. In: Leadbeater, B.S.C. & Riding, R. (eds) *Biomineralisation in Lower Plants and Animals*. Systematics Association Special Volume 30. Clarendon Press, Oxford, pp. 32–67.

Brasier, M.D. 1988. Foraminiferid extinction and ecological collapse during global biological events. In: Larwood, G.P. (ed.) *Extinction and Survival in the Fossil Record*. Systematics Association Special Volume 34. Clarendon Press, Oxford, pp. 37–64.

Brasier, M.D. 1995a. Fossil indicators of nutrient levels. 1: Eutrophication and climate change. In: Bosence, D.W.J. & Allison, P.A. (eds) *Marine Palaeoenvironmental Analysis from Fossils*. Geological Society Special Publication, No. 84, 113–133. The Geological Society, London.

Brasier, M.D. 1995b. Fossil indicators of nutrient levels. 2: Evolution and extinction in relation to oligotrophy. In: Bosence, D.W.J. & Allison, P.A. (eds) *Marine Palaeoenvironmental Analysis from Fossils*. Geological Society Special Publication, No. 84, 133–151. The Geological Society, London.

Castagnoli, G.C., Bonino, G. & Taricco, C. 2002. Long term solar-terrestrial records from sediments: carbon isotopes in planktonic foraminifera during the last millennium. *International Solar Cycle Study (ISCS): advances in space research* 29, 1537–1549.

Cavalier-Smith, T. 1993. Kingdom Protoza and its 18 phyla. *Microbiological Review* 57, 953–994.

Cavalier-Smith, T. 1998. A revised six-kingdom system of life. *Biological Review* 73, 203–266.

Corlisss, J.C. 1994. An interim utilitarian ('user friendly') hierarchical classification and characterization of the protists. *Acta Protozoologica* 33, 1–51.

Cummings, R.H. 1955. *Nodosinella* Brady, 1876, and associated Upper Palaeozoic genera. *Micropalaeontology* 1, 221–238.

Curtis, D.M. (ed.) 1976. *Depositional Environments and Paleoecology: Foraminiferal Paleoecology/selected papers reprinted from Journal of Paleontology and Journal of Sedimentary Petrology*. Society of Economic Paleontologists and Mineralogists, Tulsa, Oklahoma.

Cushmann, J.A. 1948. *Foraminifera, their classification and economic use*, 4th edn. Cambridge, MA. Harvard University Press.

Darling, K.F., Wade, C.M., Kroon, D., Leigh Brown, A.J. & Bijma, J. 1999. The diversity and distribution of modern planktic foraminiferal small unit ribosomal RNA genotypes and their potential tracers of present and past ocean circulation. *Paleoceanography* 14, 3–12.

D'Hondt, S. & Arthur, M.A. 2002. Deep water in the Late Maastrichtian ocean. *Paleoceanography* 17, article no. 1008.

Dill, H.G. & Dultz, S. 2001. Chemical facies and proximity indicators of continental marine sediments (Triassic to Liassic, SE Germany). *Neues Jahrbuch für Geologie und Paläontologie-Abhandlungen* 221, 289–324.

Ellis, B.F. & Messina, A.R. 1940 to date. *Catalogue of Foraminifera*. Special Publication American Museum Natural History.

Geel, T. 2000. Recognition of stratigraphic sequences in carbonate platform and slope deposits: empirical models based on microfacies analysis of Palaeogene deposits in southeastern Spain. *Palaeogeography, Palaeoclimatology, Palaeoecology* 155, 211–238.

Glaessner, M.F. 1945. *Principles of Micropalaeontology*. Hafner Press, New York.

Gooday, A.J. 1993. Deep sea benthic foraminiferal species which exploit phytodetritus: characteristic features and controls on distribution. *Marine Micropaleontology* 22, 187–205.

Green, O.R. 2002. *A Handbook of Palaeontological Techniques*. Kluwer, Rotterdam.

Grigelis, A.A. 1978. Higher foraminiferid taxa. *Paleontological Journal* 121, 1–9.

Hallock, P. 1985. Why are larger foraminifera large? *Paleobiology* 11, 195–208.

Hansen, H.J. 1979. Test structure and evolution in the Foraminifera. *Lethaia* 12, 173–182.

Hart, M. & Carter, D.J. 1975. Some observations on the Cretaceous Foraminiferida of southeast England. *Journal of Foraminiferal Research* 5, 114–126.

Haslett, S.K. (ed.) 2002. *Quaternary Environmental Micropalaeontology*. Arnold, London.

Hass, H.C. & Kaminski, M.A. (eds) 1997. *Contributions to the Micropaleontology and Paleoceanography of the Northern North Atlantic (collected results from the GEOMAR Bungalow Working Group*. The Gryzbowski Foundation, Krakow.

Haynes, J.R. 1981. *Foraminifera*. Macmillan, London.

Haynes, J.R. 1990. The classification of the Foraminifera – a review of historical and philosophical perspectives. *Palaeontology* 33, 503–528.

Hedley, R.H. & Adams, C.G. 1974. *Foraminifera*, vol. 1. Academic Press, London.

Hemleben, C., Spindler, M. & Anderson, O.R. 1989. *Modern Planktonic Foraminifera*. Springer-Verlag, New York.

Hermelin, J.O. & Shimmield, G.B. 1990. The importance of the oxygen minimum zone and sediment geochemistry in the distribution of Recent benthic foraminifera in Northwest Indian Ocean. *Marine Geology* 91, 1–29.

Holzmann, M. & Pawlowski, J. 1997. Molecular, morphological and ecological evidence for species recognition in

Ammonia (Foraminifera, Protozoa). *Journal of Foraminiferal Research* 27, 311–318.

Huber, B.T., Bijma, J. & Darling, K. 1997. Cryptic speciation in the living planktonic foraminifer *Globigerinella siphonifera* (d'Orbigny). *Paleobiology* 23, 33–62.

Jenkins, D.G. & Murray, J.W. (eds) 1989. *Stratigraphical Atlas of Fossil Foraminifera*. Ellis Horwood, Chichester.

Jones, R.W. 1996. *Micropalaeontology in Petroleum Exploration*. Oxford Science Publications, Clarendon Press, Oxford.

Kafescioglu, I.A. 1971. Specific diversity of planktonic foraminifera on the continental shelves as a paleobathymetric tool. *Micropalaeontology* 17, 455–470.

Keeling, P.J. 2001. Foraminifera and cercozoa are related in actin phylogeny: two orphans find a home? *Molecular Biology and Evolution* 18, 1551–1557.

Keller, G., Adatte, T., Stinnesbeck, W. *et al.* 2002. Paleoecology of the Cretaceous-Tertiary mass extinction in planktonic foraminifera. *Palaeogeography, Palaeoclimatology, Palaeoecology* 178, 257–297.

Kucera, M. & Kennett, J.P. 2002. Causes and consequences of a Middle Pleistocene origin of the modern planktonic foraminifer *Neogloboquadrina pachyderma* sinistral. *Geology* 30, 539–542.

Kummel, B. & Raup, D. (eds) 1965. *Handbook of Paleontological Techniques*. W.H. Freeman, San Francisco.

Lee, J.J. & Anderson, O.R. (eds) 1991. *Biology of Foraminifera*. Academic Press, London.

Lipps, J. (ed.) 1993. *Fossil Prokaryotes and Protists*. Blackwell Scientific Publications, Oxford.

Loeblich Jr, A.R. & Tappan, H. 1964. Protista 2; Sarcodina, chiefly 'Thecamoebians' and Foraminiferida. In: Moore, R.C. (ed.) *Treatise on Invertebrate Palaeontology, Part C*, 2 vols. Geological Society of America and University of Kansas Press, Lawrence, Kansas.

Loeblich Jr, A.R. & Tappan, H. 1988. *Foraminiferal Genera and their Classification*. Van Nostrand Reinhold, New York.

McIlroy, D., Green, O.R. & Brasier, M.D. 2001. Palaeobiology and evolution of the earliest agglutinated Foraminifera: Platysolenites, Spirosolenites and related forms. *Lethaia* 34, 13–29.

Moguilevsky, A. & Whatley, R. (eds) 1996. *Microfossils and Oceanic Environments*. University of Wales, Aberystwyth Press, Aberystwyth.

Morley Davies, A. 1971. *Tertiary Faunas*, 2nd edn (revised by F.E. Eames & R.J.G. Savage), vol. 1. *The composition of Tertiary faunas*. George Allen & Unwin, London.

Mulitza, S., Durkoop, A., Hale, W., Wefer, G. & Niebler, H.S. 1997. Planktonic foraminifera as recorders of past surface-water stratification. *Geology* 25, 335–338.

Murray, J.W. 1979. *British Nearshore Foraminiferids: keys and notes for the identification of the species*. Published for the Linnean Society of London and the Estuarine and Brackish-water Sciences Association by Academic Press, London.

Murray, J.W. 1991. *Ecology and Palaeoecology of Benthic Foraminifera*. Longman, Harlow, Essex.

Naidu, P.D. & Malmgren, B.A. 1996. Relationship between Late Quaternary upwelling history and coiling properties of *Neogloboquadrina pachyderma* and *Globigerina bulloides* in the Arabian Sea. *Journal of Foraminiferal Research* 26, 64–70.

Oberhänsli, H. 1992. Planktonic foraminifers as tracers of ocean currents in the eastern South Atlantic. *Paleoceanography* 7, 607–632.

Oxford, M.J., Gregory, F.J., Hart, M.B., Henderson, A.S., Simmons, M.D. & Watkinson, M.P. 2002. Jurassic planktonic foraminifera from the United Kingdom. *Terra Nova* 14, 205–209.

Paul, C.R.C., Lamolda, M.A., Mitchell, S.F., Vaziri, M.R., Gorostidi, A. & Marshall, J.D. 1999. The Cenomanian–Turonian boundary at Eastbourne (Sussex, UK): a proposed European reference section. *Palaeogeography, Palaeoclimatology, Palaeoecology* 150, 83–121.

Pawlowski, J. 2000. Introduction to the molecular systematics of foraminifera. *Micropalaeontology* 46, supplement 1, 1–12.

Pawlowski, J. & Holzmann, M. 2002. Molecular phylogeny of Foraminifera – a review. *European Journal of Protistology* 38, 1–10.

Pawlowski, J., Bolivar, I., Fahrni, J., De Vargas, C., Gouy, M. & Zaninetti, L. 1997. Extreme differences in rates of molecular evolution of foraminifera revealed by comparison of ribosomal DNA sequences and the fossil record. *Molecular Biology and Evolution* 14, 498–505.

Pessagno, E.A. & Miyano, K. 1968. Notes on the wall structure of the Globigerinacea. *Micropalaeontology* 14, 38–50.

Pokorny, V. 1963. *Principles of Zoological Micropalaeontology* (English translation edited by J.W. Neale), vol. 1. Pegammon Press, Oxford.

Price, G.D. & Hart, M.B. 2002. Isotopic evidence for Early to mid-Cretaceous ocean temperature variability. *Marine Micropalaeontology* 46, 45–58.

Resig, J.M., Lowenstam, H.A., Echols, R.J. & Weiner, S. 1980. *Cushman Foundation Foraminiferal Research Special Publication* 19, 205–214.

Roe, H.M., Charman, D.J. & Gehrels, W.R. 2002. Testate amoebae in coastal deposits in the UK: implications for studies of sea-level change. *Journal of Quaternary Science* 17, 411–429.

Saidova, Kh.M. 1967. Sediment stratigraphy and palaeogeography of the Pacific Ocean by benthonic Foraminifera during the Quaternary. *Progress in Oceanography* 4, 143–151.

Schafer, C.T. & Pelletier, B.R. 1976. *First International Symposium on Benthonic Foraminifera of Continental Margins*. Special Publication Maritime Sediments, no. 1 (2 vols).

Schmiedl, G. & Mackensen, A. 1997. Late Quaternary paleoproductivity and deep-water circulation in the eastern South Atlantic Ocean. Evidence from benthic foraminifera. *Palaeogeography, Palaeoclimatology, Palaeoecology* 130, 43–80.

Schmitz, B., Keller, G. & Stenvall, O. 1992. Stable isotope and foraminiferal changes across the Cretaceous–Tertiary boundary at Stevns Klint, Denmark – arguments for long-term oceanic instability before and after bolide-impact event. *Palaeogeography, Palaeoclimatology, Palaeoecology* 96, 233–260.

Schnitker, D. 1994. Deep sea benthic foraminifers: food and bottom water masses. In: Zahn, R. *et al.* (eds) *Carbon Cycling in the Glacial Ocean: constraints on the ocean's role in global change*. Spriner-Verlag, Berlin, pp. 539–553.

Sen Gupta, B.K. 1988. Water mass relation of the benthic foraminifer *Cibicides wuellerstorfi* in the eastern Caribbean Sea. *Bulletin de l'Institut de Géologie du Bassin d'Aquitaine (Bordeaux)* 44, 23–32.

Sen Gupta, B.K. (ed.) 1999. *Modern Foraminifera*. Kluwer Academic Publishers, Boston.

Shackleton, N.J. 2000. The 100,000-year ice-age cycle identified and found to lag temperature, carbon dioxide, and orbital eccentricity. *Science* 289, 1897–1902.

Shackleton, N.J. & Kennett, J.P. 1975. Paleotemperature history of the Cenozoic and the initiation of Antarctic glaciation: oxygen and carbon isotope analyses in Deep Sea Drilling Project sites 277, 279 and 281. *Initial Reports of the Deep Sea Drilling Project* 29, 743–755.

Sliter, W.V. 1972. Upper Cretaceous planktonic foraminiferal zoogeography and ecology – eastern Pacific margin. *Palaeogeography, Palaeoclimatology, Palaeoecology* 12, 15–31.

Stehli, F.G. & Creath, W.B. 1964. Foraminiferal ratios and regional environments. *Bulletin. American Association of Petroleum Geology* 48, 1810–1827.

Swain, F.M. (ed.) 1977. *Stratigraphic Micropalaeontology of Atlantic Basin and Borderlands*. Elsevier, Amsterdam.

Tappan, H. & Loeblich Jr, A.R. 1988. Foraminiferal evolution, diversification, and extinction. *Journal Paleontology* 62, 695–714.

Waelbroeck, C., Labeyrie, L., Michel, E., Duplessy, J.C., McManus, J.F., Lambeck, K., Balbon, E. & Labracherie, M. 2002. Sea-level and deep-water temperature changes derived from benthic foraminifera isotopic records. *Quaternary Science Reviews* 21, 295–305.

CHAPTER 16

Radiozoa (Acantharia, Phaeodaria and Radiolaria) and Heliozoa

Cavalier-Smith (1987) created the phylum Radiozoa to include the marine zooplankton Acantharia, Phaeodaria and Radiolaria, united by the presence of a central capsule. Only the Radiolaria including the siliceous Polycystina (which includes the orders Spumellaria and Nassellaria) and the mixed silica–organic matter Phaeodaria are preserved in the fossil record. The Acantharia have a skeleton of strontium sulphate (i.e. celestine $SrSO_4$). The radiolarians range from the Cambrian and have a virtually global, geographical distribution and a depth range from the photic zone down to the abyssal plains. Radiolarians are most useful for biostratigraphy of Mesozoic and Cenozoic deep sea sediments and as palaeo-oceanographical indicators.

Heliozoa are free-floating protists with roughly spherical shells and thread-like pseudopodia that extend radially over a delicate silica endoskeleton. Fossil heliozoans occur as scales or spines, less than 500 µm in size. They are found in marine or freshwater habitats from the Pleistocene to Recent.

Phylum Radiozoa

The living radiolarian

The individual single-celled radiolarians average between 50 and 200 µm in diameter, with colonial associations extending to metres in length. The cytoplasm of each cell is divided into an outer **ectoplasm** (extracapsulum) and an inner **endoplasm** (intracapsulum), separated by a perforate organic membrane called the **central capsule** (Fig. 16.1a), a feature that is unique to the Radiolaria. The nucleus or nuclei in multinucleate species are found within the endoplasm.

Radiating outwards from the central capsule are the pseudopodia, either as thread-like **filopodia** or as **axopodia**, which have a central rod of fibres for rigidity. The ectoplasm typically contains a zone of frothy, gelatinous bubbles, collectively termed the **calymma** and a swarm of yellow symbiotic algae called **zooxanthellae**. The calymma in some spumellarian Radiolaria can be so extensive as to obscure the skeleton.

A mineralized **skeleton** is usually present within the cell and comprises, in the simplest forms, either radial or tangential elements, or both. The radial elements consist of loose **spicules**, external **spines** or internal **bars**. They may be hollow or solid and serve mainly to support the axopodia. The tangential elements, where present, generally form a porous **lattice** shell of very variable morphology, such as spheres, spindles and cones (Fig. 16.1b,c). Often there is an arrangement of concentric or overlapping lattice shells.

Skeletal composition differs within the Radiozoa, being of strontium sulphate (i.e. celestine, $SrSO_4$) in the class Acantharia, opaline silica in the class Polycystina (orders Spumellaria and Nassellaria) and organic with up to 20% opaline silica in the order Phaeodaria. Radiolarians are able to repair broken elements and grow by adding to their skeletons. The absence of gradational forms between adults and juveniles in plankton samples indicates this process is not a simple addition of material alone.

Radiolarians reproduce by fission and possibly sexually by the release of flagellated cells, called **swarmers**. In the family Collosphaeridae (Spumellaria), the cells remain attached to form colonies. Individual radiolarians are thought to live no longer than 1 month. As marine zooplankton, radiolarians occupy a wide range of trophic types including bacterivores, detritivores, omnivores and osmotrophs (Casey 1993, in Lipps 1993, pp. 249–285). With increasing size there is a trend from herbivory to omnivory (Anderson 1996). Many species use their sticky radiating axopodia to trap and paralyse passing organisms (e.g. phytoplankton and bacteria). Food particles are digested in vacuoles within the calymma and nutrients are passed through the perforate central capsule to the endoplasm. Those living in the photic zone may also contain zooxanthellae and can survive by symbiosis.

Buoyancy is maintained in several ways. The specific gravity is lowered by the accumulation of fat globules or gas-filled vacuoles. Frictional resistance is increased by the development of long rigid axopods borne on skeletal spines. Holes in the skeleton allow the cytoplasm to pass through and also reduce weight. The spherical and discoidal skeletal shapes are further devices to reduce sinking, as in foraminifera, coccolithophores and diatoms. The turret- and bell-like skeletons of the Nassellaria appear to be adaptions for areas of ascending water currents, the mouth being held downwards and the axis held vertically, much as in silicoflagellates.

Radiolarian distribution and ecology

Living radiolarians prefer oceanic conditions, especially just seaward of the continental slope, in regions where divergent surface currents bring up nutrients from the depths and planktonic food is plentiful. Although most diverse and abundant at equatorial latitudes, where they may reach numbers of up to 82,000 m^{-3} water, they also thrive with diatoms in the subpolar seas (Fig. 16.2). Radiolarians tend to bloom seasonally in response to changes in food and silica content, currents and water masses.

Different trophic types live in different parts of the ocean; herbivores are restricted to the upper 200 m of the ocean whereas symbiotrophs are found to dominate the subtropical gyres and warm shelf areas. Detritivores and bacterivores dominate high latitude shallow subsurface waters. Different species may also occur in vertically stratified assemblages, each approximately corresponding to discrete water masses with certain physical and chemical characteristics (Fig. 16.3). Assemblage boundaries at 50, 200, 400, 1000 and 4000 m

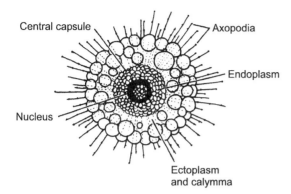

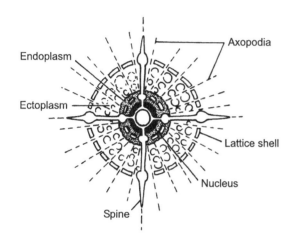

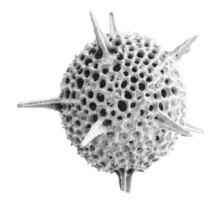

Fig. 16.1 (a) Cross-section through a naked radiolarian cell (*Thalassicola*). (b) Cross-section through a spumellarian showing the relationship of the nucleus, endoplasm and ectoplasm to three concentric lattice shells and radial spines. (c) SEM photomicrograph of a Neogene spumellarian radiolarian. ((b) After Westphal 1976.)

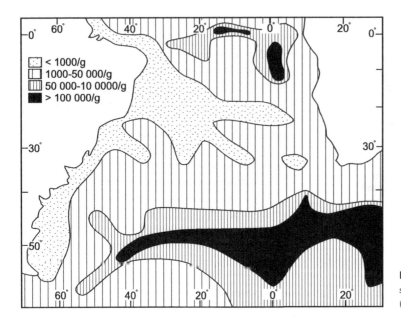

Fig. 16.2 Abundance of Radiolaria in surface sediments of the South Atlantic. (Modified from Goll & Bjørklund 1974.)

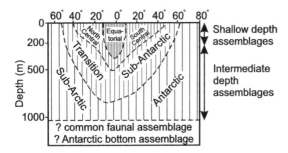

Fig. 16.3 Latitudinal and vertical assemblages of polycystine radiolarians in the pacific along 170° W transect. (Modified from Casey, in Funnell & Reidel 1971, figure 7.1.)

are reported, though these depths vary with latitude. Acantharia and Spumellaria generally dominate the photic zone (<200 m) and Nassellaria and Phaeodaria dominate in depths below 2000 m. Some radiolarian species occupy a wide depth range, with juveniles and small adults thriving at the shallower end of the range and the larger adults living in the deeper waters.

Radiolaria zoogeography is directly related to ocean circulation and water mass distribution patterns. The boundaries of radiolarian provinces thus correspond to major current convergences in the subtropical and tropical regions and have been used to plot the changing history of currents and water masses through the Cenozoic (see Casey et al. 1983; Casey 1989) and hence as a proxy for palaeotemperature. Gradients of temperature, silica and other macronutrient concentrations probably influence the latitudinal abundance of living Radiolaria (Abelmann & Gowing 1996). In the modern oceans eight shallow-water and seven deep-water provinces have been defined (Casey et al. 1982; Casey 1989). Of these the Subtropical Anticyclonic Gyre Province has the highest radiolarian diversity, specimen density and species endemism, probably reflecting to the presence of algal symbionts in the majority of taxa inhabiting this province. Deep-water provinces appear to be related to water masses found at depth. As with the Foraminiferida, some cold-water species that live near the surface in subpolar waters occur at greater depths near the Equator (Fig. 16.3).

Radiolarians and sedimentology

Both the $SrSO_4$ skeletons of the Acantharia and the weakly silicified tubular skeletons of the Phaeodaria are very prone to dissolution in the water column after death and on the deep sea floor, and they are therefore rare as fossils. Conversely, the solid opaline skeletons of the Spumellaria and Nassellaria tend to be more resistant even than in silicoflagellates and diatoms, although all are susceptible to dissolution because sea water is very undersaturated relative to silica. Below the calcium carbonate compensation depth (usually 3000–5000 m) nearly all $CaCO_3$ enters into solution so that siliceous radiolarian or diatomaceous oozes tend to accumulate. Radiolarian oozes are mostly found in the equatorial Pacific below zones of high productivity at 3000–4000 m depth and can contain as many as 100,000 skeletons per gram of sediment, but they may also occur abundantly in marine diatomaceous oozes or in *Globigerina* and coccolith oozes.

With increasing depth and dissolution, the abundance of Radiolaria in deep sea sediments decreases, through a progressive loss of the more delicate skeletons (Fig. 16.4). If the settling or sedimentation rate is slow, the chances for solution of skeletons will also increase, eventually lending a bias to the composition of fossil assemblages. Consequently, red muds of the abyssal plains mainly consist of volcanic and meteoritic debris barren of all but the most resistant parts of radiolarian skeletons and fish debris. The best preserved of radiolarians are those that have sunk rapidly to the ocean floor, usually within the faecal pellets of copepod crustaceans (Casey 1977, in Swain 1977, p. 542).

Fossil radiolarians are frequently found in chert horizons. Nodular cherts found interbedded with calcareous pelagic sediments of Mesozoic and Cenozoic age are probably deep-water deposits formed below belts of upwelling plankton-rich waters, as at the present day (see Casey 1989). The massive and ribbon-bedded cherts (**radiolarites**) found in Palaeozoic successions are interbedded with black shales and basic volcanic rocks in settings that have been interpreted as ancient oceanic crust. Ancient radiolarite deposits have been compared to those in the modern Owen and Somalia basins, both narrow, partially restricted basins with active monsoonal upwelling. However, the best radiolarian assemblages of Palaeozoic age come from continental shelf facies (Holdsworth 1977, in Swain 1977, pp. 167–184) and Bogdanov & Vishnevskaya (1992) proposed a shift in radiolarian habitats from the shallow, carbonate shelves of the Palaeozoic to the exclusively oceanic realm today.

Like other microfossils, Radiolaria are very prone to exhumation and reburial in younger sediments. These

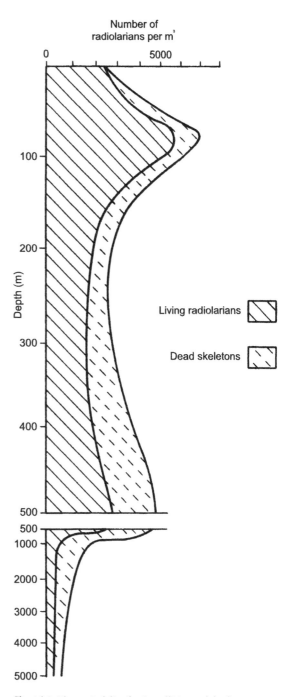

Fig. 16.4 The vertical distribution of living and dead radiolarians through the water column at a station in the central Pacific. (After Petrushevskaya, in Funnell & Reidel 1971, figure 21.4.)

and other aspects of radiolarians in sediments are reviewed more fully in Anderson (1983), Sanfilippo et al. (in Bolli et al. 1985) and Casey (1993, in Lipps 1993, pp. 249–285).

Classification of radiolarians

The classification of the radiolarians is in a state of flux. Living radiolarians are subdivided on morphology of the unmineralized (and therefore unfossilized) central capsule as well as on the composition and geometry of the skeleton. Fossil Radiolaria are classified using skeletal morphology. Separate schemes have been devised for the taxa present in the different eras and to date little attempt has been made to rationalize the many schemes. The scheme followed herein (Box 16.1) follows that proposed by Hart & Williams (1993 in Benton 1993, pp. 66–69) with modifications as recommended by Cavalier-Smith (1993). Suprageneric categories are probably best regarded as informal.

    Kingdom PROTOZA
    Parvkingdom ACTINOPODA
    Phylum RADIOZOA
    Subphylum RADIOLARIA
    Class POLYCYSTINEA

Polycystine Radiolaria are generally spherical. The Palaeozoic spherical Radiolaria (order Archaeospicularia) may not be closely related to the younger Spumellaria and comprise several but as yet little-studied groups (see Holdsworth 1977, in Swain 1977, pp. 167–184). For example, *Entactinosphaera* (U. Dev.-Carb., Fig. 16.5a) has a six-rayed internal spicule supporting two or more concentric lattice shells.

Order Spumellaria comprises skeletons in the form of a spherical or discoidal lattice, with several concentric shells bearing radial spines and supporting bars. In *Thalassicola* (Rec., Fig. 16.1a) a skeleton is either lacking or consists merely of isolated spicules. *Actinomma* (Rec., Fig. 16.5c) has three concentric, spherical lattice shells with large and small radial spines and bars. *Dictyastrum* (Jur.-Rec., Fig. 16.5d) has a flattened skeleton with three concentric chambers leading to three radiating chambered arms. Related genera also have radial beams and subdivide the chambers into chamberlets. *Albaillella* (Carb., Fig. 16.5b) belongs to a group of radiolarians with bilaterally symmetrical, triangulate skeletons that flourished in Silurian to Carboniferous times. Their systematic position is uncertain, with Holdsworth (1977, in Swain 1977, p. 168) placing them in a separate suborder Albaillellaria; but they have also been compared with the later Nassellaria.

Order Nassellaria have skeletons usually comprising a primary spicule, a ring or a lattice shell. The primary spicule comprises three, four, six or more rays that may be simple, branched or anastomosing. In *Campylacantha* (Rec., Fig. 16.6a), for example, the skeleton comprises a three-rayed spicule, each ray bearing similar but smaller branches. Evolutionary modifications of these rays led in certain stocks to a **sagittal ring** that may bear spines, sometimes in the form of tripod-like basal feet. In *Acanthocircus* (Cret., Fig. 16.6b) the ring bears three simple spines, two of them projecting inwards.

The phylogeny of taxa with more elaborate lattice shells can be traced from the study of the form of the primary spicule or ring elements (see Campbell 1954). The lattice may be spherical, discoidal, ellipsoidal or fusiform and constructed of successive chambers (segments) that partially enclose earlier ones. The skeletons differ from those of Acantharia and Spumellaria in having a wide aperture (basal shell mouth) at the **terminal pole**. This may be open or closed by a lattice. The initial chamber (**cephalis**) is closed and contains the primary spicule elements referred to above. The cephalis may also bear diagnostic features such as an apical horn. The second chamber is called the thorax and the third the abdomen with, sometimes, many more post-abdominal segments, each separated by a 'joint' or constriction. *Bathropyramis* (Cret.-Rec., Fig. 16.6c) has a conical lattice with rectangular pores and about nine radial spines around the open basal shell mouth. *Podocyrtis* (Cret.-Rec., Fig. 16.6d) has a conical, segmented skeleton with an apical horn and a tripod of three radial spines around the open mouth. Successive chambers of the fusiform *Cyrtocapsa* (Jur.-Rec., Fig. 16.6e) form prominent segments and the mouth is closed by a lattice.

Class Phaeodaria have skeletons that comprise 95% organic and 5% opaline silica constructed in the form of a lattice of hollow or solid elements, often with complex dendritic spines called **styles**. The central

Box 16.1 The classification of Radiolaria with diagrammatic representatives of a typical form (after Casey, in Lipps 1993, figure 13.5)

Class POLYCYSTINA
Order ARCHAEOSPICULARIA: Includes Lower Palaeozoic Radiolaria previously included in the Spumullaria and Collodaria characterized by spherical forms with a globular shell of several spicules. Members of this order are some of the earliest radiolaria and may have provided the ancestors to the Spumullaria and the Albiaillellidae.
Order SPUMELLARIA

Actinommidae Spongy cylindrical forms. Cosmopolitan. ?Trias.-Rec.	*Actinommidium*	**Phacodiscidae** Lens or biconvex disc-shaped forms. Warm water, ?Palaeo./Meso.-Rec.	
Coccodiscidae Lens shaped with a latticed centre and spongy chambered girdle or arms. Meso.-Eoc.	*Lithocyclia*	**Pseudoaulophacidae** Lens-like commonly triangular, usually with a few marginal spines. Cret. (Val.-Mass.)	
Collosphaeridae Single spheres, usually more interpore area than pore area; weakly developed external projections. Commonly colonial and possessing symbionts. Warm water in oligotrophic anticyclonic gyres, Mioc.-Rec.	*Collospaera*	**Pyloniidae** Skeleton comprises an ellipsoid of girdles and holes (gates). Warm water, Eoc.-Rec.	
Entactiniidae Spherical or ellipsoidal; latticed wall structure, bars running to the centre of the skeleton. L. Sil.-Carb.		**Spongodiscidae** Polyphyletic grouping of discoid, spongy forms. Dev.-Rec.	
Hagiastriidae Spongy 'rectangular' mesh and two to four large radial arms Palaeo.-Meso./Rec.		**Sponguridae** A polyphyletic group containing many subgroups of discoidal and 'spongy' forms. Cosmopolitan. Meso.-Rec.	
Litheliidae Coiled and latticed forms; tightly coiled morphotypes are cold water and loosely coiled morphotypes warm water. Cosmopolitan Carb.-Rec.		**Tholoniidae** Outer shell elliptical with bulb-like extensions. Deep cold water. Mioc.-Rec.	
Orosphaeridae Spherical or cup shapes with coarse polygonal lattice. Usually large specimens. Eoc.-Rec.			

Box 16.1 (cont'd)

Order NASSELLARIDA: Cone-shaped polycystine radiolaria
Suborder SPYRIDA
Suborder CYRTIDAE

Acanthodesmiidae D-shaped ring or latticed bilobed chamber with an internal D-shaped sagittal ring. Mainly warm shallow water; containing symbionts. Ceno.

Eucyrtinidae Usually more than one postcephalic chamber. The Spongocapsidae, Syringocapsidae and Xitidae can be included here. Mainly warm water, Meso.-Rec.

Amphipyndacidae Small usually poreless cephalis with several postcephalic joints. Warm water, Cret.-Eoc.

Plagoniidae Walls of the thorax and sometimes the cephalis can be extremely reduced to a spicular form. Spines are typically faceted. Cosmopolitan, Mio.-Rec. Similar spines are known from the Palaeo. and Meso.

Artostrobiidae Lobate or tubular cephalis with latitudinally arranged pores. Cold- and deep-water forms are more robust than warm-water forms. Cret.-Rec.

Pterocorythidae Large, elongate, lobate, porate cephalis and commonly more than one post-cephalic chamber. Bear a long spine that emerges from the side of the cephalis. Warm-water forms, Eoc.-Rec.

Cannobotrythiidae Lobate and randomly porate cephalis that may extend as tubes. Cosmo., warm- and cold-water forms Cret. or L. Palaeogene-Rec.

Rotaformiidae Lens-shaped with central area enclosing a nasselarian cephalis. Cret.

Carpocaniidae Cephalis small and recessed into a commonly elongte thorax. Warm water, Eoc.-Rec.

Theoperidae Small spherical forms with one or more post-cephalic chambers; lacking pores in the cephalis but bearing a single apical spine. Cold water, Trias.-Rec.

Class PHAEODAREA: Low preservation potential make this group rare in the fossil record. Two families are reported from the Tort.-Rec., the Challengeridae and the Getticellidae.
Subphylum SPASMARIA
Class ACANTHARIA: Rarely occur as fossils but comprises the Holacantharia and Euacanthia.

Radiolaria Incertae Sedis

Albaillellidae Sil.-Carb.
Anakrusidae, U. Caradoc-Ashgill.
Archeoentactiniidae M. Camb.
Ceratoikiscidae M. Sil.
Haplentactiniidae L. Ord./Sil.-Perm.

Inaniguttidae = Palaeoactinommids L. Ord.-U. Sil.
Palaeoscenidiidae?Ord./Dev.-Carb.
Palaeospiculumidae M. Camb.
Pylentonemiidae Ord.

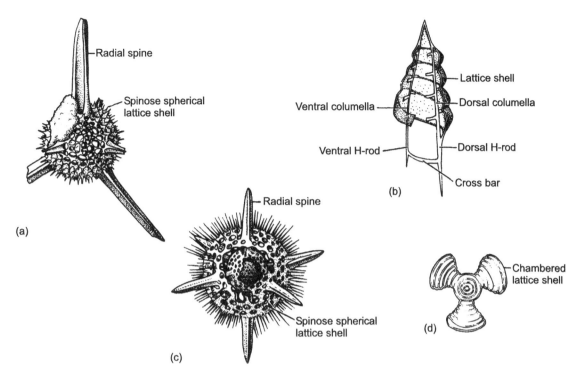

Fig. 16.5 Polycystine Radiolaria. (a) *Entactinosphaera* ×195. (b) *Albaillella* (scale unknown). (c) *Actinomma* (scale unknown). (d) *Ditryastrum* × 66. ((a) After Foreman 1963; (b) after Holdsworth 1969; (c) after Campbell 1954; (d) after Campbell 1954 from Haeckel ((c), (d) from the *Treatise on Invertebrate Paleontology*, courtesy of and © 1954, Part C, The Geological Society of America and The University of Kansas).)

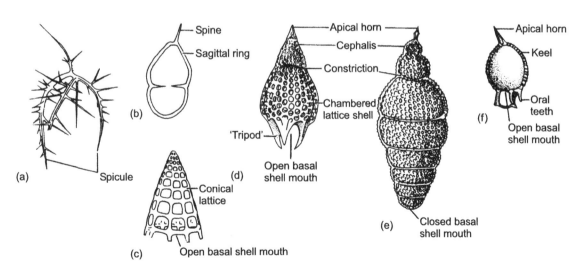

Fig. 16.6 Nasellarian and phaeodarian Radiolaria. (a) *Campylacantha* ×200. (b) *Acanthocircus* ×40. (c) *Bathropyramis* ×133. (d) *Podocyrtis* ×100. (e) *Cyrtocapsa* ×200. (f) *Challengerianum* ×187. ((a) After Campbell 1954 from Jorgensen; (b) after Campbell 1954 from Squinabol; (c), (d), (e) after Campbell 1954 from Haeckel; (f) redrawn after Reshetnjak, in Funnell & Riedel 1971, figure 24.19b. ((a), (c)–(e) from the *Treatise on Invertebrate Paleontology*, courtesy of and © 1954, Part C, The Geological Society of America and The University of Kansas).)

capsule also has a double wall rather than the single wall found in the former groups, and a **basal shell mouth** as in the Nassellaria. Only the more robust shells are known as fossils, such as *Challengerianum* (Mioc.-Rec., Fig. 16.6f). This has an ovate shell with an **apical horn**, a **marginal keel**, an open basal shell mouth surrounded by **oral teeth** and a skeleton wall with a fine hexagonal, diatom-like mesh.

Subphylum SPASMARIA
Class Acantharia

These have skeletons generated at the cell centre rather than peripherally as is usual in the other groups. This skeleton generally comprises 20 spines of $SrSO_4$ joined at one end (in the endoplasm) and arranged like the four spokes of five wheels in different planes and of varying diameters (e.g. *Zygacantha*, ?Mioc.-Rec., Fig. 16.7a).

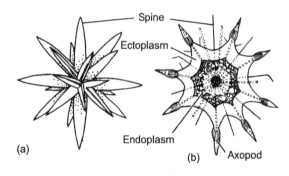

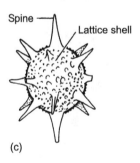

Fig. 16.7 Acantharian radiolarians. (a) *Zygacantha* skeleton ×160. (b) *Acanthometra* cell with spicules ×71. (c) *Belonaspis* skeleton ×100. ((a) After Campbell 1954 from Popofsky; (b) redrawn after Westphal 1976; (c) after Campbell 1954 from Haeckel ((a), (c) from the *Treatise on Invertebrate Paleontology*, courtesy of and © 1964, Part C, The Geological Society of America and The University of Kansas).)

Acanthometra (Rec., Fig. 16.7b) has thin radial spines embedded in cytoplasm that invariably disarticulate after death. *Belonaspis* (Rec., Fig. 16.7c) has an ellipsoidal lattice formed by fused spine branches (or **apophyses**) with 20 projecting radial spines.

General history of radiolarians

Radiolaria first appeared in the Cambrian and were one of the first groups to change from a benthic to free-floating mode of life (Knoll & Lipps 1993, in Lipps 1993, pp. 19–29). The earliest well-preserved examples are spicules, cones and the closed spheres of spherical Archeoentactiniidae and spicules of the Palaeospiculumidae from the Middle Cambrian of the Georgina Basin, Australia (Won & Below 1999) and the Upper Cambrian and Lower Ordovician of Kazakhstan (Nazarov 1975). Cold- and warm-water types can be distinguished in the Cambrian and a deep-water radiolarian fauna was present by the Silurian. A variety of distinct spumellarians flourished in the Palaeozoic, joined by the first deep, cold-water albaillellarians in the Late Devonian to Early Permian (Holdsworth 1977, in Swain 1977, pp. 167–184).

The dramatic reduction of cold- and warm-water species during the Permian and Triassic periods (Tappan & Loeblich 1973; Kozur 1998) has been attributed to the tectonic closure of some Late Palaeozoic ocean basins, the reorganization and reduction in the number of surface currents and eutrophication due to Late Permian glaciation (Hallam & Wignall 1997; Martin 1998). The first unequivocal nassellarians appeared in the Triassic. About half of the extant groups of Radiolaria appeared in the Mesozoic. The earliest unequivocal Phaeodaria are of Cretaceous age, with equivocal records from the Permian or even older.

From the Cretaceous Radiolaria had to share their niches with the rapidly radiating planktonic foraminifera (Anderson 1996). The fossil record suggests that, unlike diatoms and silicoflagellates, the Radiolaria did not flourish in the cooler Cenozoic Era (Fig. 16.8), as the equatorial belt in which they achieve their highest diversity contracted steadily during this time. Through the Cenozoic, radiolarians also show a progressive

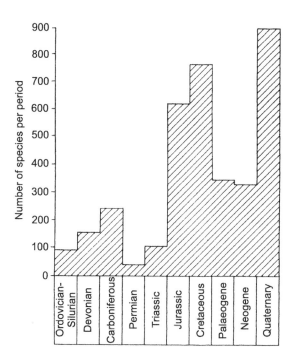

Fig. 16.8 Apparent changes in the species diversity of polycystine Radiolaria through time. (Based on Tappan & Loeblich 1973 with modifications after Vishnevskaya 1997.)

decrease in the amount of silica used to build the skeleton, particularly in silica-depleted, warm surface waters (Casey et al. 1983).

The Eocene–Oligocene boundary is marked by greatly enhanced siliceous ooze accumulation, Horizon A^c, comprising radiolarian-diatom deposits that occur in a broad belt across the northern Atlantic, equatorial Pacific and Mediterranean region. This event has been correlated with a large hiatus associated with volcanogenic deposits and vigorous deep-water currents and upwelling. Berger (1991) in his chert-climate hypothesis noted the Eocene 'opal revolution' corresponded with a declining volcanic silica source, increased oceanic mixing and progressive oxygenation, colonization of upwelling zones by diatoms and the more effective recirculation of biogenic silica.

A substantial change in the marine siliceous plankton occurred in the Early Oligocene, with a marked decline in many thickly silicified radiolarians (e.g. Conley et al. 1994; Khokhlova 2000). This decline in diversity has been interpreted as an increase in competitive pressure for dissolved silica by the diatoms and silicoflagellates, a pattern also seen during the coolers periods of the Cenozoic (Harper & Knoll 1975). Hence many living Nassellaria and Phaeodaria are delicately constructed and do not occur as fossils. The apparent drop in diversity may be misleading, with the data merely recording a decrease in the preservation potential of radiolarians, of oceanic environments, or both.

Fossil Acantharia have been reported from Paleocene and younger strata. The last major radiolarian radiation occurred at the Paleocene–Neogene boundary in response to the development of new intermediate and circumpolar water masses and the oligotrophic subtropical gyres.

Applications of radiolarians

Most of the studies of fossil Radiolaria emphasize their value to biostratigraphical correlation of oceanic sediments, particularly where the calcareous microfossils have suffered dissolution. Sanfilippo et al. (in Bolli et al. 1985) provides a review of Mesozoic and Cenozoic radiolarian biozonations, with tropical Cenozoic biozonations the best developed. The Late Paleocene to Recent has been divided into 29 biozones that can be recognized around the world and have been related directly to the well-dated magnetostratigraphy. The complex nature of the radiolarian skeleton and an almost complete Mesozoic to Recent geological record makes this group ideal for charting microevolutionary changes (see Moore 1972; Foreman 1975; Knoll & Johnson 1975).

Radiolaria have an increasing value as depth, palaeoclimate and palaeotemperature indicators and changes in radiolarian provinciality through the Cenozoic are highlighted (Casey et al. 1990). They have also been used to indicate palaeogeographic and tectonic changes in ocean basins. For example, radiolarian stratigraphy gave early support to the hypothesis of sea-floor spreading (Riedel 1967), and the closure of the Panama isthmus about 3.5 Myr is reflected in changing radiolarian assemblages in the Atlantic (Casey & McMillen in Swain 1977, pp. 521–524). The resistant nature of radiolarian chert to tectonism and diagenesis means Radiolaria are often the only common fossils preserved in orogenic belts and within accreted terranes (e.g. Murchey 1984; De Wever *et al.* 1994; Nokleberg *et al.* 1994; Cordey 1998).

Phylum Heliozoa

The Heliozoa closely resemble the Radiolaria but they lack the distinctive central capsule membrane between ectoplasm and endoplasm. Their skeletons may comprise a spherical lattice of chitinous matter weakly impregnated with silica, or isolated siliceous spicules and plates embedded in the mucilage near the outer ectoplasm. A few can agglutinate a skeleton of sand grains or diatom frustules or even survive without a skeleton at all. These delicate structures tend to fall apart after death, thereby obscuring their heliozoan origin. Heliozoans are, none the less, known as fossils from a few Pleistocene lake sediments (Moore 1954).

Further reading

Anderson's book on Radiolaria (1983) provides an excellent review of the biology of living radiolarians plus other aspects on the research into Radiolaria up to that time. Casey (in Lipps 1993, pp. 249–285) provides a good general review with sections on oceanographic applications and biostratigraphy. Sanfilippo *et al.* (in Bolli *et al.* 1985) provides a detailed review of radiolarian biostratigraphy plus many illustrations. Case studies of the application of Radiolaria in orogenic belts can be found in a special issue of *Palaeogeography, Palaeoclimatology, Palaeoecology* 1996, 96, 1–161). Identification of specimens should be assisted by reference to Foreman & Riedel (1972 to date). Racki & Cordey (2000) review radiolarian palaeoecology in the context of the evolution of the marine silica cycle.

Hints for collection and study

Fossil Radiolaria can be extracted from mudstones, shales and marls using methods A to E (see Appendix), from limestones using method F and from cherts using method F using HF. The residues should then be washed over a 125- and 68-µm sieve, dried and concentrated with CCl_4 (methods I and J) and viewed with reflected light (method O) or with well-condensed transmitted light, as with diatoms (q.v.). Radiolarian cherts can be studied in relatively thick petrographic thin sections, viewed with transmitted light at about 400× or higher. Further information on preparatory techniques is given by Riedel & Sanfilippo (in Ramsay 1977, pp. 852–858).

REFERENCES

Abelmann, A. & Gowing, M.M. 1996. Horizontal and vertical distribution pattern of living radiolarians along a transect from the Southern Ocean to the South Atlantic Subtropical Region. *Deep Sea Research, Part I* 43, 361–382.

Anderson, O.R. 1983. *Radiolaria*. Springer Verlag. New York.

Anderson, O.R. 1996. The physiological ecology of planktonic sarcodines with application to palaeoecology: patterns in space and time. *Journal of Eukaryotic Microbiology* 43, 261–274.

Benton, M.J. (ed.) 1993. *The Fossil Record 2*. Chapman & Hall, London.

Berger, W.H. 1991. Produktivität des Ozeans aus geologischer Sicht: Denkmodelle und Beispiele. *Zeitschrift. Deutsche Geologische gesellschaft* 42, 149–178.

Bogdanov, N.A. & Vishnevskaya, V.S. 1992. Influence of evolutionary changes in Radiolaria on sedimentary processes. *Doklady Akademii Nauk SSSR* 324, 162–166 [in Russian].

Bolli, H.M., Saunders, J.B. & Perch-Nielsen, K. 1985. *Plankton Stratigraphy*. Cambridge University Press, Cambridge.

Campbell, A.S. 1954. Radiolaria. In: Moore, R.C. (ed.) *Treatise on Invertebrate Paleontology. Part D, Protista 3: Protozoa (chiefly Radiolaria and Tintinnina)*. Geological Society of America and University of Kansas Press, Lawrence, Kansas, pp. 11–163.

Casey, R.E. 1989. Model of modern polycystine radiolarian shallow-water zoogeography. *Palaeogeography, Palaeoclimatology, Palaeoecology* 74, 15–22.

Casey, R.E., Spaw, J.M. & Kunze, F.R. 1982. Polycystine radiolarian distribution and enhancements related to oceanographic conditions in a hypothetical ocean. *Bulletin. American Association of Stratigraphic Palynologists*, 66, 1426.

Casey, R.E., Wigley, C.R. & Perez-Guzmann, A.M. 1983. Biogeographic and ecologic perspective on polycystine radiolarian evolution. *Paleobiology* 9, 363–376.

Casey, R.E., Weinheimer, A.L. & Nelson, C.O. 1990. Cenozoic radiolarian evolution and zoogeography of the Pacific. *Bulletin. Marine Science* 47, 221–232.

Cavalier-Smith, T. 1987. The origin of eukaryote and archaeobacterial cells. *Annals of the New York Academy of Sciences* 503, 17–54.

Cavalier-Smith, T. 1993. Kingdom Protozoa and its 18 phyla. *Microbiological Reviews* 57, 953–994.

Conley, D.J., Zimba, P.V. & Theriot, E. 1994. Silica content of freshwater and marine benthic diatoms. In: Kociolek, J.P. (ed.) *Proceedings of the 11th International Diatom Symposium, San Francisco, 1990. Memoir. Californian Academy of Science* 17, 95–101.

Cordey, F. 1998. Radiolaires des complexes d'accretion de la Cordillere Canadienne (Columbie-Britannique). *Bulletin. Geogical Survey of Canada* 207, 1–209.

De Wever, P., Azéma, J. & Fourcade, E. 1994. Radiolaires et radiolarites: production primaire, diagenése et paléogéographie. *Buletin. Centres des Recherches Exploration-Production ELF-Aquitaine* 18, 315–379.

Foreman, H.P. 1963. Upper Devonian Radiolaria from the Huron member of the Ohio Shale. *Micropalaeontology* 9, 267–304.

Foreman, H.P. 1975. Radiolaria from the North Pacific, Deep Sea Drilling Project, Leg 32. *Initial Reports of the Deep Sea Drilling Project* 32, 579–673.

Foreman, H.P. & Riedel, W.R. 1972 to date. *Catalogue of Polycystine Radiolaria*. Micropaleontology Press, American Museum of Natural History, New York.

Funnell, B.M. & Riedel, W.R. (eds) 1971. *The Micropalaeontology of Oceans*. Cambridge University Press, Cambridge.

Goll, R.M. & Bjørklund, K.R. 1974. Radiolaria in the surface sediments of the South Atlantic. *Micropalaeontology* 20, 38–75.

Hallam, A. & Wignall, P. 1997. *Mass Extinctions and their Aftermath*. Oxford University Press, Oxford.

Harper, H.E. & Knoll, A.H. 1975. Silica, diatoms and Cenozoic radiolarian evolution. *Geology* 3, 175–177.

Holdsworth, B.K. 1969. The relationship between the genus *Albaillella* Deflandre and the ceratoikiscid Radiolaria. *Micropalaeontology* 15, 230–236.

Khokhlova, I.E. 2000. Changes in generic composition of Cenozoic radiolarians in tropical realm of the World ocean: correlation with abiotic events. *Byulletin' Moskovskogo Obshchestva Ispytatelei Prirody Otdel Geologicheskii* 75, 34–40 [in Russian].

Knoll, A.H. & Johnson, D.A. 1975. Late Pleistocene evolution of the collosphaerid radiolarian *Buccinosphaera invaginata*. *Micropalaeontology* 21, 60–68.

Kozur, H.W. 1998. Some aspects of the Permian–Triassic boundary (PTB) and the possible causes for the biotic crisis around this boundary. *Palaeogeography, Palaeoclimatology, Palaeoecology* 143, 227–272.

Lipps, J.H. (ed.) 1993. *Fossil Prokaryotes and Protists*. Blackwell, Boston.

Martin, R.E. 1998. Catastrophic fluctuations in nutrient levels as an agent of mass extinction: upward scaling of ecological processes? In: McKinney, M.L. & Drake, J.A. (eds) *Biodiversity Dynamics. Turnover and populations, taxa and communities*. Columbia University Press, New York, pp. 405–429.

Moore, R.C. 1954. Heliozoa. In: Moore, R.C. (ed.) *Treatise on Invertebrate Paleontology. Part D, Protista 3: Protozoa (chiefly Radiolaria and Tintinnina)*. Geological Society of America and University of Kansas Press, Lawrence, Kansas.

Moore Jr, T.C. 1972. Mid-Tertiary evolution of the radiolarian genus *Calocycletta*. *Micropalaeontology* 18, 144–152.

Murchey, B. 1984. Biostratigraphy and lithostratigraphy of chert in Franciscan Complex, Marin headlands, California. In: Blake, M.C. (ed.) *Franciscan Geology of Northern California. SEPM Pacific Section* 43, 51–70.

Nazarov, B.B. 1975. Lower and Middle Paleozoic radiolarians of Kazakhstan. *Trudy Instituta Geologiceskih Nauk SSSR* 275, 202pp. [in Russian].

Nokleberg, W.J., Parfenov, L.M. & Monger, J.W.H. 1994. Circum-North Pacific Tectono-Stratigraphic Terrane Map. *US Geological Survey Open-File* 94.

Racki, G. & Cordey, F. 2000. Radiolarian palaeoecology and radiolarites: is the present the key to the past? *Earth-Science Reviews* 52, 83–120.

Ramsay, A.T.S. (ed.) 1977. *Oceanic Micropalaeontology*, 2 vols. Academic Press, London.

Riedel, W.R. 1967. Radiolarian evidence consistent with spreading of the Pacific floor. *Science* 157, 540–542.

Swain, F.M. (ed.) 1977. *Stratigraphic Micropaleontology of Atlantic Basin and Borderlands*. Elsevier, Amsterdam.

Tappan, H. & Loeblich Jr, A.R. 1973. Evolution of the ocean plankton. *Earth Science Reviews* 9, 207–240.

Vishnevskaya, V.S. 1997. Development of Palaeozoic-Mesozoic Radiolaria in the Northwestern Pacific Rim. *Marine Micropalaeontology* 30, 79–95.

Westphal, A. 1976. *Protozoa*. Blackie, Glasgow.

Won, M.Z. & Below, R. 1999. Cambrian Radiolaria from the Georgina Basin, Queensland, Australia. *Micropalaeontology* 45, 325–363.

CHAPTER 17

Diatoms

Diatoms are unicellular algae with golden-brown photosynthetic pigments that differ from other chrysophytes in lacking flagella. Their cell wall is silicified to form a **frustule**, comprising two **valves**, one overlapping the other like the lid of a box. Diatoms live in almost all kinds of aquatic and semi-aquatic environments that are exposed to light, and their remains may accumulate in enormous numbers in **diatomite**. Diatoms are the dominant marine primary producers (see Nelson *et al.* 1995 for a review) and play a particularly important role in the carbon, silica and nutrient budgets of the modern ocean. Over a century of careful botanical research into living forms has resulted in a relatively clear-cut taxonomy and considerable knowledge about their biology and ecology. Living species are extremely sensitive to physical and chemical conditions, so that they provide a valuable tool for studies of modern water quality and for the reconstruction of past environments. Diatoms are also important as biostratigraphical zone fossils in marine deposits from high latitudes or at great water depth, both of which tend to lack calcareous microfossils.

The living diatom

The diatom cell ranges in size from 1 to 2000 μm in length, although most species encountered are in the size range 10 to 100 μm. The cell may be single or colonial, the latter bound together by mucous filaments or by bands into long chains. Each cell possesses two or more yellow, olive or golden-brown photosynthetic **chloroplasts**, a central vacuole and a large central diploid nucleus, although it lacks flagella and pseudopodia. **Pennate** diatoms (Fig. 17.1) can glide over the substrate by the production of a stream of mucus between the frustule and the sediment, but the planktonic **centric** diatoms (Fig. 17.2) are non-motile. To avoid sinking below the photic zone, the latter are therefore provided with low-density fat droplets or occasionally with spines and they may also construct long colonial chains of frustules.

Reproduction is primarily asexual, by mitotic division of the parent cell into two. This binary fission can take place from one to eight times per day. Because each daughter cell takes one of the parent valves for its own and adds a new valve, there is a gradual diminution in the average size of the diatom stock with each generation. This trend is eventually reversed by sexual reproduction.

The frustule

About 95% of the cell wall in diatoms is impregnated with opaline silica. The region of overlap between the **epivalve** and **hypovalve** is called the **girdle**, and a study of the valve and the girdle view aids identification (Fig. 17.1a). Frustules are usually either circular (**centric**) or elliptical (**pennate**) in valve view, these kinds also comprising the two orders of diatoms (Centrales and Pennales). From 10 to 30% of the valve surface area is covered in tiny pores called **punctae**, the arrangement of which is also significant for classification, perforate surface. The punctae, which allow connection between the cytoplasm and the external environment, can either be simple holes or are occluded by thin transverse plates with minute pores, referred to as **sieve membranes** (Fig. 17.1d). Arrangement of the punctae in lines gives rise to **striae**, usually separated by imperforate ridges called **costae**.

Members of the order Pennales, or pennate diatoms, have frustules that are elliptical or rectangular in valve view, with sculpture that is bilaterally symmetrical about a central line. In many diatoms, this central line is a longitudinal unsilicified groove down the middle of each valve face called a **raphe**, which has rows of punctae arranged at right angles on either side (Fig. 17.1a). The raphe facilitates a flow of mucus that leads to a creeping motion. Some do not have a groove but merely a similar, silicified area clear of punctae, called a **pseudoraphe** (Fig. 17.1b,c). A **central nodule** in the mid-point of the valve face divides the raphe into two, and similar **polar nodules** may occur at the extremities (Fig. 17.1a). The raphe or pseudoraphe can occur on one or both valves. Such features are used for further taxonomic subdivision of pennate diatoms. Members of the suborder Araphidineae only have a pseudoraphe and generally occur attached by mucilage pads at the apex of the cell. For example, *Fragilaria* (Fig. 17.1b) is a benthic, freshwater genus with a very narrow frustule, rectangular in girdle view and commonly united on the valve faces into long chains. The punctae are arranged in striae without intervening costae. In the suborder Monoraphidineae, a raphe is present on the hypovalve and a pseudoraphe on the epivalve. *Achnanthes* (Fig. 17.1c), for example, is solitary or united in chains and has boat-shaped (**naviculoid**) valves with punctae arranged in striae. The example in Fig. 17.1c is a brackish-water species but freshwater and marine species occur. The Biraphidineae have a true raphe on both valves, such as in the common freshwater genus *Pinnularia* (Fig. 17.1a).

The order Centrales (syn. Coscinodiscophyceae) is characterized by members that have a structural centre formed by a point. Centric diatoms have frustules which are circular, triangular or quadrate in valve view

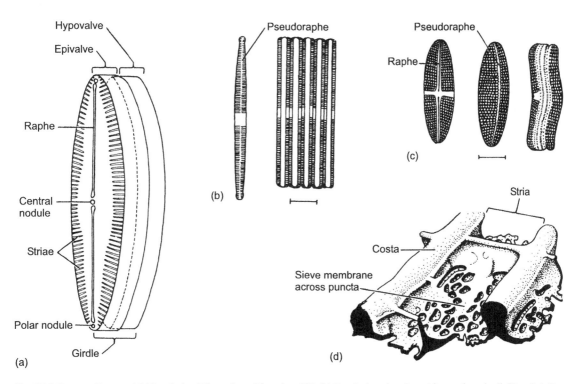

Fig. 17.1 Pennate diatoms. (a) *Pinnularia*, oblique view with raphe ×320. (b) *Fragilaria*, valve view with pseudoraphe (left) and girdle view of colony (right, about ×545). (c) *Achnanthes*, hypovalve with raphe (left), epivalve view with pseudoraphe (centre) and girdle view (right, all about ×545). (d) Detail of diatom punctae. Scale bar = 10 μm ((a) After Scagel *et al.* 1965; (b) and (c) after van der Werff & Huls 1957–1963; (d) after Chapman & Chapman 1973 from Fott.)

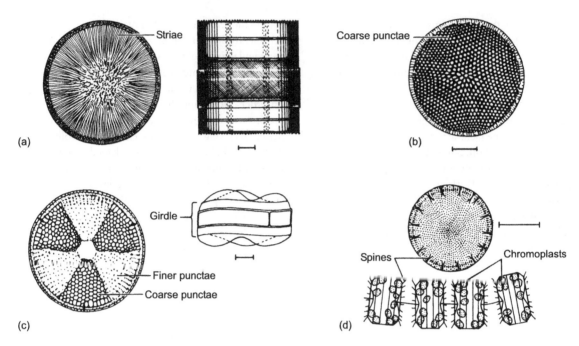

Fig. 17.2 Centric diatoms. (a) *Melosira*, valve view (left) and girdle view of colony (right, about ×342). (b) *Coscinodiscus*, valve view, about ×535. (c) *Actinoptychus*, valve view (left, about ×277) and girdle view (right, about ×340). (d) *Thalassiosira*, valve view (above) and girdle view of colony (below, both ×670). Scale bar = 10 μm. (After van der Werff & Huls, 1957–1963.)

and rectangular or ovate in girdle view. Being mostly planktonic and non-motile, they lack the raphe and pseudoraphe. *Melosira* (Fig. 17.2a) thrives in freshwater and brackish-water habitats, its pillbox-like frustules united into long filaments. The punctae are small and arranged in numerous fine striae radiating from a central region of fewer punctae. In *Coscinodiscus* (Fig. 17.2b) the frustule is also discoidal but with very large radiating punctae. This genus is typical of many inshore and outer shelf planktonic assemblages. *Actinoptychus* (Fig. 17.2c) has the valve face divided into compartments, alternately elevated and depressed, with punctae of different size and shape. It thrives in the nearshore plankton. *Thalassiosira* (Fig. 17.2d) is an open-ocean planktonic form with radial punctae and small submarginal spines, the frustules united in chains by a delicate mucus filament.

Many planktonic diatoms living in shelf seas produce a thick-walled siliceous resting cyst or **statospore** when temperature and nutrients fall below a critical level. The statospore may sink down to the sea floor until favourable conditions return as, for example, during seasonal upwelling. Unlike normal frustules, the sculpture of the epivalve and hypovalve of the statospore is different, and it also lacks a girdle.

Diatom distribution and ecology

Diatoms are autotrophic and form the basis of food chains in many aquatic ecosystems. Different species occupy benthic and planktonic niches in ponds, lakes, rivers, salt marshes, lagoons, seas and oceanic waters, while some thrive in the soil, in ice, or attached to trees and rocks.

Pennate diatoms dominate the freshwater, soil and epiphytic niches although they also thrive in benthic marine habitats. Centric diatoms thrive as plankton in marine waters, especially at subpolar and temperate latitudes. Distinct planktonic assemblages are known to dwell in nearshore, neritic and oceanic environments. They can also occur as plankton in freshwater bodies.

Diatoms require light and are therefore limited to the photic zone (<200 m) during life. Each species

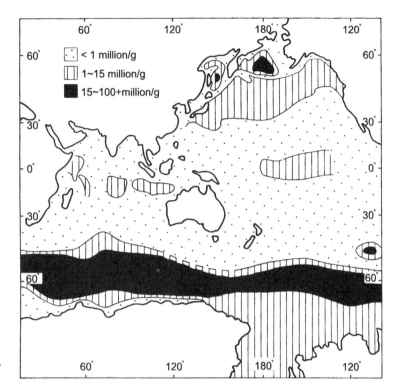

Fig. 17.3 The distribution of diatom frustules in surface sediments of the Indian and Pacific oceans, in millions per gram of sediment. (Based on Lisitzin, in Funnell & Riedel 1971, figure 10.11.)

tends to have a preference for a particular water mass, with distinctive ranges of temperature, salinity, acidity, oxygen and mineral concentrations. Seasonal fluxes in these factors at high latitudes may lead to spring and late summer blooms, particularly among the plankton where diatoms may number as much as 1000 million cells per m^3 of water. Diatoms are especially abundant in regions of upwelling caused by current divergences, as in those of the Antarctic divergence (Fig. 17.3) and off the coast of Peru. These waters are favoured because of their high silica, phosphate, nitrate and iron content. Diatoms living at times of high nutrient availability often face an acute shortage of dissolved silica, which they overcome by the production of weakly silicified frustules (Conley et al. 1994; Baron & Baldauf 1995). After death, the thin and highly porous skeleton dissolves rapidly, making more silica available for the next generation of diatoms. Where the concentrations of biolimiting nutrients are low, as in the centres of oceanic gyres, then diatoms tend to be rare.

Diatomaceous sediments

Diatom productivity is high where nutrient levels are high. These conditions can lead to the accumulation of diatomites on the deep sea floor. Diatomites are forming in three main areas at the present time: beneath sub-Arctic waters of the northern hemisphere; beneath sub-Antarctic waters of the southern hemisphere; and in an equatorial belt around the Indian and Pacific Oceans, related to a belt of equatorial upwelling (Fig. 17.3). These equatorial deposits are some 4–6 m thick and may contain over 400 million valves per gram (largely of *Ethmodiscus* sp.). Such vast accumulations of diatoms also require conditions of low terrigenous influx (e.g. away from coastlines) and high $CaCO_3$ solubility (e.g. at abyssal depths).

Modern sea water is undersaturated with respect to silica, largely because of the huge amount of silica removed from solution by diatom biomineralization. This means that diatom frustules are prone to dissolution by pressure at depth or under alkaline conditions;

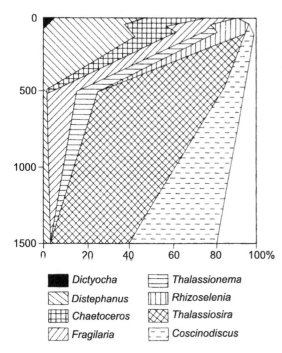

Fig. 17.4 Changes in the silicoflagellate and diatom flora with depth, mainly through dissolution. *Dictyocha* and *Distephanus* are silicoflagellates; the rest are diatoms. (Based on Lisitzin, in Funnell & Riedel 1971, figure 10.8.)

especially the less robust or weakly silicified forms (Fig. 17.4). This selective dissolution in marine environments and the non-preservation of many freshwater forms leads to fossil assemblages rarely being representative of the living assemblage. Less than 5% of the living assemblage at the ocean surface reaches the sea floor to form a death assemblage. The latter mainly comprises robust frustules and statospores, plus more delicate forms that have reached the bottom through incorporation into zooplankton faecal pellets. Furthermore, planktonic diatoms may travel far before coming to rest on the sea bed; even freshwater diatoms are not uncommon in deep sea sediments. The latter are mainly blown off the land by strong winds.

Classification

Kingdom CHROMISTA
Subkingdom EUCHROMISTA
Infrakingdom DIATOMEA

The classification of diatoms has been traditionally based on frustule form and sculpture. Hustedt (1930) gave the group the status of a division, Bacillariophyta, but Hendey (1964) and many others regard diatoms as a class within the Chrysophyta, which also includes the coccolithophores. Cavalier-Smith (1993) placed the diatoms (along with the coccolithophores) in the kingdom Chromista based on the location of the chloroplasts in the lumen of the endoplasmic reticulum. This organelle arrangement also distinguishes the diatoms from other photosynthetic Protozoa including the dinoflagellates. Two orders are widely recognized, namely the Pennales and the Centrales (Table 17.1).

Evolutionary history

The fossil record of marine diatoms is still incompletely known due to dissolution and taphonomic effects (e.g. Hesse 1989; De Wever *et al.* 1994; Martin 1995; Schieber *et al.* 2000). The ancestor of the diatoms may have been a spherical chrysophyte provided with thin siliceous scales, such as are known from Proterozoic cherts. The earliest unequivocal recorded diatom frustules are centric forms from the Early Jurassic although very few remains are known before the Campanian (Late Cretaceous).

Diatoms were only moderately affected by events at the Cretaceous–Tertiary boundary (c. 23% extinction). A major radiation took place among centric diatoms in the Paleocene when the first pennate types also appeared (Fig. 17.5), expanding their numbers gradually through time.

Periods of turnover in diatom species have coincided with steps in global cooling leading to the increasing of latitudinal thermal gradients through the Cenozoic. High- and low-latitude diatom assemblages began to differentiate in the Late Eocene to Oligocene, and provincialism increased again in the latest Miocene. Within the Pleistocene, diatom assemblages closely resemble modern ones but they show a marked increase in abundance during glacial maxima owing to increased surface water circulation, upwelling and raised nutrient levels.

Prior to the Oligocene, diatom assemblages are dominated by robust genera such as *Hemiaulus* (Jur.-Oligo., Fig. 17.5). From the Oligocene onward these robust forms are progressively replaced by more finely

Table 17.1 Suprageneric classification of the diatoms. (Redrawn after Barron in Lipps 1993, after Simonsen 1979.)

Order	Suborder	Family
Centrales Central point formed by a point, auxospore formation by oogamy	Coscinodiscineae Valves with a ring of marginal pores, symmetry primarily without development of polarities, e.g. *Coscinodiscus*	Thalassiosiraceae Melosiraceae Coscinodiscaceae Hemidiscaceae Asterolampraceae Heliopeltaceae
	Rhizosoleniineae Valves primarily unipolar, strongly elongated in the direction perpendicular to the plane at which the two valves are joined in the frustule, e.g. *Pyxilla*	Pyxillaceae Rhizosoleniaceae
	Biddulphineae Valves primarily bipolar, secondarily tri- to multipolar to cicular, e.g. *Triceratium*	Biddullphiaceae Chaetoceraceae Lithodemiaceae Eupodiscaceae
Pennales Structural centre normally formed by a line, auxospore formation not by oogamy	Araphidineae – valves without a raphe, e.g. *Thalassiothrix*	Diatomaceae Protoraphidaceae
	Raphidineae – valves with a raphe, e.g. *Nitzchia*	Eunotiaceae Achanthaceae Naviculaceae Auriculaceae Epithemiaceae Nitzsciaceae Surirellaceae

silicified genera such as *Coscinodiscus* (Eoc.-Rec., Figs 17.2b, 17.5), *Thalassiosira* (?Eoc.-Rec., Figs 17.2d, 17.5) and *Thalassionema* (Oligo.-Rec., Fig. 17.5). By the Late Miocene, very finely silicified forms such as *Nitschia* (Mioc.-Rec., Fig. 17.5) and *Denticulopsis* (Mioc.-Rec., Fig. 17.5) are abundant. Very delicate, small and elongate forms such as *Chaetoceras* and *Skeletonema* dominate blooms in modern coastal upwelling zones. The high number of living species (Fig. 17.6) reflects the contribution made by such small, weakly silicified forms with low preservation potential. This trend towards more weakly silicified forms through the Cenozoic has accompanied global cooling, more vigorous circulation and raised nutrient levels. It therefore seems that diatoms, along with the Radiolaria, have adapted to the increasing competition for scarce silica in surface waters by reducing their demand for silica in the skeleton.

The fossil record of freshwater diatoms is also very incomplete owing to dissolution of their delicate frustules. Pennate diatoms had certainly colonized freshwater habitats by the Paleocene, while a radiation of both centric and pennate diatoms during the Middle Miocene may have been enhanced by added supplies of silica from widespread volcanism.

Applications of diatoms

Few microfossil groups can rival diatoms for the breadth of their potential applications and these have been reviewed by Stoermer & Smol (1999). Planktonic diatoms provide the primary means of correlating high-latitude, and deep-water deposits where calcareous microfossils tend to be sparse and of low diversity. Their value as biozonal indices for the Cretaceous and Tertiary successions

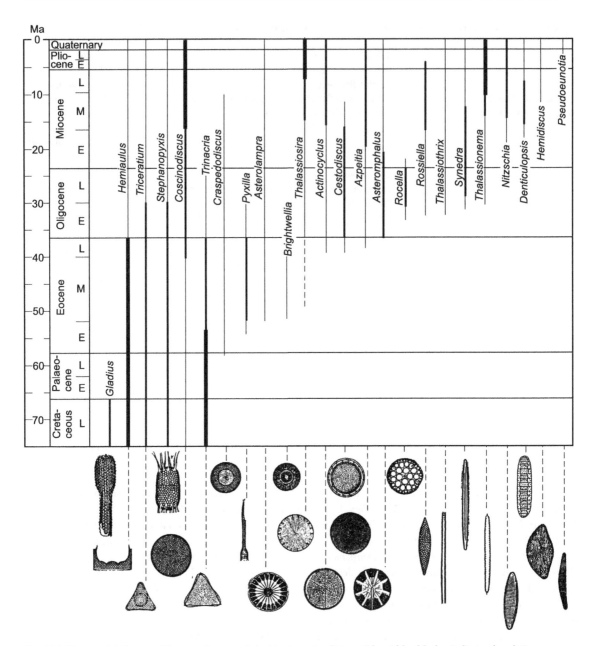

Fig. 17.5 Stratigraphical ranges of important genera of planktonic marine diatoms. The width of the bar indicates the relative abundance of the genus during its range. (Reproduced from Barron in Lipps 1993, figure 10.11.)

is outlined in Barron (in Lipps 1993, pp. 155–169) and detailed by Fenner (in Bolli *et al.* 1985) and Barron (in Bolli *et al.* 1985). After the Eocene, both high- and low-latitude biozonations are needed and correlation between these can be problematic. Although the northern and southern high-latitude assemblages may share species, few of these show synchronous appearances and extinctions. Before the Miocene, the retrieval of biostratigraphical information is also hampered by the adverse effect of burial upon frustule preservation.

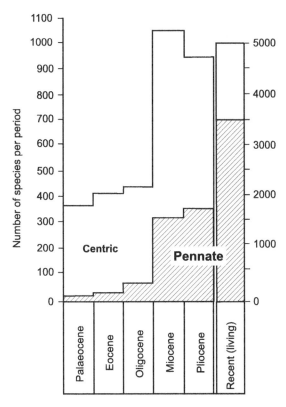

Fig. 17.6 Changes in species diversity of diatoms through the Cenozoic Era. (Based on Tappan & Loeblich 1973.)

The palaeoecological value of diatoms is well established, particularly for evidence of climatic cooling and changing sedimentation rates in the Arctic and Antarctic oceans (e.g. Retallack et al. 2001; Bianchi & Gersonde 2002; Shemesh et al. 2002; Wilson et al. 2002; Whittington et al. 2003) and to provide sea surface temperature estimates (e.g. Birks & Koc 2002). Gardner & Burckle (1975) have shown that the *Ethmodiscus* oozes of the equatorial Atlantic were deposited during glacial maxima.

Quaternary diatoms are useful indicators of local habitat changes from terrestrial to deep marine environments and provide insight into relative lake level and sea-level changes and water chemistry. Diatom assemblages have been classified – the halobian system (Hustedt 1957) – according to their salinity preferences. Mapping these assemblages through cores from salt marsh, lake and estuaries allows the construction of diatom diagrams and plotting of sea-level change (e.g. Shennan et al. 1994). Freshwater diatoms have been used to study the history of lakes since the last glaciation, revealing the effects of changing pH and climate (e.g. Battarbee 1984; Battarbee & Charles 1987; Mackay et al. 1998; Leng et al. 2001; Marshall et al. 2002) and the effects of human pollution (e.g. Jones et al. 1989; Stewart et al. 1999; Joux-Arab et al. 2000; Ek & Renberg 2001).

The ratios between the oxygen isotopes ^{18}O and ^{16}O in the silica of fossil diatom frustules can also be used to indicate absolute temperatures in Quaternary deposits (e.g. Mikkelsen et al. 1978; Shemesh et al. 1992, 2002), though vital effects cannot be fully discounted (Schmidt et al. 1997, 2001). Carbon isotopic records from diatoms have been used to model whether the Southern Ocean was a source or sink for carbon dioxide during the last glacial (Rosenthal et al. 2000).

Mention should be made here of the economic value of diatomites, a porous and lightweight sedimentary rock resulting from the accumulation of diatom frustules. Deposits in California are marine ranging in age from Late Cretaceous to Late Pliocene. In the Miocene diatomites occur in units up to 1000 m thick with over six million frustules per cubic centimetre. Diatomites have been deposited in freshwater environments since at least the Eocene. Although rarely greater than 1 m thick, they are still of economic interest. The silica is graded and used for filtering, sugar refining toothpaste, insulation, abrasive polish, paint and lightweight bricks. The association of many oil fields with diatom-bearing shale indicates the high lipid-oil content in diatoms is a likely source of petroleum (Harwood 1997, in Stoermer & Smol 1999, pp. 436–447).

Further reading

The biology and ecology of living diatoms are outlined by Round et al. (1990) and Barron (in Lipps 1993, pp. 155–169) and more detailed aspects and taxonomy can be found in Hendey (1964) and Simonsen (1979). Sieburth (1975) provides many beautiful photographs of living forms in their natural habitat. The distribution and significance of diatoms in oceanic sediments are also reviewed in Funnell & Riedel (1971), whilst Stoermer & Smol (1999) clearly sets forth the geological value of diatoms and contains a useful bibliography.

Recent and fossil genera and species can be identified with the aid of the catalogue by van Landingham (1967 to date). Hartley (1996) provides a useful guide to British living diatoms. Reviews of the importance of terrestrial diatoms can be found in Clarke (2003) and Conley (2002).

Hints for collection and study

Recent diatoms can be easily collected by scraping up the green scum from the floors of ponds or from the surfaces of mud, pebbles, shells and vegetation in shallow marine waters. Temporary mounts in distilled water can be prepared on glass slides and viewed with well-condensed transmitted light at about 400× magnification or higher.

Fossil diatoms are readily studied in freshwater or marine diatomites. Any reputable aquarium shop or pet shop will sell the 'diatom powder' used for aquarium filters. These diatomites usually require little in the way of disaggregation or concentration, but diatoms in shales and limestones will need to be treated with methods B, C, D or E to release them and method J to concentrate them (see Appendix). Treat the sample with formic (or even concentrated hydrochloric) acid if calcareous shells are not wanted (see method F). Temporary mounts can be prepared with distilled water and strewn on glass slides. For permanent mounts dry the residue on a glass slide, add a blob of Canada Balsam to the cover slip and place this over the residue. When dry, examine with transmitted light. Some more sophisticated techniques are given by Setty (1966).

REFERENCES

Baron, J.A. & Baldauf, J.G. 1995. Cenozoic marine diatom biostratigraphy and applications to paleoclimatology and paleoceanography. Siliceous Microfossils. *Paleontological Society Short Course, Paleontology* 8, 107–118.

Battarbee, R.W. 1984. Diatom analysis and the acidification of lakes. *Philosophical Transactions of the Royal Society of London* B305, 451–477.

Battarbee, R.W. & Charles, D.F. 1987. The use of diatom assemblages in lake sediments as a means of assessing the timing, trends and causes of lake acidification. *Progress in Physical Geography* 11, 552–580.

Bianchi, C. & Gersonde, R. 2002. The Southern Ocean surface between Marine Isotope Stages 6 and 5d: shape and timing of climate changes. *Palaeogeography, Palaeoecology, Palaeoclimatology* 187, 151–177.

Birks, C.J.A. & Koc, N. 2002. A high-resolution diatom record of late-Quaternary sea-surface temperatures and oceanographic conditions from the eastern Norwegian Sea. *Boreas* 31, 323–344.

Bolli, H.M., Saunders, J.B. & Perch-Nielsen, K. 1985. *Plankton Stratigraphy*. Cambridge University Press, Cambridge.

Cavalier-Smith, T. 1993. Kingdom Protozoa and its 18 phyla. *Microbiological Reviews* 57, 953–994.

Chapman, V.J. & Chapman, D.J. 1973. *The Algae*. Macmillan, London.

Clarke, J. 2003. The occurrence and significance of biogenic opal in the regolith. *Earth Science Reviews* 60, 175–194.

Conley, D.J. 2002. Terrestrial ecosystems and the global biogeochemical silica cycle. *Global Biogeochemical Cycles* 16, article no. 1121.

Conley, D.J., Zimba, P.V. & Theriot, E. 1994. Silica content of freshwater and marine benthic diatoms. In: Kociolek, J.P. (ed.) *Proceedings of the 11th International Diatom Symposium, San Francisco, 1990. Memoir. Californian Academy of Science* 17, 95–101.

De Wever, P., Azéma, J. & Fourcade, E. 1994. Radiolaires et radiolarites: production primaire, diagenése et paléogéographie. *Buletin. Centres des Recherches Exploration-Production ELF-Aquitaine* 18, 315–379.

Ek, A.S. & Renberg, I. 2001. Heavy metal pollution and lake acidity changes caused by one thousand years of copper mining at Falun, central Sweden. *Journal of Paleolimnology* 26, 89–107.

Funnell, B.M. & Riedel, W.R. (eds) 1971. *The Micropalaeontology of Oceans*. Cambridge University Press, Cambridge.

Gardner, J.V. & Burckle, L.H. 1975. Upper Pleistocene *Ethmodiscus rex* oozes from the eastern equatorial Atlantic. *Micropalaeontology* 21, 236–242.

Hartley, B. 1996 (ed.). *An Atlas of British Diatoms*. Biopress Ltd, Bristol.

Hendey, N.L. 1964. *An Introductory Account of the Smaller Algae of British Coastal Waters, Part v: Bacillariophyceae (Diatoms)*. Fisheries Investigations Series, IV. HMSO, London.

Hesse, R. 1989. Origin of chert diagenesis of biogenic siliceous sediments. *Geoscience Canada* 15, 171–192.

Hustedt, F. 1930. Bacillariophyta (Diatoms). In: Pascher A. (ed.) *Die Sussltwasserflora Mitteleuropas*. G. Fischer, Jena.

Hustedt, F. 1957. Die Diatomeenflora des Fluss-systems der Weser im Gebiet der Hansenstadt Bremen. *Abhandlungen herausgegeben vom naturwissen schaflichen Verein zu Bremen* 34, 181–440.

Jones, V.J., Stevenson, A.C. & Batterbee, R.W. 1989.

Acidification of lakes in Galloway, south-west Scotland: a diatom and pollen study of the post-glacial history of the Round Loch of Glenhead. *Journal of Ecology* 77, 1–23.

Joux-Arab, L., Berthet, B. & Robert, J.M. 2000. Do toxicity and accumulation of copper change during size reduction in the marine pennate diatom *Haslea ostrearia*? *Marine Biology* 136, 323–330.

van Landingham, S.L. 1967 to date. *Catalogue of the Fossil and Recent Genera and Species of Diatoms and their Synonyms. (A revision of F.W. Millsa, An index to the genera and species of the Diatomaceae and their synonyms.)* J. Cramer Verlag, Germany.

Leng, M., Barker, P., Greenwood, P. et al. 2001. Oxygen isotope analysis of diatom silica and authigenic calcite from Lake Pinarbasi, Turkey. *Journal of Palaeolimnology* 25, 343–349.

Lipps, J. (ed.). 1993 *Fossil Prokaryotes and Protists*. Blackwell Scientific Publications, Oxford.

Mackay, A.W., Flower, R.J., Kuzmina, A.E. et al. 1998. Diatom succession trends in recent sediments from Lake Baikal and their relation to atmospheric pollution and to climate change. *Philosophical Transactions of the Royal Society of London* B353, 1011–1055.

Marshall, J.D., Jones, R.T., Crowley, S.F. et al. 2002. A high resolution Late-Glacial isotopic record from Hawes Water, Northwest England Climatic oscillations: calibration and comparison of palaeotemperature proxies. *Palaeogeography, Palaeoecology, Palaeoclimatology* 185, 25–40.

Martin, R.E. 1995. Catastrophic fluctuations in nutrient levels as an agent of mass extinction: upward scaling of ecological processes? In: McKinney, M.L. & Drake, J.A. (eds) *Biodiversity Dynamics. Turnover and populations, taxa and communities*. Columbia University Press, New York, pp. 405–429.

Mikkelsen, N., Labeyris Jr, L. & Berger, W.H. 1978. Silica oxygen isotopes in diatoms, a 20000 yr record in deep sea sediments. *Nature* 271, 536–538.

Nelson, D.M., Tréguer, P., Brzezinski, M.A., Leynaert, A. & Quéguiner, B. 1995. Production and dissolution of biogenic silica in the ocean-revised global estimates, comparison with regional data and relationship to biogenic sedimentation. *Global Biogeochemical Cycles* 9, 359–372.

Retallack, G.J., Krull, E.S. & Bockheim, J.G. 2001. New grounds for reassessing palaeoclimate of the Sirius Group, Antarctica. *Journal of the Geological Society, London* 158, 923–935.

Rosenthal, Y., Dahan, M. and Shemesh, A. 2000. Southern Ocean contributions to glacial-interglacial changes of atmospheric P_{CO_2}: an assessment of carbon isotope records in diatoms. *Paleoceanography* 15, 65–75.

Round, F.E., Crawford, R.M. & Mann, D.G. 1990. *The Diatoms: biology and morphology of the genera*. Cambridge University Press, Cambridge.

Scagel, R.F.R.J., Bandoni, G.E., Rouse, W.E. et al. 1965. *An Evolutionary Survey of the Plant Kingdom*. Blackie, London.

Schieber, J., Krinsley, D. & Riciputi, L. 2000. Diagenetic origin of quartz silt in mudstones and implication for silica cycling. *Nature* 406, 981–985.

Schmidt, M., Botz, R., Stoffers, P., Anders, T. & Bohrmann, G. 1997. Oxygen isotopes in marine diatoms: a comparative study of analytical techniques and new results on the isotope composition of recent marine diatoms. *Geochimica et Cosmochimica Acta* 61, 2275–2280.

Schmidt, M., Botz, R., Rickert, D., Bohrmann, G., Hall, S.R. & Mann, S. 2001. Oxygen isotopes of marine diatoms and relations to opal-A maturation. *Geochimica et Cosmochimica Acta* 65, 201–211.

Setty, M.G.A.P. 1966. Preparation and method of study of fossil diatoms. *Micropalaeontology* 12, 511–514.

Shemesh, A., Charles, C.D. & Fairbanks, R.G. 1992. Oxygen isotopes in biogenic silica-global changes in ocean temperature and isotopic composition. *Science* 256, 1434–1436.

Shemesh, A., Hodell, D., Crosta, X. et al. 2002. Sequence of events during the last deglaciation in Southern Ocean sediments and Antarctic ice cores. *Palaeoceanography* 17, article no. 1056.

Shennan, I., Innes, J., Long, A.J. & Zong, Y. 1994. Late Devensian and Holocene sea-level at Loch nan Eala near Arisaig, northwest Scotland. *Journal of Quaternary Science* 9, 261–284.

Sieburth, J.M. 1975. *Microbial Seascapes. A pictorial essay on marine microorganisms and their environments*. University Park Press, Baltimore.

Simonsen, R. 1979. The diatom system: ideas on phylogeny. In: *Bacillaria*, vol. 2. J. Cramer, Brauschweig, pp. 9–71.

Stewart, P.M., Butcher, J.T. & Gerovac, P.J. 1999. Diatom (Bacillariophyta) community response to water quality and land use. *Natural Areas Journal* 19, 155–165.

Stoermer, E.F. & Smol, J.P. (eds) 1999. *The Diatoms: applications for the environmental and earth sciences*. Cambridge University Press, Cambridge.

Tappan, H. & Loeblich Jr, A.R. 1973. Evolution of the ocean plankton. *Earth Science Reviews* 9, 207–240.

van der Werff, A. & Huls, H. 1957–1963. *Diatomeeeriflora van Nederland* (in 7 parts). Published privately.

Whittington, G., Buckland, P., Edwards, K.J. et al. 2003. Multiproxy Devensian Late-glacial and Recent environmental records at an Atlantic coastal site in Shetland. *Journal of Quaternary Science* 18, 151–168.

Wilson, G.S., Barron, J.A., Ashworth, A.C. et al. 2002. The Mount Feather Diamicton of the Sirius Group: an accumulation of indicators of Neogene Antarctic glacial and climatic history. *Palaeogeography, Palaeoecology, Palaeoclimatology* 182, 117–131.

CHAPTER 18

Silicoflagellates and chrysophytes

Silicoflagellates have been traditionally referred to the Chrysophyceae, the golden algae, due to the colour imparted by their photosynthetic pigments (chlorophylls *a* and *c*, β-carotene, fucoxanthin and carotenoids) and the Phytomastigophora. However, Cavalier-Smith (1993) placed them in the kingdom Chromista based on their chloroplast structure and 18sRNA phylogenetics. He considered them a separate class based on their silica skeleton. Silicoflagellates have been minor components of marine phytoplankton since Early Cretaceous times. They are only well preserved in siliceous rocks such as diatomites and have been little used except in deep oceanic strata where they are now widely employed both for correlation and for estimation of palaeoclimatic conditions.

The living silicoflagellate

The unicellular organism is usually from 20 to 100 μm in diameter and contains golden-brown photosynthetic pigments, a nucleus, several **pseudopodia** and a single **flagellum** at the anterior end of the cell (Fig. 18.1a). The cytoplasm is supported internally by a skeleton of hollow rods, composed of opaline silica. Reproduction appears to be predominantly asexual, beginning with the secretion of a daughter skeleton and followed by simple cell division. Silicoflagellates are **photoautotrophic**; that is, they feed both by photosynthesis and the function of the tentacles is unknown (Moestrup, in Sandgren *et al.* 1995, pp. 75–93). They are restricted to the shallow photic zone of the ocean (0–300 m), thriving in the silica-enriched waters associated with current upwelling in equatorial and high-latitude waters along the western margins of continents. In these cooler waters they may also bloom seasonally. For these reasons the silicoflagellates are commonest as fossils in biogenic silica deposits formed during cool periods with marked seasonality or with strong upwelling. They are unknown in freshwater habitats.

The silicoflagellate skeleton

The basic skeleton is built upon a **basal ring** which may be elliptical, circular or pentagonal with from two to seven **spines** on the corners of the outer margins (Fig. 18.1b). This basal ring is usually traversed by one or, rarely, more **apical bars** arched upward in an apical direction and connected to the basal ring by shorter lateral bars. In some genera these features are elaborated into a complex hemispherical lattice (Fig. 18.1). The skeleton is thought to function as a mechanism for improvement of buoyancy, spreading out the cytoplasm and increasing resistance to downward sinking. To reduce weight further, the skeleton is also hollow. In life, the domed apical portion is orientated upwards towards the light.

Classification

Botanists generally consider silicoflagellates members of the Chrysophyceae whilst protozoologists consider them to be an order of the Phytomastigophora. Palaeontologists emphasize skeletal morphology as the primary taxonomic criterion and many form species have been described. Biologists prefer a broad species concept. Six basic morphological groups (placed in the

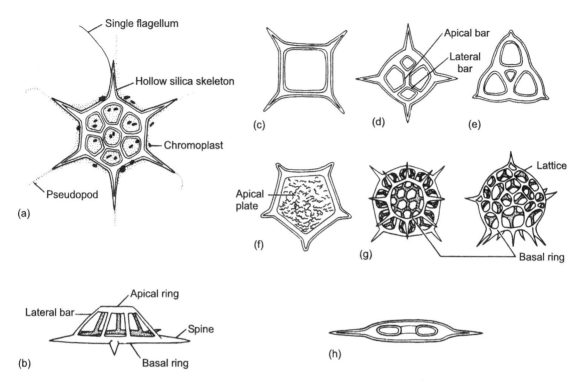

Fig. 18.1 Silicoflagellates. (a) Living cell and skeleton of *Distephanus* ×267. (b) Side view of *Distephanus* skeleton ×267. (c) *Mesocena* ×533. (d) *Dictyocha* ×400. (e) *Corbisema* ×533. (f) *Vallacerta* ×446. (g) *Cannopilus* ×500. (h) *Naviculopsis* ×375. ((a) Modified from Marshall 1934.)

family Dictyochaceae) are recognized in the Cenozoic with an additional four groups in the Cretaceous.

Corbisema group

This group has a basal ring with three sides and three **struts**. In *Corbisema* (Cret.-Rec., Fig. 18.1e) the symmetry is trigonal. Most species are roughly equilateral but some have one side shorter that the others which gives the impression of bilateral symmetry. Rare two- and four-sided variants have been reported. This group dominates the Cretaceous but declines in importance from the Oligocene.

Dictyocha group

Silicoflagellates with a four-sided basal ring bearing a spine at each corner. The struts generally meet to form an **apical bridge** but this is absent in *D. medusa* and forms of Neogene age. In *Dictyocha* (Cret.-Rec., Fig. 18.1d) the quadrate basal ring has corner spines and a diagonal **apical bar** with bifid ends (**lateral bars**). This group is common throughout the Cenozoic.

Distephanus group

This group contains species with three to eight sides, the apical and basal rings are typically the same size.

Cannopilus group

This group contains multiwindowed forms that resemble Radiolaria in appearance. *Cannopilus* (Olig.-Rec., Fig. 18.1g) resembles a radiolarian but has a hemispherical lattice with spines both on the basal ring and on the lattice.

Bachmannocena group

Three- to many-sided forms that only develop a basal ring comprise this group (Fig. 18.2a). Some workers consider these ecophenotypic variants of the *Corbisema* and *Dictyocha* groups.

Naviculopsis group

These are elongate forms, lacking struts and with major axis spines. *Naviculopsis* (Palaeoc.-Mioc., Fig. 18.1h) has a long and narrow ring with an arched cross bar and a spine at each end. An apical bridge can vary in width and appears to be a useful taxonomic discriminator and is biostratigraphically significant. Members of this group are locally abundant from the Eocene but declined through the Oligocene to become extinct in the Miocene.

Vallacerta group

A unique group of silicoflagellates with basal rings that have apical domes, lacking windows and portals. *Vallacerta* (Cret., Fig. 18.1f) has a pentagonal basal ring with corner spines and a convex, sculptured disc (**apical plate**) of silica. This group is known from the Cretaceous.

Lyramula Group

This group includes Y-shaped forms and some workers question a silicoflagellate affinity. The skeleton typically has two relatively long limbs and rarely a shorter third limb (Fig. 18.2b). This group can be extremely common in siliceous sediments of Late Cretaceous age.

Cornua group

This group is characterized by three radiating skeletal elements that bifurcate distally (Fig. 18.2c). Some consider these are aberrant corbisemids. Others have suggested *Cornua* is an evolutionary intermediate between *Corbisema* and the more primitive *Variramus*. *Cornua* is only found in shallow-water sediments.

Variramus Group

Silicoflagellates with branching skeletons and lacking basal rings are included in this group (Fig. 18.2d). The skeleton incorporates spines and spikes and is extremely variable in morphology.

Geological history of silicoflagellates

Lyramula and *Vallacerta* are the earliest silicoflagellates, found in the lower Cretaceous of the southern hemisphere high latitudes. *Corbisema* and *Dictyocha* survived the Mesozoic and are ancestral to the Cenozoic lineages. *Distephanus* (Eoc.-Rec.), *Dictyocha* (U. Cret.-Rec.) and *Octactis* (Pleisto.-Rec.) are the only living genera. Silicoflagellates have been both numerous and diverse during periods of climatic cooling, i.e. the Late Cretaceous, Late Eocene, Miocene and

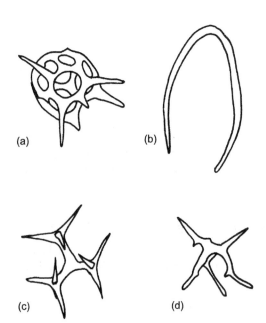

Fig. 18.2 Diagrammatic representation of silicoflagellate genera from the Cretaceous and Cenozoic, all about ×500. (a) *Bachmannocea* (L. Plio.). (b) *Lyramula* (U. Cret.). (c) *Cornua* (Cret.). (d) *Variramus* (L. Cret.). (Redrawn from illustrations in Lipps & McCartney, in Lipps 1993.)

Chapter 18: Silicoflagellates and chrysophytes

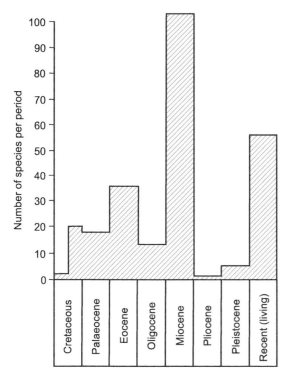

Fig. 18.3 Species diversity of described silicoflagellates through time. (Based on Tappan & Loeblich 1973.)

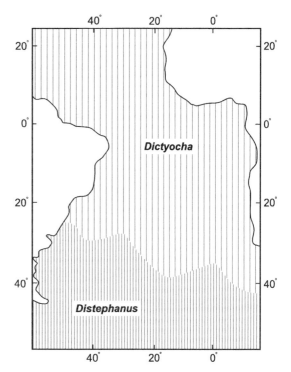

Fig. 18.4 Distribution of Recent *Dictyocha* and *Distephanus* in the South Atlantic waters. (Based on Lipps 1970.)

Quaternary (Fig. 18.3). At these times oceanic current circulation is thought to have been more rapid leading to more vigorous upwelling of mineral-rich waters and blooms of siliceous phytoplankton (see Lipps, 1970; McCartney, in Lipps 1993, pp. 143–155).

Applications of silicoflagellates

Silicoflagellate biostratigraphy is particularly well developed for tropical and subtropical settings and has been reviewed in SEPM Special Publication 32 (1981) and Perch-Nielsen (in Bolli *et al.* 1985, pp. 811–847). Because silicoflagellates have evolved slowly, silicoflagellate biozones are comparatively few and long ranging, with reliance on the relative abundance of species (or assemblages of species). The high degree of ecophenotypic variation means biozonations are only of local use and at higher latitudes where other groups are rare. Emphasis has also been placed on their value as palaeoclimatic indicators, especially from ratios of the warm-water *Dictyocha* to the cool-water *Distephanus* in sediments (Fig. 18.4). Warm- and coolwater species of *Dictyocha* have been used by Cornell (1974) to indicate fluctuations in the Miocene climate of California. Aspects of palaeoclimatology are reviewed by Louse (pp. 407–421) and Muhina (pp. 423–431) in Funnell & Riedel (1971). Although only a minor fraction of biogenic silicates, their role in sedimentology has also been outlined in the above volume by Kozlova (in Funnell & Riedel 1977, pp. 271–275).

Chrysophyte cysts

Chrysophyte cysts are commonly abundant in marine, freshwater and damp terrestrial habitats. The siliceous cyst encloses a granular cytoplasm. For the most part they are unicellular, non-marine and phytoplanktonic, others can be colonial with a **coccoid** or filamentous

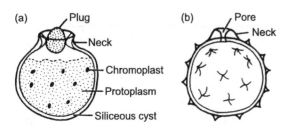

Fig. 18.5 Chrysophyte cysts. (a) Recent statocyst ×2670. (b) Fossil *Archaeomonas* ×3330.

habit. The living *Mallomonas* and *Synura* form benthic resting cysts called **stomatocycsts**, especially after reproduction (Duff *et al.* 1995).

Stomatocysts are from 3 to 25 μm across, usually spherical and can be distinguished by the presence of a single pore through which the germinating cell emerges (Fig. 18.5a). The pore may be surrounded by an elevated collar (Zeeb & Smol 1993). The outer surface sculpture, form of collar and pore and the overall shape may be used to distinguish cyst genera and species (e.g. fossil *Archaeomonas*, Fig. 18.5b).

Fossil chrysophyte cysts are mainly known from Late Cretaceous and younger diatomites, shales and silts (Cornell 1970; Tynan 1971), but similar structures are reported from the Late Precambrian Beck Spring Chert, about 1300 Myr (Cloud 1976). It appears that some of the more distinctive species may have potential as guide fossils in deep sea strata (Gombos 1977).

Further reading

Tappan (1980), McCartney (in Lipps 1993, pp. 143–155) and Sandgren *et al.* (1995) provide useful reviews of biology, ecology and evolution. Perch-Nielsen (in Bolli *et al.* 1985, pp. 811–847) summarizes the biostraigraphical utility and illustrates many species.

Hints for collection and study

Silicoflagellates and chrysomonads are most readily obtained from marine diatomites and prepared and studied in the same way as diatoms. Disaggregated residues in water can be smeared on a glass slide and covered with a cover slip for viewing in transmitted light. For more permanent mounts allow the residue to dry on the slide, add a drop of Caedax or Canada Balsam to the cover slip and place over the residue. Allow to dry before examining with transmitted light.

REFERENCES

Bolli, H.M., Saunders, J.B. & Perch-Nielsen, K. 1985. *Plankton Stratigraphy*. Cambridge University Press, Cambridge.

Cavalier-Smith, T. 1993. Kingdom Protozoa and its 18 phyla. *Microbiological Reviews* 57, 953–994.

Cloud, P. 1976. Beginnings of biospheric evolution and their biochemical consequences. *Paleobiology* 2, 351–387.

Cornell, W.C. 1970. The chrysomonad cyst-families Chrysostomataceae and Archaeomonadaceae: their status in paleontology. *Proceedings North American Paleontological Convention 1969* Part a, 958–994.

Cornell, W.C. 1974. Silicoflagellates as paleoenvironmental indicators in the Modelo Formation. *Journal of Paleontology* 48, 1018–1029.

Duff, K.E., Zeeb, B.A. & Smol, J.P. 1995. Atlas of chrysophyte cysts. *Developments in Hydrobiology* 99, 1–189.

Funnell, B.M. & Riedel, W.R. (eds) 1971. *The Micropalaeontology of Oceans*. Cambridge University Press, Cambridge.

Gombos Jr, A.M. 1977. Archaeomonads as Eocene and Oligocene guide fossils in marine sediments. *Initial Reports of the Deep Sea Drilling Project* 36, 689–695.

Lipps, J.H. 1970. Ecology and evolution of silicoflagellates. *Proceedings. North American Paleontological Convention 1969*, Part G, 965–993.

Lipps, J.H. (ed.) 1993. *Fossil Prokaryotes and Protists*. Blackwell Scientific Publications, Oxford.

Marshall, S.M. 1934. The Silicoflagellata and Titininoinea. *British Museum (Natural History) Great Barrier Reef Expedition 1928-1-29, Scientific Reports* 4, 623–624.

Sandgren, C.D., Smol, J.P. & Kristiansen, J. 1995. *Chrysophyte Algae: ecology, phylogeny and development*. Cambridge University Press, Cambridge.

Tappan, H. 1980. *The Palaeobiology of Plant Protists*. W.H. Freeman, San Fransisco.

Tappan, H. & Loeblich Jr, A.R. 1973. Evolution of the ocean plankton. *Earth Science Reviews* 9, 207–240.

Tynan, E.J. 1971. Geologic occurrence of the archaeomonads. *Proceedings. 2nd International Plankton Conference: Rome 1970*. Edizioni Tecnoscienza, Rome, pp. 1225–1230.

Zeeb, B.A. & Smol, J.P. 1993. Chrysophycean cyst record from Elk Lake, Minnesota. *Canadian Journal of Botany* 71, 737–756.

CHAPTER 19

Ciliophora: tintinnids and calpionellids

The Ciliophora contains protozoa that are covered within an outer layer, the **pellicle**, which bears rows of tiny **cilia**. These serve both for locomotion and food gathering by beating together in waves. Internally cells have a large, irregular-shaped macronucleus for normal cell functions and a tiny micronucleus for reproductive purposes. A distinct cell **mouth** and a **buccal cavity** are further distinctive features of these active protists (Fig. 19.1).

Only the suborder Tintinnina are of much geological interest. These comprise about 14% of the 7200 known ciliate species and form an important component of the microzooplankton. Unfortunately, however, few of these become fossilized and the fossil record (L. Ord.-Rec.) is very patchy. Calcareous forms, known as **calpionellids**, and **pseudacellids**, may be related to the tintinnids and are sufficiently abundant in Mesozoic pelagic limestone facies to be useful in biostratigraphy.

The living tintinnid

The tintinnid cell is generally tubular, conical or cup-shaped with the posterior end drawn out into a stalk or **peduncle** for attachment to the external shell, called a **lorica** (Fig. 19.1). The anterior end of the cell is broader and fringed by a feathery crown of tentacle-like **membranelles**, which are actually bundles of fused cilia. Beneath these membranelles occur the buccal cavity and the cell mouth, whilst the remainder of the pellicle is traversed by spiralling rows of cilia.

The cell is able to rotate freely within the lorica and the attachment by the peduncle may only be temporary. A crown of beating membranelles project from the **aperture** of the lorica and serve to propel the whole structure backwards with a spiral motion. Captured food such as bacteria, green algae, coccolithophores, dinoflagellates and diatoms are passed by the cilia to the mouth for ingestion, leading to internal digestion within **food vacuoles**. Chlorophyll in the diet imparts a green colour to the cell.

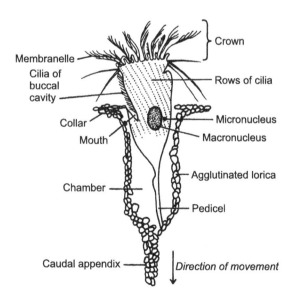

Fig. 19.1 Recent *Tintinnopsis*, about ×400. (Modified from Colom 1948, after Fremiet.)

The lorica

The lorica may be from 10 to 1000 μm long although most are 120–200 μm in length. In outline, the loricae vary from globular, through conical, cup- and

215

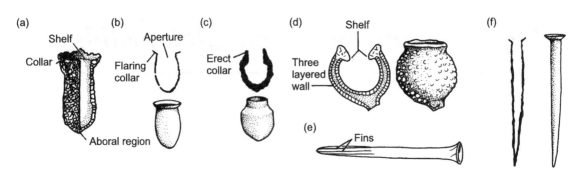

Fig. 19.2 Tintinnid and calpionellid loricae. (a) *Tintinnopsis*, external and longitudinal section ×133. (b) *Tintinnopsella*, longitudinal section and external views ×133. (c) *Calpionella* longitudinal section and reconstruction, ×333. (d) *Tytthocorys* longitudinal section and external view, ×150. (e) *Salpingella*, ×133. (f) *Salpingellina* longitudinal section and reconstruction, ×166. ((a) After Remane, in Brönnimann & Renz 1969, figure 2.14; (b), (c), (f) partly after Colom 1948; (d) after Tappan & Loeblich 1968; (e) after Kofoid & Campbell 1939 ((b), (c), (d), (f) with the permission of The Paleontological Society; (e) with the permission of the Museum of Comparative Zoology, Harvard).)

bottle-shaped to bullet- or nail-shaped. All have an aperture at the **oral** end and most have a closed **aboral** region of rounded or pointed form at the opposite end (Fig. 19.2). Within is a very spacious, single **chamber** that can be up to 10 times as voluminous as the cell itself.

Sculpture takes the form of spines, costae, fins, transverse grooves, longitudinal grooves or spiral grooves, reticulate patterns or fenestrate (i.e. window-like) structures. Minute cavities within the wall, called **alveoli**, are filled with low-gravity fluids that no doubt help to keep the lorica more buoyant. The relatively large surface area and the development of **collars** (Figs 19.1, 19.2a) may help to retard sinking in some forms, but streamlining for efficient motility and protection from UV radiation appears to be the major function of the tintinnid lorica.

The tintinnid wall is a delicate organic structure of chitin or xanthoprotein but it may be strengthened by the agglutination of tiny particles of quartz, coccoliths and diatom frustules. At certain times in the past when oceans were supersaturated in calcium carbonate, tropical offshoots of primitive tintinnid stocks developed a calcareous lorica. Calcareous walls are unknown in living tintinnids. By contrast, fossil calpionellids of the Jurassic and Cretaceous, most Palaeozoic representatives and the pseudocellids of the Tertiary have a primary calcite lorica. Agglutinated calpionellids are known from the Mesozoic.

Distribution and ecology of tintinnids

Tintinnids feed on the nannoplankton, and provide an important trophic link to the larger zooplankton and fish. Tintinnina occur in the photic zone in all seas and oceans, but are rarely abundant except in the Antarctic where they are exceeded in number only by the diatoms on which they feed (Wasik 1998). Sensitivity to temperature and salinity gives rise to Recent assemblages typical of subtropical and tropical, boreal and Austral-Asian seas. Neritic species occur in brackish water. Freshwater forms comprise only 10 of the 840 described species and these are mostly isolated relict populations left by retreating seas of the Tertiary, such as in the Caspian Sea and Lake Baikal. Cold-water assemblages are low in diversity and high in abundance. In tropical waters diversity is greater but cells are smaller and are less numerous. Tintinnid 'blooms' in the modern ocean appear to be strongly seasonally influenced.

Only a few organic loricae have been reported as fossils. The agglutinated loricae of fossil tintinnids may be found in neritic limestones and glauconitic clays or in estuarine and lacustrine deposits. Fossil calpionellids (L. Tithonian-E. Valanginian) are more abundant and occur in pelagic fine-grained limestones deposited in the subtropical, Mesozoic Tethys Ocean, occurring in high abundance along with coccoliths (including *Nannoconus*), planktonic foraminifera and radiolarians. Calpionellids have also been reported from DSDP and ODP sites in the North Atlantic as far north as the Scotia Shelf and Grand Banks. None has been described from the Boreal realm. The fragility of the lorica in both groups means they rarely survive into the fossil record and require careful preparation if they are to be extracted from rocks.

Classification

The phylogenetic relationship between the tintinnids and the calpionellids is uncertain as mineralized tests are completely unknown among ciliates in general, and in living tintinnids in particular. Remane (in Brönnimann & Renz 1969, vol. 2, pp. 574–587) amongst others argued that the calpionellids may not even have been tintinnids or ciliates at all.

The classification of the tintinnids has varied little in 60 years. They have been classified within the class Ciliata, order Spirotrichida. Cavalier-Smith (1993), however, elevated this order to a class, the Spirotrichea within the phylum Ciliophora. Tappan (in Lipps 1993, pp. 285–303) provides a review to family level.

Order TINTINNINA

Taxonomy is based mainly on the shape, composition, wall structure and sculpture of the lorica. Caution is necessary, though, because the size, shape and composition of the lorica can vary with ecological conditions. In addition, the living species *Favella ehrenbergi* develops three morphologically distinct loricae, previously considered separate species.

As most fossil assemblages occur in limestones, they are usually studied from randomly orientated thin sections, a process requiring considerable practice. *Tintinnopsis* (Rec.; Figs 19.1, 19.2a) has an organic and agglutinated wall with a slightly constricted aperture surrounded by a flaring collar.

The superfamily Calpionelloidea (class '*Incertae sedis*') includes two families of calpionellids, the Colomiellidae (Aptian-Albian) and the Calpionellidae (Tithonian-Hauterivian). The fossil calpionellid *Tintinnopsella* (U. Jur.-Cret., Fig. 19.2b) is very similar to *Tintinnopsis* but has a calcareous wall of radially arranged fibres. In *Calpionella* (U. Jur.-L. Cret., Fig. 19.2c) the wall is calcareous and the collar is short and erect. *Tytthocorys* (U. Eoc., Fig. 19.2d) has a three-layered calcareous lorica, the aperture constricted by a shelf just below the short, flaring collar. *Salpingella* (Rec., Fig. 19.2e) has a wholly organic, nail-shaped lorica with a flaring collar and longitudinal fins. The superficially similar calpionellid *Salpingellina* (Fig. 19.2f) occurs in lower Cretaceous rocks and has a calcareous lorica.

Geological history of tintinnids

The fossil record provides a very patchy view of tintinnid history. They are rare in both Palaeozoic and Tertiary rocks and have not yet been reported from Cambrian, Carboniferous, Permian, upper Cretaceous, Palaeocene, Miocene and Pliocene sediments (Tappan & Loeblich 1968). Even the rare Pleistocene records do little justice to the number of living species, indicating the group has a poor preservation potential. In the Late Jurassic and Early Cretaceous, the more readily preserved calpionellids bloomed from Mexico to the Caucasus in the Tethys Ocean, building deep sea limestones together with the coccoliths. Their dramatic decline in the Late Cretaceous and Eocene may be related to global cooling or to vigorous competition from the thriving planktonic foraminifera, radiolarians and dinoflagellates.

The Late Cretaceous and Cenozoic record of the tintinnids is sparse, though well-preserved calcareous loricae are known from the Eocene and rare lower Oligocene. These pseudarcellids are associated with diverse foraminifera, invertebrates and detrital sediment indicating neritic facies.

Applications

Calpionellids can be used to correlate those Tethyan limestones in which they abound (e.g. Remane, in

Brönnimann & Renz 1969, vol. 2, pp. 559–573). The poor fossil record limits the palaeoecological and biogeographical usefulness of tintinnids and calpionellids. Wide geographic range and rapid evolution make the calpionellids a useful biostraigraphic tool in the Upper Tithonian and Berriasian. Both groups have potential in defining ocean water masses and current patterns (e.g. Dolan 2000). Echols & Fowler (1973) have investigated the potential of brackish-water species for palaeosalinity and shoreline studies in Pleistocene sediments of the North Pacific. Benest (1981) examined salinity controls on Tithonian species from Algeria. Some stratigraphically useful calpionellids are widely distributed and hyaline forms have been used to erect a standard biozonation for the Jurassic and Cretaceous of the Tethyan Realm (Allemann et al. 1971; Remane, in Bolli et al. 1985, pp. 555–573).

Further reading

An introduction to the ecology, classification and geological applications of tintinnids and calpionellids can be found in chapters by Remane (in Haq & Boersma 1978, pp. 161–170) and Tappan (in Lipps 1993, pp. 285–303). Stratigraphically useful species are described and illustrated by Remane (in Bolli et al. 1985, pp. 555–573).

Hints for collection and study

Calpionellids are best studied in thin sections or peels of Mesozoic Tethyan limestones of deep-water origin (see method N in Appendix). The morphology of the species must then be reconstructed from the various unorientated cross-sections, an axial (longitudinal) section being the most helpful for identification.

REFERENCES

Allemann, F., Catalano, R., Farès, F. & Remane, J. 1971. Standard calpionellid zonation (Upper Tithonian-Valanginian) of the western Mediterranean province. *Proceedings. 2nd International Plankton Conference, Roma 1970*, 2, 1337–1342.

Benest, M. 1981. Calpionellid facies interbedded in rhythmic platform deposits showing a deficient salinity – example of carbonate upper Tithonian in Chellala Mountains (Tellian Foreland, West Algeria). *Comptes Rendus des Seances de l'Academie des Sciences Serie II – Mechanique Physique Chimie Sciences de L'Univers Sciences de la Terre* 292, 1287–1290.

Bolli, H.M., Saunders, J.B. & Perch-Nielsen, K. (eds) 1985. *Plankton Stratigraphy*. Cambridge University Press, Cambridge.

Brönnimann, P. & Renz, H.H. (eds) 1969, *Proceedings. First International Conference on Planktonic Microfossils, Geneva 1967*, vol. 1, 422pp.; vol. 2, 745pp. E.J. Brill, Leiden.

Cavalier-Smith, T. 1993. Kingdom Protozoa and its 18 phyla. *Microbiological Reviews* 57, 953–994.

Colom, G. 1948. Fossil tintinnids – 1 loricated infusoria of the order Oligotricha. *Journal of Paleontology* 22, 233–263.

Dolan, J.R. 2000. Tintinnid ciliate diversity in the Mediterranean Sea: longitudinal patterns related to water column structure in late spring–early summer. *Aquatic Microbiology Ecology* 22, 69–78.

Echols, R.J. & Fowler, G.A. 1973. Agglutinated tintinnid loricae from some Recent and Late Pleistocene shelf sediments. *Micropalaeontology* 19, 431–443.

Haq, B.L. & Boersma, A. 1978. *Introduction to Marine Micropalaeontology*. Elsevier, New York.

Kofoid, C.A. & Campbell, A.S. 1939. The Tintinnoinea. *Bulletin. Museum of Comparative Zoology Hart* 84, 1–473.

Lipps, J.H. (ed.) 1993. *Fossil Prokaryotes and Protists*. Blackwell Scientific, Oxford.

Tappan, H. & Loeblich Jr, A.R. 1968. Lorica composition of modern and fossil Tintinnida (ciliate Protozoa), systematics, geologic distribution and some new Tertiary taxa. *Journal of Paleontology* 42, 1378–1394.

Wasik, A. 1998. Antarctic tintinnids: their ecology, morphology, ultrastructure and polymorphism. *Acta Protozoology* 37, 5–15.

CHAPTER 20

Ostracods

Ostracods are one of the most diverse groups of living crustaceans, they are the most abundant of fossil arthropods and are represented by some 33,000 living and fossil species (Cohen *et al.* 1998). Ostracods are small, bivalved Crustacea, with two chitinous or calcareous valves that hinge above the dorsal region of the body form their carapace. Ostracods were originally marine and probably benthic, but by the Silurian had expanded into reduced salinity and pelagic environments (Siveter, in Bassett & Lawson 1984, pp. 71–85). Some ostracods are adapted to a semi-terrestrial life living in damp soil and leaf litter. The class is subdivided into two subclasses, the weakly calcified, marine Myodocopa and the Podocopa. The podocopans include a high diversity of ecologically widespread forms and have the better fossil record. The vast majority of living forms are podocopans.

Ostracods are widely used in biostratigraphy, in determining palaeoenvironments and plaeoclimates and are indispensable as indicators of ancient shorelines and plate distributions. Ostracods have a long and well-documented fossil record from the Ordovician to the present day; the affinities of putative Cambrian forms are hotly disputed.

Soft body structure

The soft parts are rarely preserved in fossils, although there are some spectacular exceptions (e.g. Bate 1972; Smith 2000). As in other arthropods the soft parts are covered by a rigid, jointed exoskeleton of **chitin**. The head is large and bears a centrally placed mouth and a dorsal, usually single, eye. The anus is at the posterior end of the body. The head and the thorax are fused to form a **cephalothorax** and it is difficult to homologize the segments and appendages with other crustaceans (Fig. 20.1a). On either side of the head/thorax junction arise large, flap-like outgrowths, the **duplicatures**, which totally enclose the rest of the animal and form the **carapace** (Fig. 20.1b).

Ostracods have commonly seven, but up to eight pairs of jointed limbs in the adult stage, borne on the ventral side of the body (Fig. 20.1c). In addition they have a **furca** (a pair of **caudal rami**) near the posterior end of the body, not generally considered as limbs, which may be homologous with the telson in other arthropods. As in other Crustacea, the limbs are basically **biramous**, comprising two distinct branches: an outer **exopodite** and an inner **endopodite**. In many instances, however, the exopodite has become reduced or lost during evolution, resulting in a uniramous limb. These ostracod appendages bear fine chitinous bristles called **setae** (which usually arise from just below the joints) and terminate in claws.

In the Myodocopa, five pairs of limbs arise from the head, whilst the Podocopa may only have four pairs. The first and second, the **antennula** (which is uniramous) and the **antenna** (biramous), are long, tapering, limbs attached to the forehead and are employed variously in walking, swimming and feeding. The upper lip (or **labrum**) forms the front and the **hypostome** the back of the mouth. A pair of biramous **mandibula** and **maxillula** are attached to the hypostome and aid mastication of the food. In the Podocopa, but not the Myodocopa, the exopodites of mandibula and first maxillula are modified into a large **brancial plate**, which stirs up the water to provide feeding currents, improve water circulation around the animal or to take up oxygen. Oxygen uptake occurs over the entire

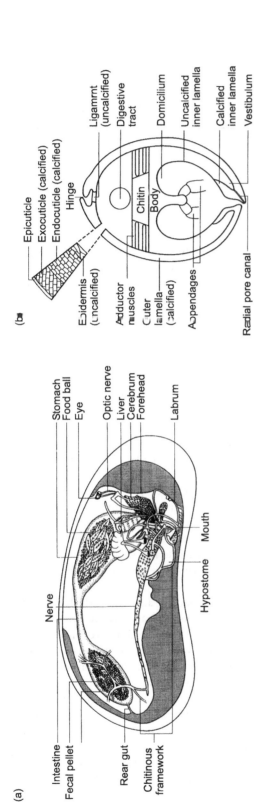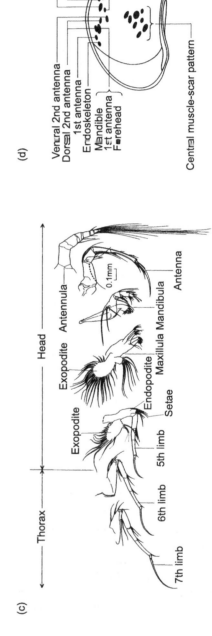

Fig. 20.1 (a) Soft-part anatomy of an ostracod. (b) Diagrammatic transverse section through an ostracod. (c) Appendage morphology of *Bairdia* (order Podocopida). (d) Details of muscle scars. (Figures redrawn from Kaesler in Boardman *et al.* 1987, figures 13.1, 13.2, 13.3, with permission from Blackwell Scientific Publications.) (Terminology from Horne *et al.* (in Holmes & Chivas 2002, pp. 5–37).)

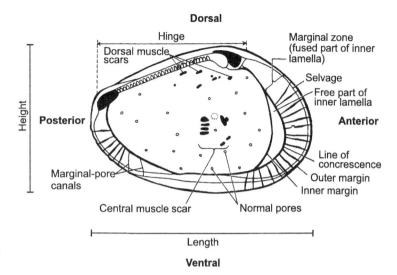

Fig. 20.2 The internal features of a podocopid left valve. (Modified from van Morkhoven 1962–1963.)

body surface. The fifth to seventh limbs are essentially similar and mostly take the form of walking legs in which the endopodite has a well-developed claw and the exopodite is reduced. In some ostracods the fifth limb is variously used for walking, feeding, in aiding respiration or clasping in sexually dimorphic taxa, or a combination of these. An eighth limb is present only in the rare Puncioidea (living Palaeocopida). Paired male copulatory appendages are situated in front of the furca and in some taxa include a pair of sperm pumps called **Zenker's organs**.

The respiratory and circulatory systems are greatly reduced. Large blood vessels and heart are also lacking in all except the relatively large and planktonic Myodocopida. A pair of lateral eyes rather than the single dorsal eye typical of many benthic forms can further distinguish some of this group. Many deeper water genera are blind. Muscles that operate the appendages are attached to the chitinous endoskeleton or the central or dorsal part of the carapace where they form the **dorsal muscle-scar** pattern. The **adductor muscles** (Fig. 20.1b) close the valves and form the **central muscle-scar** pattern on the valves (Figs 20.1d, 20.2), their position is marked on the outside of the valve by a **subcentral tubercle** or an infold of the valve known as the **sulcus**. The number and arrangement of muscle scars is diagnostic for many of the higher ostracod taxa (Box 20.1), but they are rarely seen in Archaeocopida and Palaeocopida. In the latter, the position of the muscles is marked by a prominent **median sulcus**, running from dorsum to venter.

The ostracod carapace

The ostracod carapace is usually ovate, kidney-shaped or bean-shaped with a hinge along the dorsal margin. Most adult carapaces measure only 0.5–3 mm long though some species can reach up to 30 mm long. The bivalved carapace is secreted by the epidermis and forms a continuous sheet covering the whole body and limbs. The carapace is formed by two lateral folds of **epidermis**, the **duplicatures**, which originate in the head region and extend forwards, backwards and downwards to enclose the body and limbs. The duplicatures have an **outer** and **inner lamella**. The inner lamella that may either be fused or free from the outer lamella (Fig. 20.2) and can have calcified and uncalcified parts (Fig. 20.3). In the latter case a space between these lamellae, a **vestibulum**, is an extension of the body, which in some taxa can house digestive and reproductive organs (Fig. 20.1b). Ridges on the duplicature called **selvages** may aid closure of the valves along the ventral margin. The innermost line of

222 Part 4: Inorganic-walled microfossils

> **Box 20.1** Ostracod classification based upon Whatley *et al.* (in Benton 1993, pp. 343–357) and Horne *et al.* (in Holmes & Chivas 2002, pp. 5–37). Figures show the general appearance (oriented anterior to the left) and their characteristic muscle scars. (Not to scale and based upon illustrations in Horne *et al.* (in Holmes & Chivas 2002 (with permission of the American Geophysical Union) and from the *Treatise on Invertebrate Paleontology* (courtesy of and © 1961, Part Q, The Geological Society of America and The University of Kansas).)

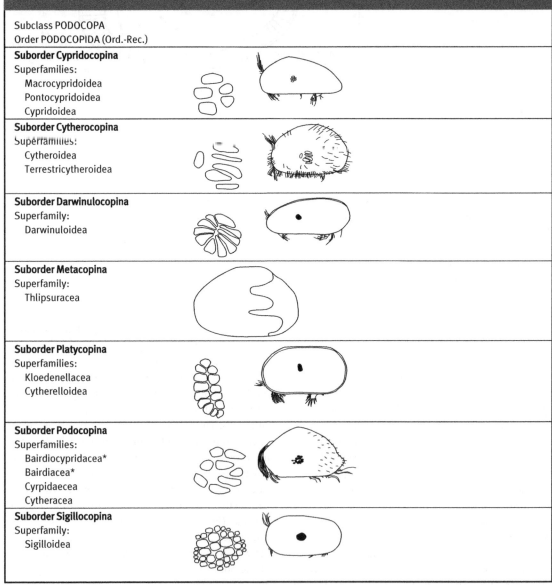

Subclass PODOCOPA
Order PODOCOPIDA (Ord.-Rec.)

Suborder Cypridocopina
Superfamilies:
  Macrocypridoidea
  Pontocypridoidea
  Cypridoidea

Suborder Cytherocopina
Superfamilies:
  Cytheroidea
  Terrestricytheroidea

Suborder Darwinulocopina
Superfamily:
  Darwinuloidea

Suborder Metacopina
Superfamily:
  Thlipsuracea

Suborder Platycopina
Superfamilies:
  Kloedenellacea
  Cytherelloidea

Suborder Podocopina
Superfamilies:
  Bairdiocypridacea*
  Bairdiacea*
  Cyrpidaecea
  Cytheracea

Suborder Sigillocopina
Superfamily:
  Sigilloidea

Chapter 20: Ostracods 223

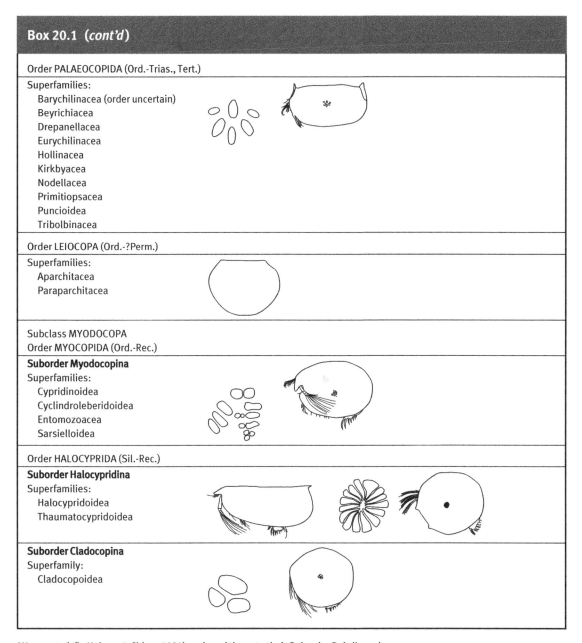

Box 20.1 (cont'd)
Order PALAEOCOPIDA (Ord.-Trias., Tert.)

Superfamilies:
   Barychilinacea (order uncertain)
   Beyrichiacea
   Drepanellacea
   Eurychilinacea
   Hollinacea
   Kirkbyacea
   Nodellacea
   Primitiopsacea
   Puncioidea
   Tribolbinacea

Order LEIOCOPA (Ord.-?Perm.)

Superfamilies:
   Aparchitacea
   Paraparchitacea

Subclass MYODOCOPA Order MYOCOPIDA (Ord.-Rec.)

Suborder Myodocopina
Superfamilies:
   Cypridinoidea
   Cyclindroleberidoidea
   Entomozoacea
   Sarsielloidea

Order HALOCYPRIDA (Sil.-Rec.)

Suborder Halocypridina
Superfamilies:
   Halocypridoidea
   Thaumatocypridoidea

Suborder Cladocopina Superfamily: Cladocopoidea

*Horne et al. (in Holmes & Chivas 2002) assigned these to their Suborder Bairdiocopina.

contact between the fused lamellae is called the **line of concrescence**, and the area between this and the outer margin is known as the **marginal zone** (Fig. 20.2).

In the Podocopa the valves are formed by the secretion of calcite by the epidermis of the outer lamella and the peripheral part of the inner lamella. In the living animal the calcareous valves are thus enclosed by a chitinous **epicuticle**. A different terminology has been used for myodocopan, in which the likely homologue of the calcified inner lamella is the **infold**, so the

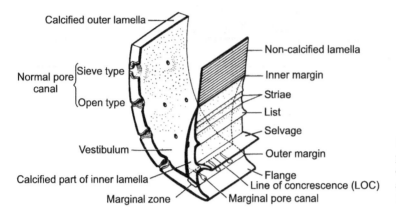

Fig. 20.3 Diagramatic section of the peripheral part of the podocopid ostracod valve, with the outer lamella and duplicature. (Modified from Kesling 1951.)

carapace comprises the outer lamella and the infold, the uncalcified inner lamella and body cuticle are the **vestment**. The valves of many myodocopans are not or weakly calcified, as fossils they are commonly secondarily calcified. Calcification is weaker in juvenile or freshwater podocopids, many marine myodocopids and archaeocopids, and for this reason they are not common as fossils.

In the Podocopa one valve is usually larger than the other and overlaps the smaller valve along all or part of the margin. In the Myodocopa this overlap is usually less obvious. The external surfaces of the inner and outer lamellae are covered with a thin layer of cuticle which is continuous across the dorsal margin and connects the valves as a **ligament** (Fig. 20.1b).

The valves are closed by the adductor muscles running through the body and fixed to the inner surface of the calcified outer lamellae, where distinctive and diagnostic muscle scars are formed (Fig. 20.1b,d). There may also be frontal scars associated with the mandibular muscles and a pair of mandibular scars which are not muscle scars but an area for the attachment of chitinous rods that help support the mandibulae.

The ostracod is kept in touch with its surroundings by tactile bristles (**sensilla**) which penetrate the outer lamella through **normal pore canals**. Sensilla that penetrate the marginal zone are called **marginal pore canals** whilst those traversing the rest of the shell are termed normal pore canals (Fig. 20.3). Their form (e.g. branched or unbranched) and arrangement can be useful to taxonomy. **Sieve pores** (Fig. 20.3), also with sensilla, are found in some podocopans and can also be useful taxonomically but their shape has also been linked to changes in salinity. Clear **eye spots** or raised **eye tubercles** may also be developed adjacent to the eyes especially in shallow-water species.

In some taxa the valves have a dorsal hinge structure of interlocking grooves and teeth and sockets. Three basic kinds of dorsal hinge structure are considered here but they may be subdivided (Fig. 20.4). Most Palaeozoic and freshwater ostracods have an **adont** hinge (Fig. 20.4a), this is the simplest, lacking teeth and sockets but often provided with a single groove along the margin of the larger valve and a corresponding ridge on the smaller valve. The **merodont** hinge (Fig. 20.4b,c) has elongate and strongly crenulated terminal elements on the right valve; the median elements may be smooth or crenulated. The entomodont hinge recognized by some authors differs only slightly from the merodont type and has not been defined as a separate category herein. The **amphidont** hinge (Fig. 20.4d) has short terminal elements that consist of well-developed teeth on the right valve that may be crenulated, divided or smooth. The median element comprises an anterior socket (can be smooth or divided) and a median groove that is usually smooth.

Dimorphism

The reproductive system of the ostracod is highly developed. The male and female are separate and often secrete carapaces of different size and shape. This **sexual dimorphism** is especially marked in the fossil

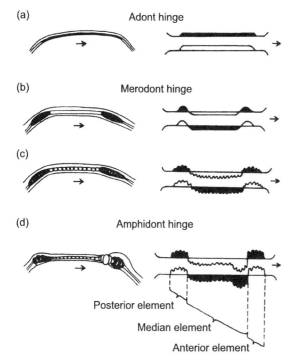

Fig. 20.4 Some ostracod hinge types, seen in lateral view of the left valve and from above. (Modified from van Morkhoven 1962–1963.)

Palaeocopida, where it has special taxonomic value (see Figs 20.10a, 20.13c,d). In this order the distinctive female forms are called **heteromorphs** and differ from the **tecnomorphs** (males and juveniles) in having a more inflated posterior region, pronounced ventral lobes, prominent hemispherical bulges called **brood pouches** or **crumina** (Fig. 20.10a), or wide frills extending beyond the free edges of the valves. Podocopid dimorphism is less obvious, but the males are generally longer and narrower and the females have greater posterior inflation (Fig. 20.9l). Imprints of the reproductive organs are sometimes found in the posterior part of the inner surface of valves.

Ostracod reproduction and ontogeny

Reproduction can take place at any time throughout the year. Some cypridinid ostracods are bioluminescent (e.g. *Vargula*, Fig. 20.13n) and are thought to use this in courtship displays (Cohen & Morin, in Whatley & Maybury 1990, pp. 381–401). The testes of the male produce spermatozoa of unique proportions, being up to 10 times the length of the male carapace. Copulation with the female results in fertilization of the eggs which are then either brooded in the carapace, shed into the water or laid amongst water weeds and stones. **Parthenogenesis**, the ability of females to lay fertile female eggs, is relatively common in freshwater species. Even in normal marine populations, the females may greatly outnumber the males, the ratio between them varying with environmental conditions. Ostracod eggs of freshwater species are resistant to desiccation and cold, and hence this stage can help survival through severe winters and prolonged droughts or even allow dispersal on the feet or feathers of birds.

As in many crustaceans, young ostracods grow in discontinuous stages called **instars** (Fig. 20.5a). When the body of an instar has grown too large for its exoskeleton, the rigid chitinous and calcareous layers are moulted. Rapid growth and development follow, together with the hardening of a new carapace. There are usually eight or nine such instars between the egg and the adult stage.

Podocopid ontogeny usually consists of eight juvenile and one adult instars. The first instar (**metanauphilus**) possesses a thin bivalved carapace but the body lacks maxillulae and thoracic legs. Maxillulae generally develop at instar two, and the legs appear between instars four and six. Muscle scars are not usually seen before instar six, genital impressions before instar seven, and sexual dimorphism before the adult stage. By instar eight, all the limbs have developed.

Modocopan ontogeny consists of between four and seven juvenile and a single adult instar. Females brood embryos in the postero-dorsal part of the carapace and release the first instar, which already has five or six legs and a furca. With each moult some or all of the limbs acquire additional setae, claws and occasionally segments. A seventh limb usually appears in the seventh instar.

The valves of ostracod instars increase progressively in size and become thicker and more heavily calcified. These changes are accompanied by modifications in shape (Fig. 20.5b) and sculpture, and in the

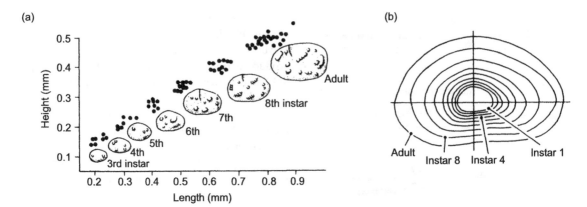

Fig. 20.5 Ostracod growth. (a) Discontinuous size distribution and changes in shape of *Neocyprideis colwellensis* from the Eocene, Lower Headon Beds. (b) Lateral outlines of successive instars of *Cypridopsis vidua*. ((a) Modified from Keen 1977; (b) modified from Kesling 1951.)

Podocopida by the increasing complexity of hinge, duplicature and marginal pore canals. It is important to distinguish morphological variations that result from ontogeny and those that result from evolution or dimorphism. For these reasons most taxonomic work is based on adult specimens.

Ostracod distribution and ecology

Substrate and food

Living ostracods are predominantly benthic or pelagic throughout their life cycle. Benthic ostracods occupy freshwater and marine habitats (Fig. 20.6). Members of the Terrestricytheroidea (e.g. *Mesocypris*) are adapted to living in damp soils and leaf litter. Freshwater ostracods tend to have smooth, thin, weakly calcified carapaces of a simple bean shape (e.g. *Halocypris*, Fig. 20.12c). Many of these consume detritus or living organisms (e.g. diatoms, protists, bacteria) stirred up by the antennae or mandibulae. *Cypridopsis* (Fig. 20.15,2) is a scavenger that holds dead plant and animal particles with its mandibulae or antennae and tears at these with its maxillulae. One species is known to predate the gastropod vector of the sickness 'Bilharzia' and hence has some medical interest.

Whereas the freshwater ostracods may spend much time swimming several centimetres above the substrate, marine benthic forms are heavier and tend to be either crawlers, burrowers or interstitial, feeding on detritus or predate diatoms, foraminifera and small polychaete worms. Such ostracods thrive best in muddy sands and silts or on seaweed and seagrasses. They are scarcer in *Globigerina* oozes and scarcest in euxinic black muds, evaporites, well-sorted quartz sands and calcareous sands.

In the Palaeozoic there were numerous different groups of filter-feeding ostracods (Lethiers & Whatley 1994) including in the Triassic the Metacopina and Platycopina, the former becoming extinct in the Toarcian in Britain. From this time the Platycopina have been the sole filter-feeding ostracods and this is reflected in their limb morphology. None of the thoracic appendages is used for walking, but they are provided with abundant setae which act as sieves to extract particles from the water. The seventh limb (the third thoracic limb) is absent. Locomotion is enabled by the enlarged postero-terminal furca and a large number of brancial plates assist the circulation of water over the ventral surface.

It has often been observed that the size, shape and sculpture of benthic ostracods broadly reflects the stability, grain size and pore size of the substrate on, or in which, they live. For example, crawling forms dwelling on soft, relatively fine-grained substrates tend to have a flattened ventral surface perhaps with weight-distributing projections called **alae** or frills, keels and

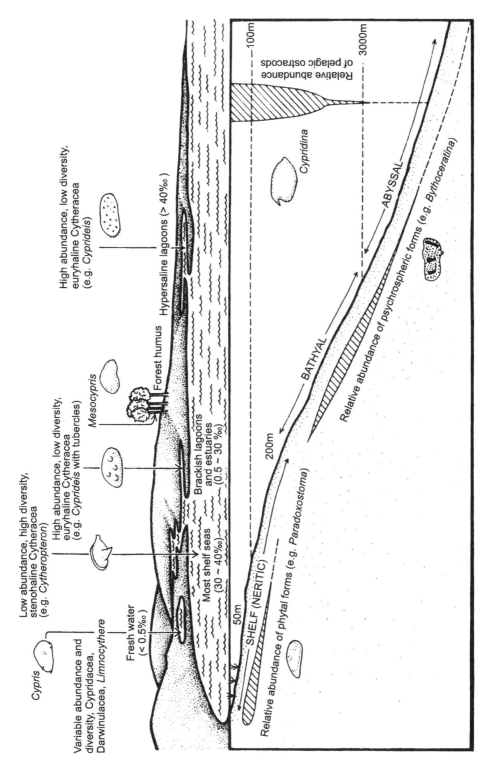

Fig. 20.6 Distribution of living Ostracoda with some typical forms represented.

lateral spines. Ostracods dwelling on coarser substrates from the more turbulent, nearshore habitats are commonly thicker shelled with a coarse sculpture of ribs, reticulations or robust spines housing sensory setae. Infaunal ostracods, which live within or burrow through the pore spaces of sandy substrates, tend to be small, smooth and robust (e.g. *Polycope*, Figs 20.12f, 20.13o); those that burrow through silts and muds need more streamlined carapaces and are usually smooth and elongated (e.g. *Krithe*, Fig. 20.9j). Burrowing is achieved with the assistance of short stout spines on the antennules. The Paradoxostomatidae contains many ostracods that are generally smooth with slim and elongate valves and have modified, tubular mandibulae for feeding on plants or animals (e.g. *Paradoxostoma*, Fig. 20.9lt).

Nektonic ostracods (particularly in the order Myocopida) spend their lives swimming in the oceans. They do this primarily by means of pairs of the hairy exopodites on the antenna. Food particles in the water are moved towards the maxillulae and thoracic limbs by water currents produced from the beating of **epipodites** on the modified first thoracic legs. In *Conchoecia* this mode of feeding is supplemented by a carnivorous diet; *Gigantocypris* subsists largely on copepods, chaetognaths and small fish caught with its antennae. As with other plankton, pelagic ostracods thrive in regions of current upwelling, rich in phosphates and nitrates. They sometimes grow very large, with *Gigantocypris* reaching lengths of up to 30 mm. Their carapaces are smooth, thin-shelled and ovate to subcircular in lateral profile. The long and active antennules and antennae have in some cases led to the formation of **rostral incisures** and projecting **rostra** at the anterior end of the carapace (e.g. *Cypridina*, Fig. 20.12e).

Many ostracods, such as *Entocythere* (order Podocopida), are **commensal**. These live attached to the appendages of larger crustaceans such as crayfish, isopods and amphipods, taking advantage of the feeding currents of their hosts. They are not common as fossils.

Salinity

Ostracods are ubiquitous in aquatic environments with species and genera living under well-defined salinity ranges within the freshwater to hypersaline range. Chlorinity is a good measure of salinity in the marine realm, but in inland saline lakes other solutes contribute more to the salinity, in these settings **athalassic** is used instead of brackish. For example, the distribution of species of *Limnocythere* (Figs 20.9m) in lakes throughout the USA is controlled by variations in Ca^{2+}, Mg^{2+}, Na^+, SO_4^{2-} and Cl^- ion concentration (Forester 1983). It is the podocopid ostracods that inhabit the greatest variety of environments, from terrestrial forms living in wet peat, all freshwater habitats to brackish through marine to hypersaline. Three main salinity assemblages are distinguishable: freshwater (<0.5‰), brackish-water (0.5–30‰) and marine (30–40‰). Hypersaline assemblages (>40‰) mainly contain euryhaline marine and brackish-water forms (e.g. Fig. 20.15). The majority of living species are adapted to a normal marine salinity of around 35‰ (i.e. stenohaline). Ostracod assemblages and species adundance can be used to plot rapid or cyclic changes in environment. Examples can be found from the Jurassic and Lower Cretaceous in southern England (Anderson, in Anderson & Bazley 1971, pp. 27–138; Anderson 1985).

Many living ostracods are remarkably tolerant of a wide range of salinities. For example, in nature *Darwinula* (Fig. 20.9d) is essentially a freshwater genus, though in culture will survive a range of salinities. Similarly *Mytilocypris*, found in Lake Bathurst, Australia, at 11‰ can also live in hyposaline water (Martens 1985). The superfamilies Darwinuloidea and Cypridoidea are essentially freshwater, whilst the Cytheroidea are essentially marine, and the Limnocytheridae (Cytheroidea) are found in fresh water. In brackish water there is a marked reduction in diversity in favour of high abundances of specialized species though this relationship breaks down at approximately 10‰ where there is a reduction in the number of individuals per species. Commonly the distinction between brackish and hypersaline faunas in the fossil record may therefore depend on the nature of the associated biota and sediments (e.g. Wakefield 1994; Knox & Gordon 1999).

Hypersaline ostracod asemblages are less well known. In the inner Scammon Lagoon, Baja, California (37–47‰) six characteristic podocopid species occur whilst myodocopids only appear in the outer lagoon (34–38‰). The even higher salinity environments of the lagoons of the Persian Gulf are inhabited by species of *Loxoconcha* (Bate & Gurney 1981).

Salinity can also have a dramatic effect on the morphology of the carapace and salinity stress can often induce greater polymorphism. Ducasse (1983) noted that freshwater incursions into the Aquitaine Basin, during the Upper Eocene and Oligocene, induced bathyal species of *Cytherella* (Fig. 20.8b) and *Argilloecia* (Fig. 20.9g) to produce 'plumper' morphs thought to be better adapted to stress. Brackish-water species tend to be thick-shelled, weakly ornamented forms with prominent normal pore canals and a merodont or amphidont hinge. Polymorphism can also be reflected in shell ornament, node and tubercle development and size. A number of species develop hollow nodes or tubercles in low-salinity habitats though other factors may have an influence. For example, *Cyprideis torosa*, a widely distributed species through Eurasia, develops nodes on specific (genetically controlled) areas of the shell, though other factors including pH may also be important (Aparecido do Carmo *et al.* 1999; van Harten 2000). Euryhaline marine species may also react to lowered salinity by developing hollow tubercles on the valves (e.g. *Cyprideis*, Fig. 20.9l). As salinity decreases, these tubercles become more evident, appearing first in the juvenile instars and even developing in adults at salinities of 5‰ or less. Because such tubercles develop with environmental changes and the character is not transferred to the offspring, they are referred to as **ecophenotypic** characters. Sieve pores can show a greater number of circular openings at lower salinity. Carapace length decreases with decreasing salinity. For example, size reduction with increasing salinity has been reported in the American genera *Hemicytherura* and *Xestoleris* (Hartmann 1963). *Loxoconcha impressa* and *Leptocythere castanea* have been shown to decrease in size with increasing distance up the Tamar Estuary (Barker 1963) though this relationship has since been disputed. Shell composition can also vary with salinity, particularly Sr/Ca and Mg/Ca ratios. In non-marine species for example the Sr/Ca ratio is independent of temperature, within the 10–25°C range.

Depth

Depth in itself does not affect ostracod distributions. However, a number of important ecological factors including hydrostatic pressure, temperature, salinity and dissolved oxygen change with depth and are paralleled by changes in ostracod faunas and diversity (Brouwers, in DeDeckker *et al.* 1988, pp. 55–77). Ostracods are therefore sensitive indicators of bottom-water conditions and the geographical distribution of ostracod assemblages are effective tracers of different benthic environments and disctinct water masses (e.g. Corrège 1993). Similarly fossil ostracod biofacies can be used in palaeoceanographic (e.g. Benson *et al.*, in Hsu & Weissert 1985, pp. 325–350; Benson, in Whatley & Maybury 1990, pp. 41–58; Coles *et al.*, in Whatley & Maybury 1990, pp. 287–305) and palaeoclimatic (e.g. Brouwers *et al.* 2000) reconstructions and oceanic events (Jarvis *et al.* 1988 and papers in Whatley & Maybury 1990). It has been shown that high levels of platycopids, the sole remaining filter-feeding ostracods, indicate low palaeo-oxygen levels (Whatley *et al.* 2003).

In shallow freshwater bodies ostracods reveal little variation with depth. In deep inland lakes which may become stratified or saline the distribution of ostracods, as in the marine realm, become indicative of distinct water masses. Relationships between limnic ostracods and their chemical/physical environment and morphological responses to changing environments have been reviewed by Carbonel *et al.* (1988).

Benthic marine ostracod depth assemblages may be categorized broadly as inner-shelf, outer-shelf and bathyal-abyssal. The shelf (or **neritic**) assemblages live between 0 and 200 m depth, and include many of the marginal marine forms mentioned above. Whereas the densest populations are found in the marginal areas, the highest diversities tend to occur in shallow-shelf seas. The presence of thick valves with eye spots, strong sculpture, amphidont hinges and conspicuously branched pore canals are features common in extant shallow-water ostracods from coarse-grained substrates. Deeper-water neritic substrates, which tend also to be finer grained, support forms with smooth, thin, often translucent carapaces with relatively weak hinges and no eyes or eye spots (e.g. *Krithe*, Fig. 20.9j).

Bathyal and abyssal assemblages, or the **psychrospheric** fauna, occur mostly at depths of 1000–1500 m and at temperatures of 4–6°C but inhabit shallow water at high latitudes. At depths greater than 600 m blind forms with relatively large carapaces (>1 mm long) and thin, highly sculptured walls (e.g. *Bythoceratina*, Fig. 20.9f) commonly occur. Deep sea species tend to exhibit convergence in carapace morphology. Both ornate and smooth forms are known. Increased number of spines in psychrospheric species is probably for protection rather than strength. Morphologies tend to be stable over long periods with marked 'punctuated' changes corresponding with oceanographic events, for example a major morphological shift occurred in *Poseidonamicus* from the South Atlantic at 14 Ma, associated with a major intensification of Antarctic glaciation. Psychrospheric ostracods have adapted to conditions of darkness, constant salinity and temperature and fine grain size. These conditions stretch more or less uniformly throughout the abyssal plains and these ostracods have a cosmopolitan distribution. Specialist ostracod communities appear to have inhabited chemosynthetic mounds since at least the Carboniferous. Modern Pacific deep sea vent communities include eucytherurine and pontocypridid Podocopa (van Harten 1992). Whatley & Ayress (1988) demonstrated that many more ostracods were pan-abyssal than previously thought and documented 65 species common to the Quaternary of the North Atlantic, Indian and South West Pacific oceans, further suggesting most species entered the deep sea in the Neogene. Psychrospheric ostracods inhabit shallower waters at high latitudes.

Pelagic ostracods can develop daytime depth associations. A surface assemblage (<250 m) of rich diversity that may overlie an impoverished layer at 300–400 m, with further rich assemblages at 450–625 m and at 720 m downwards (Angel 1969). These daytime zonations with their distinct species are partly disrupted by upward migrations at night, but in general appear to correspond with different water masses found at different depths.

Temperature

Latitudinal temperature control of shallow-water species has given rise to numerous localized (**endemic**) assemblages ranging from high latitudes (at temperatures below 0°C) to the subtropics and tropics (where they may live in waters up to 51°C). This endemism is enhanced in benthic ostracods by the lack of a planktonic larval stage for dispersal. As with most groups, tropical assemblages tend to be more diverse than those in higher latitudes. Some of the latter are, however, of relatively large body size, explained by their slower metabolism and the longer time it takes them to reach maturity. As well as affecting the metabolic rate, maturation and food supply, temperature can also control the breeding season and, in some freshwater species, the incidence of parthenogenesis. Heip (1976) found that in *Cyprideis torosa* the rate of population increase could be correlated with mean temperature rise. However, other freshwater species can show the opposite relationship with the development of juveniles slowing in the summer.

Classification

Kingdom ANIMALIA
Phylum CRUSTACEA
Class OSTRACODA

The classification of the ostracods is in a state of flux. The classification used here is based upon Whatley *et al.* (in Benton 1993, pp. 343–357) and Horne *et al.* (in Holmes & Chivas 2002, pp. 5–37). Students should be aware that other classification schemes do exist. The Ostracoda form a distinct class because of their bivalved, perforate carapace into which the entire animal can be withdrawn when the carapace is closed. Biologists subdivide the extant members of the group on differences in their soft parts, particularly their limbs. Generally, these taxa correspond with the carapace-based taxa of palaeontologists (Box 20.1). This biological approach, however, cannot be extended to the extinct Palaeozoic orders, which are diagnosed entirely on carapace features.

The Ostracoda are divided into eight orders, four of these (the Archaeocopida, Bradoriida, Leperditicopida and Eridostracoda) can only be regarded as putative ostracods. In summary, the following carapace features are of value in the classification of fossil taxa:

1 basic carapace shape;
2 muscle scar position and arrangement;
3 degree of development and fusion of the duplicature with the outer lamella;
4 structure, shape, size and arrangement of normal and marginal pore canals;
5 nature, location and degree of valve overlap;
6 hinge elements;
7 nature of sexual dimorphism, if present;
8 nature of surface sculpture, and presence of eye spots;
9 nature of marginal zone;
10 form of selvages and flanges.

Obviously it is essential to be able to distinguish dorsal from ventral and posterior from anterior in fossil ostracod valves. In the extant Podocopida and Myodocopida, orientation presents no problem, but in the extinct Archaeocopida, Leperditicopida and Palaeocopida the correct orientation is less certain. Guidelines for orientation, following the currently accepted practices, are therefore included below.

Putative ostracods

Order Archaeocopida This order is characterized by forms with a weakly calcified or phosphatized carapace (e.g. *Vestrogothia*, Fig. 20.7a). Also the hinge line is straight; ventral margin convex; prominent eye tubercles; dimorphism and muscle-scar pattern unknown. Specimens of the phosphatic archaeocopids have been discovered with preserved appendages unlike any found in the other ostracod orders and suggests their exclusion from the class.

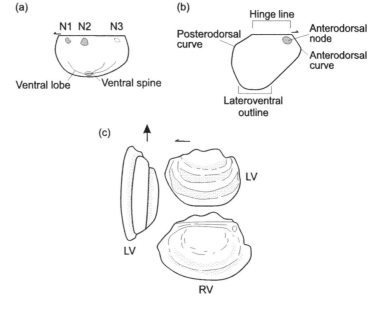

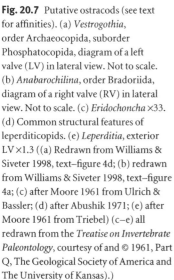

Fig. 20.7 Putative ostracods (see text for affinities). (a) *Vestrogothia*, order Archaeocopida, suborder Phosphatocopida, diagram of a left valve (LV) in lateral view. Not to scale. (b) *Anabarochilina*, order Bradoriida, diagram of a right valve (RV) in lateral view. Not to scale. (c) *Eridoconcha* ×33. (d) Common structural features of leperditicopids. (e) *Leperditia*, exterior LV ×1.3 ((a) Redrawn from Williams & Siveter 1998, text–figure 4d; (b) redrawn from Williams & Siveter 1998, text–figure 4a; (c) after Moore 1961 from Ulrich & Bassler; (d) after Abushik 1971; (e) after Moore 1961 from Triebel) (c–e) all redrawn from the *Treatise on Invertebrate Paleontology*, courtesy of and © 1961, Part Q, The Geological Society of America and The University of Kansas).)

Order Bradoriida Bradoriids are small bivalved arthropods found in Cambrian and Lower Ordovician rocks (e.g. Siveter & Williams 1997), typically adults are 1–18 mm long (Fig. 20.13a). The order includes what are now recognized as two distinct groups, the Bradoriida *sensu stricto* and the Phosphatocopida, and are described in Siveter & Williams (1997). *Anabarochilina* (L.-M. Camb., Fig. 20.7b) has a smooth or wrinkled, subquadrate carapace with prominent antero-dorsal node. Rare specimens of *Kunmingella*, with soft parts preserved, indicate bradoriids belong outside the Crustacea. Based on soft-part anatomy (e.g. Müller 1979) phosphatocopids are now considered a sister group to the Crustacea (e.g. Waloszek 1999). This has profound implications for the stratigraphical and evolutionary history of the Ostracoda; much if not all of their Cambrian record may be spurious.

Order Eridostracoda The taxonomic position and rank of this order are controversial; some hold that 'Eridostraca' may be an extinct group of marine branchiopods, others that they are part of the order Palaeocopida. *Eridochoncha* (Fig. 20.7c) has a straight hinge line curved ventral margin and concentric ridged sculpture.

Order Leperditicopida This order includes forms in which the carapace is large, well calcified and usually long. Some leperditicopids range up to 5 cm in length. Other characters include a straight hinge line and prominent eye tubercle; inner lamella uncalcified; complex muscle-scar pattern with up to 200 subsidiary small scars. Orientation of the valves should be helped by the guidelines below. *Leperditia* (L. Sil.-U. Dev., Figs 20.7e, 20.13b) was a widespread genus with a purse-shaped, smooth or punctate carapace bearing distinct eye tubercles and adductor muscle scars. Typically they are found as large abundance, monospecific assemblages in facies typical of shallow, marginal habitats. Most were probably epibenthic, detritus feeders (Vannier *et al.* 2001). Though morphological similarities with the ostracoda are an important consideration, the lack of evidence from soft parts means taxonomic and phylogenetic relationships remain inconclusive.

Class Ostracoda ('true ostracods')

Order Podocopida The Podocopida comprise the bulk of the Mesozoic and Cenozoic fossil ostracods, although they have a longer history (L. Ord.-Rec.). Living forms (e.g. Fig. 20.13e–i) are largely diagnosed from their soft parts. The antenna exopodite is greatly reduced, the maxillula has a large branchial plate and the eighth limb is usually absent. Fossil taxa have been erected on carapace morphology. Podocopid valves are well calcified, of unequal size and have a convex dorsal margin and a weakly convex, straight or concave ventral margin. Lobes and sulci are uncommon and muscle scars and duplicature are prominent. Podocopid valves may be orientated using the following guidelines.

1 The dorsal margin is convex or straight but less than the total length of the carapace. It bears adont, merodont or amphidont hinge elements. Eye spots and eye tubercles, if present, occur in an antero-dorsal position.

2 The ventral margin is often concave but may be straight or convex. The duplicature, where present, is narrow in the suborders Metacopina and Platycopina and wider in the Podocopina, with marginal pore canals in the marginal zone. The ventral region may also be provided with prominent spines, frills, flanges and wing-like alae.

3 Adductor muscle scars are variable in number and arrangement but are invariably situated just anterior of the valve centre. Their position may be marked on the outer surface by a subcentral tubercle.

4 Viewed laterally, the more pointed end is generally posterior whilst the higher, blunter end is anterior.

5 In dorsal or ventral view, the broadest region occurs near the posterior end in adults and is often more swollen in female carapaces.

6 The more complex terminal elements of the hinge line are developed towards the anterior end of the hinge.

7 Major spines, tubercles and alae generally point to the posterior.

8 The marginal area of the suborder Podocopina tends to be broader and with more marginal pore canals at the anterior end.

A large number of podocopids have adapted to crawling and burrowing niches in marine sediments or on seaweeds. However, this order also includes the terrestrial and freshwater Cypridacea and fresh- and brackish-water genera of the Cytheridacea.

The suborder Cypridocopina (Dev.-Rec.) includes many freshwater ostracods and a few marine forms. Because of the low salinity they secrete smooth, thin, chitinous or weakly calcified shells, often with a rather low preservation potential. The hinge is adont or rarely merodont. The adductor muscle-scar pattern consists generally of one large dorsal element, three anterior elements and two posterior elements, all elongated and more or less aligned (Box 20.1). The duplicature is incompletely fused to the outer lamella, leaving a prominent vestibule and a relatively narrow marginal zone. Living cypridocopeans are distinguished by their appendages, the more so because their carapaces are very similar. The palaeontologist may therefore face problems with the taxonomy of fossil specimens, making necessary the accurate measurement of all carapace features. *Halocypris* (?Jur., Pleist.-Rec., Fig. 20.12c) thrives in freshwater ponds. It has a smooth, subtriangular carapace of relatively large size, up to 2.5 mm long. *Carbonita* (?L. Carb., U. Carb.-Perm., Fig. 20.9i) was typical of freshwater or slightly brackish-water facies around coal swamps. It has a smooth, elongate carapace with a somewhat larger right valve. *Cypridea* (U. Jur.-L. Cret., Fig. 20.9h) is another fossil form typical of fresh to slightly brackish waters. Both dorsal and ventral margins are relatively straight and the antero-ventral margin is provided with a beak and notch. The hinge is merodont and the surface is usually pitted and pustulose. *Argilloecia* (Cret.-Rec., Fig. 20.9g) has adapted to outer shelf and bathyal marine conditions, and has often been found in *Globigerina* oozes. The carapace is smooth and elongate with a blunt anterior and a pointed posterior end, the right valve slightly larger than the left one.

The suborder Cytherocopina (M. Ord.-Rec.) is morphologically the most varied of this order. They have three pairs of legs adapted for locomotion, distinctive adductor muscle scars of four elements aligned in a near vertical row, anterior of which are found three mandibular and one or two frontal muscle scars. The hinge of Cytheroidea (e.g. *Celtia*, Rec., Fig. 20.13k) is usually merodont or amphidont and the duplicature with its marginal zone is prominent, often with branched marginal pore canals. Ecologically the Cytheroidea are a varied group, as the following examples will show.

The suborder Darwinulocopina (Carb.-Rec.) comprises freshwater ostracods sporting a distinctive muscle-scar pattern, i.e. an almost symmetrical rosette of 9–12 elongate scars (Box 20.1). The carapace of *Darwinula* (Fig. 20.9d) is smooth, thin-shelled, elongate and ovate, provided with an adont hinge but lacks a duplicature. The suborder Metacopina were marine and are known only from fossil carapaces (?Sil.-M. Jur.). They were ancestoral to the Platycopina and probably to certain Podocopina. The muscle scars are numerous (>25) and assembled in a compact group (Box 20.1). The hinge elements are either adont or differentiated into merodont form; the duplicature is indistinct and of narrow width. Typically the left valve is slightly larger, overlapping the free margins of the right valve. In *Kloedenella* (Sil.-Dev., Fig. 20.8a) the left valve overlaps the right one and both bear two prominent antero-dorsal sulci. The heteromorph has a higher posterior region than the tecnomorph. For example, *Healdia* (Dev.-Perm., Fig. 20.8c) has a smooth, rounded carapace with backwards directed 'shoulders' near the posterior end. The dorsal hinge is adont.

The suborder Platycopina is a still-living marine group that arose from the Metacopina in the Triassic. They differ from that group in having a larger right valve, an adont hinge and in the adductor a muscle-scar pattern, which comprises 10–18 elongate scars arranged in two slightly curved rows (Box 20.1). Platycopine and metacopine carapaces share the same ovate shape, an inequality of valve size, weakly developed duplicature and a prominent selvage with contact grooves around the free margins. Recent platycopids have biramous antennae (cf. Podocopina) and three pairs of thoracic appendages that serve as maxillulae, although the third pair are rudimentary in females. There are only a few genera, of which *Cytherella* (Jur.-Rec., Fig. 20.8b) is the most common. This has a smooth ovate carapace with the rear end more inflated in dorsal view, especially in the larger female specimens. *Cytherelloidea* (Jur.-Rec., Fig. 20.13m) is distinguished by a stronger ornament and a generally more compressed carapace. Ornamentation can vary widely within a single species and this may be ecophenotypic.

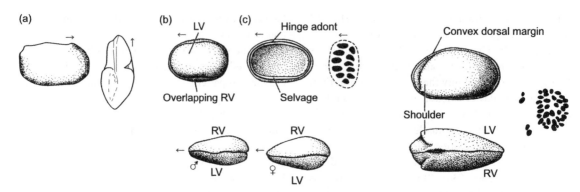

Fig. 20.8 Selected examples of the order Podocopida, suborder Platycopina, Metacopina. (a) *Kloedenella* RV exterior, dorsal view ×30. (b) *Cytherella*: exterior LV ×27; exterior RV ×27; detail of aductor muscle scar, dorsal view of male, dorsal view of female ×27. (c) *Healdia*: exterior RV of male ×67; muscle scar ×133; dorsal view of female ×67. ((a) After Moore 1961, from Ulrich & Bassler; (b) after Andreev 1971; (c) after Shaver in Moore 1961 (from the *Treatise on Invertebrate Paleontology*, courtesy of and Copyright © 1961, Part Q, The Geological Society of America and The University of Kansas).)

The suborder Podocopina (L. Ord.-Rec.) are an ancient marine stock that were common in Palaeozoic seas. Their carapaces are commonly thick and smooth with a strongly convex dorsal margin, a blunt anterior end and a pointed posterior end. Quaternary forms have a characteristic 'cocked-hat' shape and were assigned to the suborder Bairdiocopina by Horne *et al.* (in Holmes & Chivas 2002, pp. 5–37). *Bairdia* (Ord.-Rec., Figs 20.1b, 20.9b) is the longest-ranging ostracod genus. Its large left valve partly overlaps the margins of the right and the hinge is adont, with a simple ridge and groove articulation. The duplicature is wide with a prominent vestibulum, and the adductor scars consist of 6–15 elongate elements arranged radially, irregularly or aligned. The articulation is weak and adont or merodont. *Bairdiocypris* (Si.-Dev., ?Jur., Fig. 20.9a) is large and has a heavily calcified subtriangular carapace. *Limnocythere* (Jur.-Rec., Figs 20.9m, 20.13l) is one of the few freshwater cytheraceans. It has a thin, chitinous carapace with a merodont hinge and marginal zone bearing many straight marginal pore canals. The hollow tubercles may be ecophenotypic, induced by the low salinities, as in the genus *Cyprideis* (Mio.-Rec., Fig. 20.9l) which lives mostly in brackish or hypersaline waters. It has a subovate, essentially smooth carapace with an entomodont hinge. Immature valves from brackish waters may bear phenotypic hollow tubercles. Species of *Cytherura*

Fig. 20.9 (*opposite*) Selected examples of the order Podocopida, suborder Podocopina, magnifications approximate. (a) *Bairdiocypris* exterior RV, dorsal view ×20. (b) *Bairdia* interior LV, ×40 (after van Morkhoven 1963); exterior RV and dorsal view ×43 (after Andreev 1971); detail of muscle scar (after van Morkhoven 1963). L.O.C., line of concrescence; (c) *Cypris*: interior LV about ×16; exterior RV ×13; detail of *Paracypris* adductor muscle scars. (d) *Darwinula*: interior LV ×63 (after van Morkhoven 1963 from Wagner); exterior LV and dorsal view ×38 (after van Morkhoven 1963 from Sars); detail of muscle scar (after van Morkhoven 1963). (e) *Cytherura*: interior LV ×87.5; dorsal view ×55.5. (f) *Bythoceratina*: interior RV ×50; exterior LV ×50; dorsal view RV ×50. (g) *Argilloecia*: interior RV ×72; exterior LV ×63; dorsal view ×72. (h) *Cypridea*: interior LV ×30; exterior RV ×20; dorsal view ×20. (i) *Carbonita*: exterior LV ×27; dorsal view ×27. (j) *Krithe*: interior RV ×35; exterior LV ×25; dorsal view ×25. (k) *Paradoxostoma* exterior LV ×62. (l) *Cyprideis*: interior LV ×50; exterior LV ×27; dorsal view of male (above) and female with hollow tubercles (below). (m) *Limnocythere*: interior LV ×67; dorsal view ×38. ((a) After Shaver in Moore 1961; (c) redrawn after van Morkhoven 1963, from Sylvestor-Bradley; (e) after Benson *et al.* 1961, from Wagner; (f) after Moore 1961, from Hornibrook; (g) redrawn after van Morkhoven 1963, from Mueller; after Pokorny 1958 from Alexander; (h) (part) after Moore 1961, from Kesling; (i) after Moore 1961, from Jones; (k) after Pokorny 1958, from Sars; (l) after van Morkhoven 1963 from Wagner and Klie and after Moore from Goerlich; (m) after van Morkhoven 1963, from Wagner from Mueller ((a), (e), (f), (h), (i) redrawn from the *Treatise on Invertebrate Paleontology*, courtesy of and © 1961, Part Q, The Geological Society of America and The University of Kansas.)

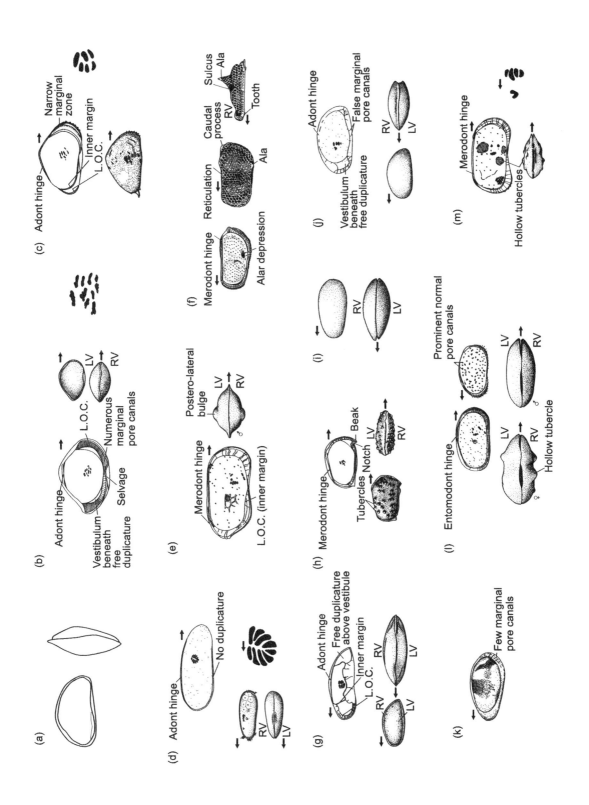

(Cret.-Rec., Fig. 20.9e) often thrive in brackish or very shallow marine waters. In these the carapace is smooth and oblong, the males more elongate and the females provided with postero-lateral bulges. The hinge is merodont and the duplicature narrow, without vestibulum. *Paradoxostoma* (?Cret., Eoc.-Rec., Fig. 20.9k) and its relatives thrive in rock pools along the intertidal zone and on subtidal seaweeds. The genus has a thin, elongate carapace that is very narrow in dorsal view and more pointed at the anterior end. Its hinge is merodont whilst the marginal zone is narrow with a few simple pore canals. *Krithe* (U. Cret.-Rec., Fig. 20.9j) is another blind mud-dweller from the outer shelf and bathyal habitats. Its smooth, thin, elongate carapace has a weak adont hinge and a duplicature of varying width that bears a broader anterior and a narrower posterior vestibulum. *Bythoceratina* (U. Cret.-Rec., Fig. 20.9f) is a typical psychrospheric ostracod, thriving best at depths between 2000 and 3000 m. Its carapace is subquadrate with a straight dorsal margin bearing a merodont hinge. The ventro-lateral margins have developed pointed alae whilst the posterior end sports a short caudal process. The outer surface of *Bythoceratina* is commonly reticulated, but it may be smooth or spinose.

The poorly represented Quaternary and living members of the Sigilliocopina contains small 'bean-shaped' ostracods, commonly less than 0.5 mm in length. They are ovate, smooth and inflated. The adductor muscle scars comprise a circular aggregation of 20–30 scars (e.g. *Saipanetta*, Box 20.1).

Order Palaeocopida (Fig. 20.10) The Palaeocopida had their acme in the Palaeozoic and can be recognized by their long straight hinge line, lobate and sulcate sculpture, and often by the distinctive sexual dimorphism, usually with a well-developed crumina in the heteromorph. The valves do not overlap and muscle-scar patterns are poorly known. The inner lamella is not calcified. Orientation of the carapace should be helped by the guidelines below.

1 The dorsal margin is long and straight, often ending in distinct **cardinal angles**. Eye spots and eye tubercles, if present, occur in an antero-dorsal position. The **sulci** and **lobes** are more sharply defined towards the dorsal margin.

2 The ventral margin is convex and may be provided with frills, flanges, brood pouches or spines, especially in the heteromorphs. Commonly a **ventral lobe** runs parallel to the ventral margin.
3 Viewed laterally with the hinge horizontal, the carapace is highest just to the anterior of the mid-line.
4 Viewed dorsally, the greatest width is usually posterior. However, the wide brood pouches of heteromorphs in the superfamily Beyrichiacea are antero-ventral in position.
5 The median sulcus, if present, approximates to the position of the numerous adductor muscles on the inside, but muscle scars are rarely seen. Both features are generally anterior of the valve centre.
6 Major spines and alae tend to be directed to the posterior.

This diverse group is usually subdivided on the basis of general shape, the nature of dimorphism (if any), the form of lobes and sulci and on superficial sculpture (e.g. spines, striae and reticulation). There are nine superfamilies (Box 20.1).

Palaeocopids are confined to the Palaeozoic and Lower Triassic, with the exception of the superfamily Kirkbyacea whose members are known from the Tertiary of Japan and Recent sediments of the eastern South Pacific. In the living genus *Manawa* the carapace is less than 1 mm long with a straight dorsal margin, the calcified duplicatures are developed into a marginal frill; adductor muscle scars have a central spot with five others arranged radially or ventrally around it. There are eight limbs and a furca. The maxillula has a leg-like endopodite without a branchial plate. The fifth to seventh limbs are used in walking and the eighth limb incorporates the male copulatory appendage, a Zenker's organ is absent.

Beyrichia (L. Sil.-M. Dev., Fig. 20.10c,d) was a widespread genus with three distinct lobes and a granular or pitted surface. The heteromorph has globular brood pouches. In *Aechmina* (M. Ord.-L. Carb., Fig. 20.10j) the dorsal margin supports a remarkably large spine, and the ventral margin bears short spines. Dimorphism is unknown in this genus. *Hollinella* (M. Dev.-M. Perm., Fig. 20.10g) has essentially four lobes, but the anterior and posterior ones (L_1 and L_4) are united with a prominent ventral lobe. The marginal frill of the heteromorph is broader than that

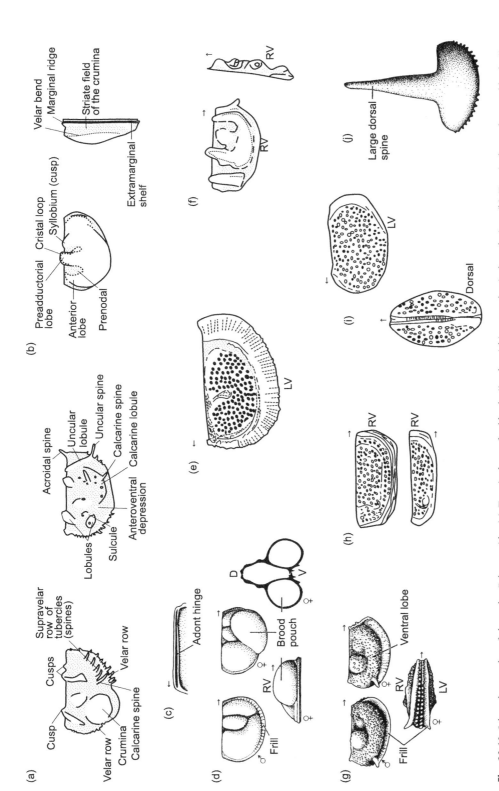

Fig. 20.10 Selected examples of the order Palaeocopida. (a) Terminology of a kloedeninine beyrichiacean valve in lateral view. (b) Terminology of a non-kloedeniine beyrichiacean valve in lateral view. (c) *Beyrichia* adont hinge line of RV. (d) *Beyrichia*: exterior RV of male ×17.5; exterior RV of female ×17.5; ventral view of RV of female ×17.5; transverse section of female. (e) *Eurychilina* LV lateral ×18. (f) *Nodella* RV lateral and dorsal views ×50. (g) *Hollinella*: exterior RV male ×20; exterior RV female ×20; ventral view of female ×20. (h) *Kirbya* RV lateral and dorsal views ×40. (i) *Primitiopsis* LV lateral and dorsal views ×25. (j) *Aechmina* exterior RV ×40. ((a), (b) Redrawn from Siveter 1980, figures 3, 4, after Martinsson; (c), (g) after Moore 1961, from Kesling; (e) after Moore 1961; (f) after Moore 1961, from Zaspelova; (h), (i) after Moore 1961, from Jones; (j) after Moore 1961, from Boucek (((c), (e)–(i) redrawn from the *Treatise on Invertebrate Paleontology*, courtesy of and © 1961, Part Q, The Geological Society of America and The University of Kansas).)

of the tecnomorph. *Eurychilina* (Ord., Fig. 20.10e) has three lobes, the anterior and posterior ones (L_1 and L_3) being very broad and the median one (L_2) virtually united with L_1. As with *Hollinella*, both the heteromorphs and the tecnomorphs possess dimorphic, radially striated frills, with the heteromorphy of *Eurychilina* amongst the largest known.

Order Leiocopida (Ord.-Perm.) Members of this order are superficially similar to palaeocopids but generally lack lobes and sulci. The carapace is inequivalved and an adductor muscle-scar is rarely visible. A velar structure can be developed as a low ridge. Dimorphism is not observed. *Aparchites* (L.-M. Ord., Fig. 20.11a) has an ovate, non-sulcate carapace in which the hinge line is shorter than the length of the carapace. Cardinal angles are obtuse and the velar ridge is smooth or tuberculate or often with small spines. Dimorphism is not known here. *Paraparchites* (Dev.-Perm., Fig. 20.11b) is ovate and smooth except for a postero-dorsal spine in a few species. The left valve usually overlaps the right along the free margin. The carapace has its greatest height medial or forward and the greatest width medial in males and posterior in females.

Order Myodocopida The Myodocopida (Ord.-Rec.) include a large number of the pelagic ostracods. They have weakly calcified carapaces with equal to unequal valves and no valve overlap; the dorsal and ventral margin may be convex and the inner lamella is only partially calcified. The muscle-scar pattern consists of numerous elongate scars. The order is classified primarily on the morphology of the antenna which is biramous, has a large basal segment and may project through a notch in the anterior margin of the carapace. This appendage bears long setae and is specialized for swimming. The carapace may reach up to 1 cm in length and larger species can develop a compound eye, heart and gills. Myodocopid ostracods are seldom preserved due to the poorly calcified carapace. Recognition and orientation of myodocopid carapaces may be helped by the following guidelines:
1 The dorsal margin is commonly convex and provided with weak, adont hinge elements. In the superfamily Cypridinoidea there is a prominent anterior

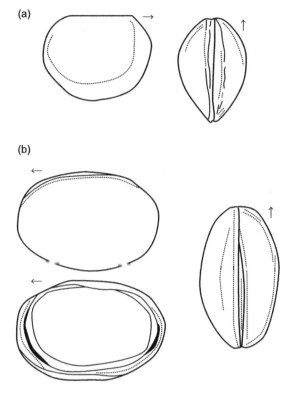

Fig. 20.11 Selected examples of the order Leiocopida, superfamilies Aparchitacea, Paraparchitacea. (a) *Aparchites*: exterior RV lateral view ×10; ventral view ×10. (b) *Paraparchites*: exterior LV; RV interior; dorsal. ((a) After Moore 1961, from Jones; (b) after Moore 1961, from Ulrich & Bassler (from the *Treatise on Invertebrate Paleontology*, courtesy of and © 1961, Part Q, The Geological Society of America and The University of Kansas).)

beak (**rostrum**) overhanging a **rostral incisure**, the former pointing antero-ventral and the latter dorsally.
2 The ventral margin is convex, occasionally furnished with a pronounced ventral spine or with ventral swellings.
3 The anterior margin of the Cypridinoidea bears a rostrum that is higher than the more pointed posterior end. The valve of the superfamily Entomozoacea bears a medial C-shaped furrow whose convex side points posteriorly. An antero-dorsal swelling is present in some of these.
4 Viewed dorsally, the broader end is posterior in many genera.

Myodocopid ostracods have a dorsal margin that may be straight or curved. The anterior margin usually has a rostral incisure and a **nuchal furrow** may be present in Palaeozoic forms. Size is highly variable with macroscopic forms reaching 2–3 cm in diameter. The antennula is modified for swimming.

Living members of the superfamily Cypridinoidea include pelagic forms, filter feeders and carnivores. They have a reduced number of appendages and are provided with two stalked, compound eyes and a median simple eye. Their carapaces can be recognized from the Silurian onwards by the prominent anterior rostrum and rostral incisure (e.g. *Cypridina*, Rec., Fig. 20.12e).

Order Halocypridina (Sil.-Rec.) This order contains ostracods in which the carapace is almost entirely uncalcified. A prominent rostrum projects as the anterior continuation of the more or less straight dorsal margin. *Halocypris* (Halocypridiina: Rec., Fig. 20.12c) is rather atypical in having a short rostrum. The Carboniferous genus *Entomoconchus* (Entomoconchidae Sil.-Rec., Fig 20.12a) has a siphonal gape at the posterior end and a distinctive set of muscle scars. Recent species of the anomalous and rare genus *Thaumatocypris* (Thaumatocypridoidea, M. Jur.-Rec., Fig. 20.12b) are thin, weakly calcified and pelagic, the fossil forms are thicker with a heavy ornament and were presumably benthic, swimming for short distances only. Both kinds bear the characteristic anterior spines.

Members of the suborder Cladocopina lack eyes, heart and the second and third thoracic limbs. They are well calcified; the muscle-scar pattern is composed of three closely often triangular juxtaposed scars. *Polycope* (?Dev., Jur.-Rec.; Figs 20.12f, 20.13o) is the principle genus and typical of the suborder. Its carapace is globular and almost circular in outline and lacks a rostrum whilst the inner surface bears the distinct cladocopine adductor muscle scars (three spots in the centre of each valve). *Polycope* as a weak swimmer prefers to live interstitially in the substrate. *Richteria* (Sil.-Perm., Fig. 20.12d) has an oblong carapace with a nuchal furrow extending downwards and a nearly straight dorsal margin. The surface of the carapace is usually ornamented with longitudinal or concentric striations.

General history of ostracods

The Early Ordovician global transgression triggered what many consider the first major radiation of ostracods (Fig. 20.14) and was probably associated with an expansion of available niches. The first Palaeocopida, Leiocopida, Podocopida and the Leperditicopida appeared at this time. The Ordovician proved to be the heyday of the Palaeocopida, their generic diversity tending to dwindle from then until their apparent extinction in the Permian. Curiously, extant palaeocopid ostracods are known from deep-water habitats in the Tertiary and Recent. Myocopids are not known before the Silurian.

Later Palaeozoic ostracod faunas were the most diverse, comprising almost as many podocopid genera as in the Jurassic Period and more fossil myodocopids than at any other time. It was also at this time that the first freshwater ostracods appear to have evolved; the Darwinulocopina, for example, flourished during the Carboniferous, Permian and Triassic but declined from the Jurassic onwards. The Late Devonian witnessed the extinction of the leperditicopids and numerous other Early Palaeozoic genera, and although new taxa appeared after this, impoverishment continued until the Jurassic.

By the Early Triassic the majority of palaeocopids, leiocopids and myodocopids appear to have become extinct. Triassic times saw the beginning of podocopid dominance amongst benthic ostracod assemblages. Many of the Jurassic lineages are of restricted time range and useful for biostratigraphy.

The Late Jurassic and Early Cretaceous heralded the worldwide expansion of non-marine deposits, including deltaic, marginal-lagoonal settings and true lacustrine with their associated limnic ostracod faunas. The faunas are often rich and highly diverse and have been used in intercontinental correlation (e.g. Anderson 1985; Horne 1995). A diverse Cretaceous fauna of marine cytheraceans suffered a minor decline at the end of the Cretaceous. Since the Paleocene, diversity of ostracod assemblages has tended to increase, although the very high numbers of Pleistocene to Recent genera (Fig. 20.14) also reflects the contribution of poorly calcified groups (e.g. myodocopids and cypridoideans) and the active interest of zoologists.

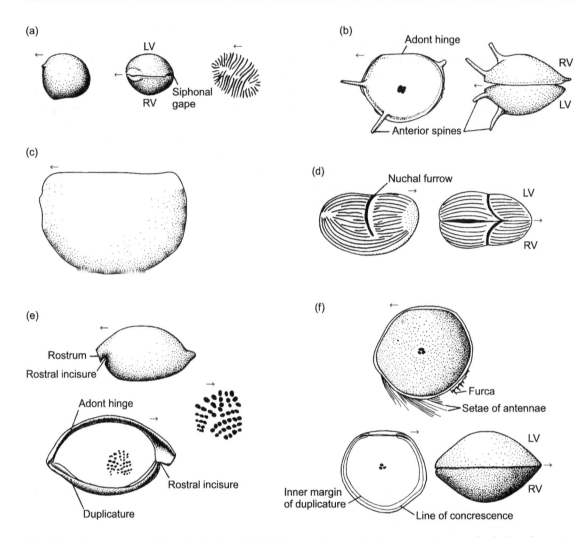

Fig. 20.12 Selected examples of the order Myodocopida. (a) *Entomoconchus*: exterior LV, ×1; ventral view ×1; detail of muscle scar ×9. (b) *Thaumatocypris*: exterior of LV living species ×20; dorsal view ×20. (c) *Halocypris*: exterior LV male; Recent species ×30. (d) *Richteria*: exterior RV ×10; dorsal view ×10. (e) *Cypridina*: exterior LV ×20; interior LV ×47; detail of muscle scar ×73.5. (f) *Polycope*: exterior LV of living species ×47; interior LV ×47; dorsal view. ((a) After Moore 1961, from Sylvester-Bradley; (b) after Moore 1961, from Mueller; (c) after Moore 1961, from Dana; (d) after Moore 1961, from Canavari; (e) after Moore 1961, from Mueller; after van Morkhoven 1963, from Keij; (f) after Pokorny 1958; after Moore 1961, from Sars; from Sylvester-Bradley in Benson *et al.* 1961 (all from Ulrich & Bassler from the *Treatise on Invertebrate Paleontology*, courtesy of and © 1961, Part Q, The Geological Society of America and The University of Kansas).)

Most deep sea genera can be traced back to Cretaceous continental shelf forms and this has led to the view that psychrospheric ostracods entered the deep sea when oceanic thermal gradients were less marked than after the formation of the psychrosphere. This led to the isolation of deep sea taxa in the Middle Eocene. Through the Cenozoic approximately 365 species are recorded from the North Atlantic and 265 species from the Pacific (Coles *et al.*, in Whatley & Maybury 1990, pp. 287–307). In the North Atlantic

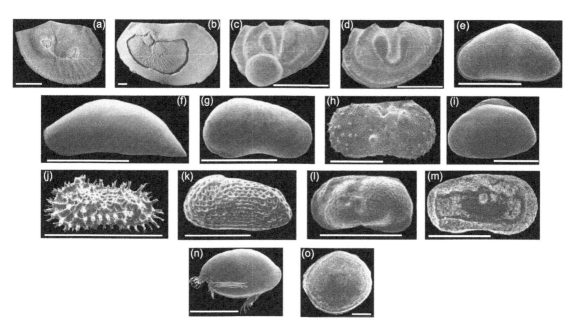

Fig. 20.13 SEM photomicrographs of selected taxa. Scale bars: (j), (o) = 100 µm; (e), (g), (h), (i), (k), (l), (m) = 500 µm; all other figures = 1 mm. (a) *Petrianna fulmenata* (Bradoriida). (b) *Leperditia* (*Hermannina*) *consobrina*, a partially exfoliated left valve showing radiate features on the internal mould. (c) *Craspedobolbina* (*Mitrobeyrichia*) *hipposiderus*, female left valve. (d) *Craspedobolbina* (*Mitrobeyrichia*) *hipposiderus*, male left valve. (e) *Propontocypris* (Podocopida), left valve. (f) *Macrocypris* (Podocopida), left valve. (g) *Potamocypris* (Podocopida), left valve. (h) *Ilyocypris* (Podocopida), right valve. (i) *Cyprinotus* (Podocopida), left side. (j) *Acanthocythereis*, left valve. (k) *Celtia*, left valve. (l) *Limnocythere*, left valve. (m) *Cytherelloidea* (Platycopina), right valve. (n) *Vargula* (Myodocopida), left side. (o) *Polycope* (Halocyprida), left valve. ((a) After Siveter *et al.* 1996, figure 6b (reproduced by permission of the Royal Society of Edinburgh from *Transactions of the Royal Society of Edinburgh: Earth Sciences* 86 (1996, for 1995), pp. 113–121; (b) after Vannier *et al.* 2001, figure 4.1 (reproduced with permission from The Paleontological Society); (c) after Siveter 1980, plate 2, figure 1 (reproduced with the permission of the Palaeontolographical Society); (d) after Siveter 1980, plate 2, figure 3 (reproduced with the permission of the Palaeontolographical Society); (e)–(o) after Horne *et al.* 2002, in Holmes & Chivas, figure 1 (reproduced with the permission of the American Geophysical Union).)

and Pacific oceans, species diversity is now known to increase in a non-uniform way throughout the Cenozoic with the largest increase in the Middle Eocene, coin-cident with the global development of the psychrospheric fauna (Coles *et al.*, in Whatley & Maybury 1990, pp. 287–307). Genera such as *Bradleya*, *Henryhowella*, *Parakrithe*, *Pedicythere*, *Pennyella* and *Thalassocythere* enter the deep ocean at this time.

Through their history the ostracods exhibit a number of generalized evolutionary trends. In the Palaeozoic, and in common with several other groups of crustaceans, the ostracods evolved towards smaller size and a simplification of muscle-scar patterns. Through time the hinge became shorter and more robust, terminal elements became more pronounced. Mesozoic families evolved curved hinge lines and the podocopids developed an ornament comprising three longitudinal ribs that can be used in phylogenetic reconstruction.

Applications of ostracods

The usefulness of ostracods in biostratigraphy has declined over the last 20 years as conodonts in the Palaeozoic and planktonic forams and calcareous nannoplankton in the Mesozoic to Recent have become more widely used. In addition, a high degree of endemism, and the often benthic niche, has restricted

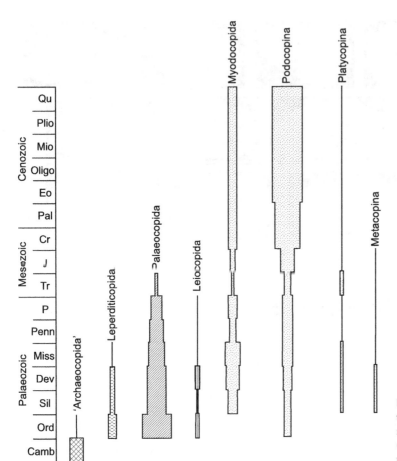

Fig. 20.14 Diversity of ostracod taxa through time. Width of scale bar equals numbers of families through time. (Data from Whatley et al. in Benton 1993, pp. 343–357.)

the use of ostracods in global correlation. However, ostracods can be useful in regional correlation (e.g. Bate & Robinson 1978). We can also note the value of ostracods in sedimentology (Brouwers, in DeDeckker et al. 1988, pp. 55–77). Krutak (1972), for example, suggested that the ostracod valve length can be used to estimate the original grain size in recrystallized sedimentary rocks. Oertli (1971, pp. 137–151) outlines how ostracod valves can be used to gauge sedimentation rate, current strength and compaction in sedimentary rocks. Two facets of ostracod population studies are useful in sedimentology: adult/juvenile ratios provide an important method of determining whether a fossil assemblage is autochthonous; trends in adult/juvenile ratio with depth can also be used to identify bathymetric gradients.

Ostracods have their widest utility in palaeoenvironmental analysis (e.g. Figs 20.15, 20.16). The majority of Recent ostracod genera are found in Miocene rocks and many have close relatives in Mesozoic assemblages – inference and uniformitarianism can therefore be used in detailed palaeoecology. Palaeoecology can also be inferred from carapace morphology. Where the ecological parameters of living species are precisely known, the history of changes in rainfall, temperature, salinity and alkalinity recorded in Quaternary lake sediments, for example, can be charted (see Delorme, in Oertli 1971, pp. 341–347; Lister, in DeDeckker et al. 1988, pp. 201–219). Many studies include both approaches supported by evidence from studies of preservation (e.g. the valve to whole-carapace ratio), sedimentology, the associated

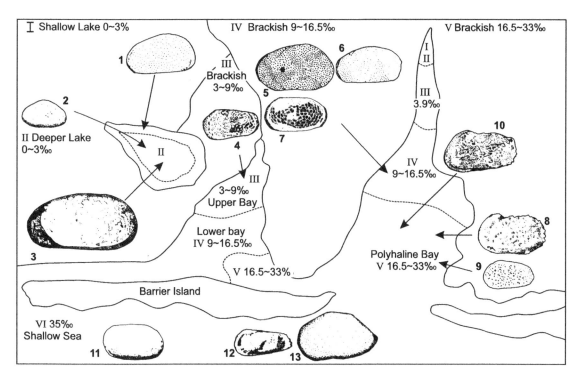

Fig. 20.15 Ostracod palaeoecology in the Late Eocene of the Hampshire basin, England. (a) The environments as reconstructed from the ostracod fauna. I, Shallow lake: 1, *Candona daleyi*; II, Deep lake: 2, *Cypridopsis bulbosa*; 3, *Moenocypris reidi*; III, Brackish 3~9‰: 4, *Cytheromorpha bulla*; IV, Brackish 16.5~33‰: 5, *Neocyprideis colwellensis*; 6, *Neocyprideis williamsoniana*; 7, *Cladarocythere hantonensis*; V, Brackish 16.5~33‰: 8, *Bradleya forbesi*; 9, *Haplocytherida debilis*; 10, *Cyamocytheridea herbertiana*; VI, Shallow sea 35‰: 11, *Cytherella* cf *C. compressa*; 12, *Idiocythere bartoniana*; 13, *Bairdia* sp. (Simplified and redrawn after Keen 1977.)

flora and fauna and stable isotope analyses (e.g. Griffiths & Holmes 2000).

Ostracods are especially useful for outlining the nature of palaeosalinities and their fluctuations in marginal marine successions (Neale, in DeDeckker *et al.* 1988, pp. 125–157) such as those of the Late Carboniferous (Pollard 1966), Middle Jurassic (Wakefield 1995) and the Cenozoic (Keen 1977; Fig. 20.15). Reviews of the palaeoecology of limnic ostracods and their applications, including carapace geochemistry, can be found in Carbonel *et al.* (1988) and papers in De Deckker *et al.* (1988).

Ostracods are valuable indicators of past climates because of their abundance and diversity in sediments and because there is a strong correlation between their distribution and temperature. Hazel (in DeDeckker *et al.* 1988, pp. 89–103) documented the palaeoclimatic controls on Pliocene to Early Pleistocene ostracod faunas from the marine sediments of the Coastal Plain of southwestern Virginia and North Carolina and of living species along the Atlantic Shelf of North America. Latitudinal control of marine ostracods is marked at the present day due to the existence of polar ice caps and narrow climate zones from the poles to the equator. Between the Late Palaeozoic and the Early Tertiary the poles were largely clear of ice and the climate zones were much less compressed. Post-Eocene faunas are much more strongly controlled by latitude than their Mesozoic counterparts.

Ostracods have also been used in the reconstruction of continental palaeoclimate records (DeDeckker & Forester, in DeDeckker *et al.* 1988, pp. 175–201). However, long-term changes in fossil assemblages associated with climate change may also reflect

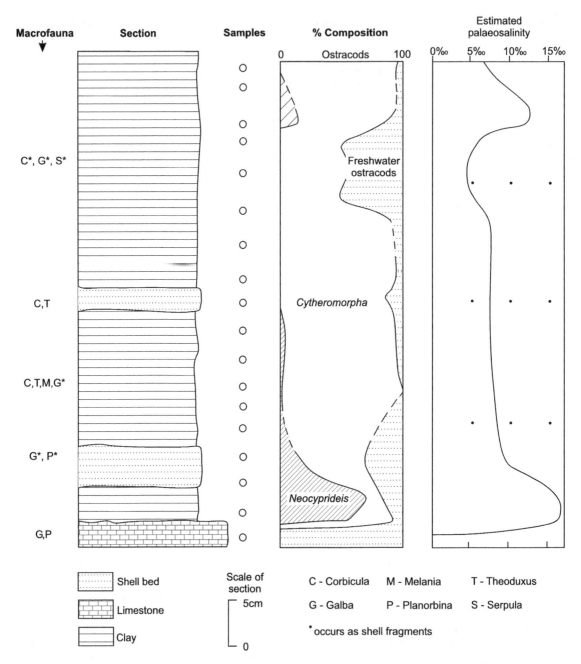

Fig. 20.16 Changes in the proportions of freshwater and brackish-water ostracoda with inferred salinity changes from part of the Lower Headon Beds. C, *Corbicula*; G, *Galba*; M, *Melania*; P, *Planorbina*; S, *Serpula*; T, *Theodoxus*. * Occurs as shell fragments. (Based on Keen 1977, with modifications from Neale in DeDeckker et al. 1988.)

changes in palaeoceanography and there is an increasing literature on the definition of ancient water masses using ostracods, particularly benthic psychrospheric forms (e.g. Ayress et al. 1997; Majoran et al. 1997; Majoran & Widmark 1998; Majoran & Dingle 2001).

The palaeoclimate signal reflected in the long-term changes in ostracod assemblages may also be in part due to continental movements from warm to cooler latitudes (e.g. the northward migration of India through the Late Jurassic to Eocene), and there are a number of studies in which ostracods have been used to determine the former position of continents. The Upper Jurassic and Lower Cretaceous non-marine ostracod faunas of north-east Brazil and West Africa are essentially the same, demonstrating these were juxtaposed if not connected at this time (Krömmelbein 1979). Schallreuter & Siveter (1985) have examined the distribution of Ordovician and Silurian ostracods across the Iapetus Ocean demonstrating that many Middle and Upper Ordovician genera are common on both sides of the ocean. Williams et al. (2003) were able to plot migration routes and rates between the rapidly converging continents of Laurentia, Baltica and Avalonia. Both studies suggest either the Iapetus Ocean was not as deep or as wide as previously thought, or that ostracods were able to migrate across a wider ocean with the help of ocean islands. Enhanced provincialism in Silurian ostracods can be related to global eustatic sea-level changes associated with the end Ordovician glaciation.

Populations of podocopid ostracods have also been described from seamounts, gyots and ocean islands where they show a high degree of endemism and both parapatric and sympatric speciation driven by both biotic (competition) and abiotic (changing summit environments) factors (Larwood et al., in Moguilevsky & Whatley 1996, pp. 385–403).

Ducasse & Moyes (in Oertli 1971, pp. 489–514) demonstrated how ostracods could be employed to plot the changing position of shorelines in the Tertiary rocks of Aquitaine. The same symposium volume contains numerous fine examples of their value to palaeoecology, as for example in the Devonian of the Eifel region where lagoon, back-reef, reef-core, fore-reef and offshore ostracod assemblages can be recognized, controlled largely by water turbulence (Becker, in Oertli 1971, pp. 801–816). Ostracod assemblages also change with sea level. Quaternary sections in North America indicate interglacial highstands are characterized by marine ostracod assemblages and low percentages of marginal marine species. Gradually, marine taxa are replaced by increasing numbers of marginal marine taxa as the glacial regression develops (Cronin, in DeDeckker et al. 1988, pp. 77–89).

Further reading

The 'Treatise on Ostracoda' by Moore (1961; under revision) gives an overview of the whole group at that time and useful morphological information with taxonomy down to genus level. Bate & Robinson (1978) compiled a useful volume on some stratigraphically important ostracods, ranging from Ordovician to Pleistocene. Quaternary ostracods and their applications have been comprehensively reviewed in a series of papers edited by Holmes & Chivas (2002). For further papers dealing with ostracod applications, ecology and evolution the reader is referred to the symposium volumes edited by Neale (1969), Oertli (1971), DeDeckker et al. (1988), Whatley & Maybury (1990), McKenzie & Jones (1993) and Horne & Martens (2000). For reviews on the salinity tolerance of ostracods the reader is directed to Neale (in DeDeckker et al. 1988, pp. 125–157) and Whatley (1983). Moguilevsky & Whatley (1996) reviewed ostracods in oceanic environments. For specific identification, palaeontologists should consult the *Catalogue of Ostracoda* by Ellis & Messina (1952 to date) or *A Stereo Atlas of Ostracod Shells*, published by the The Micropalaeontological Society and Kempf (1980, 1986, 1995, 1997). Living ostracods from Britain and Europe can be identified using Athersuch et al. (1989) and Meisch (2000). Whatley et al. (in Benton 1993, pp. 343–357) provide the most recent classification of fossil groups. Further information is available on the internet by searching under IRGO (International Research Group on Ostracoda), CYPRIS (Newsletter of IRGO), OSTRACON (ostracod listserver), ISO (International Symposium on Ostracoda) or EOM (European Ostracod Meeting).

Hints for collection and study

Living ostracods can be collected from marine and non-marine environments using the simplest of techniques. They are commonly found on seaweeds and in surface scrapes from mud flats (along with foraminifera) or in the organic detritus in freshwater ponds. Freshwater species are readily cultivated in a tank provided with pondweed and a little manure. To examine their general behaviour, study the washed muds or pond water in a glass petri dish using reflected light. The morphology and the limb movements are better seen if a specimen is placed with a blob of water under a cover slip on a glass cavity slide and viewed with transmitted light.

To extract ostracods from argillaceous rocks and marls, employ methods C to E (especially D; see Appendix). Method B is useful for hard chalks and limestones and method F where the carapace is phosphatic or silicified. Wash and dry the disaggregated sample as in methods G and I and mount as in method O.

Isolated ostracod valves should be examined in reflected light on both the internal and external surfaces. To see the muscle-scar patterns, pore canals and duplicature more clearly, place the specimen on a glass slide, cover with a drop of water (or glycerine, immersion oil or Canada Balsam) and a cover slip and view with transmitted light. Further suggestions for collection, preparation and study are given by Athersuch *et al.* (1989).

REFERENCES

Abushik, A.F. 1971. Orientation in Leperditiida. In: Vyalov O.S. (ed.) *Fossil Ostracoda*. Israel Program for Scientific Translations, Jerusalem, pp. 102–105.

Andreev, Yu.N. 1971. Sexual dimorphism of the Cretaceous ostracods of the Gissaro-Tadzhik region. In: Vyalov O.S. (ed.) *Fossil Ostracoda*. Israel Program for Scientific Translations, Jerusalem, pp. 102–105.

Angel, M.V. 1969. Planktonic ostracods from the Canary Island region; their depth distributions diurnal migrations, and community organization. *Journal of the Marine Biological Association UK* 49, 515–533.

Anderson, F.W. 1985. Ostracod faunas in the Purbeck and Wealden of England. *Journal of Micropalaeontology* 4, 1–67.

Anderson, F.W. & Bazley, R.A.B. 1971. The Purbeck Beds of the Weald (England). *Bulletin. Geological Survey of Great Britain* 34, 1–138.

Aparecido do Carmo, D., Whatley, R.C. & Timberlake, S. 1999. Variable noding and palaeoecology of a Middle Jurassic limnocytherid ostracod: implications for modern brackish water taxa. *Palaeogeography, Palaeoclimatology, Palaeoecology* 148, 23–35.

Athersuch, J., Horne, D.J. & Whittaker, J.E. 1989. *Marine and Brackish Water Ostracods*. In: Kermack, D.M. & Barnes, R.S.K. (eds) *Synopses of the British Fauna (New Series)*, no. 43. The Linnean Society.

Ayress, M., Neil, H., Passlow, V. *et al.* 1997. Benthonic ostracods and deep watermasses: a qualitative comparison of Southwest Pacific, Southern and Atlantic oceans. *Palaeogeography, Palaeocliamtology, Palaeoecology* 131, 287–302.

Barker, D. 1963. Size in relation to salinity in fossil and euryhaline ostracods. *Journal of the Marine Biological Association UK* 43, 785–795.

Bassett, M.G. & Lawson, J.D. 1984. Autecology of Silurian organisms. *Special Papers in Palaeontology* no. 32.

Bate, R.H. 1972. Phosphatized ostracods with appendages from the lower Cretaceous. *Palaeontology* 15, 379–393.

Bate, R.H. & Gurney, A. 1981. The ostracod genus *Loxoconcha* Sars from Abu Dhabi lagoon and the neighbouring near-shore shelf, Persian Gulf. *Bulletin. British Museum Natural History (Zoology)* 41, 235–251.

Bate, R.H. & Robinson, J.E. 1978. A stratigraphical index of British Ostracoda. *Special Issue, Geological Journal* no. 8.

Benson, R.H. *et al.* 1961. Ostracoda. In: R.C. Moore (ed.) *Treatise on Invertebrate Paleontology. Part Q Arthropoda 3: Crustacea*. Geological Society of America and University of Kansas Press, Lawrence, Kansas.

Benton, M.J. (ed.) 1993. *The Fossil Record 2*. Chapman & Hall, London.

Boardman, R.S., Cheetham, A.H. & Rowell, A.J. 1987. *Fossil Invertebrates*. Blackwell Scientific Publications, Oxford.

Brouwers, E.M., Cronin, T.M., Horne, D.J. & Lord, A.R. 2000. Recent shallow marine ostracods from high latitudes: implications for Late Pliocene and Quaternary palaeoclimatology. *Boreas* 29, 127–143.

Carbonel, P., Colin, J.-P., Danielopol, D.L., Loffler, H. & Neustrueva, I. 1988. Paleoecology of limnic ostracods: a review of some major topics. *Palaeogeography, Palaeoclimatology, Palaeoecology* 62, 431–461.

Cohen, A.C., Martin, J.W. & Kornicker, L.S. 1998. Homology of Holocene ostracode biramous appendages with those of other crustaceans: the protopod, epipod, exopod and endopod. *Lethaia* 31, 251–265.

Corrège, T. 1993. The relationship between water masses and benthic ostracod assemblages in the western Coral Sea, Southwest Pacific. *Palaeogeography, Palaeoclimatology, Palaeoecology* 105, 245–266.

DeDeckker, P., Colin, J.-P. & Peypoupet, J.-P. 1988. *Ostracoda in the Earth Sciences*. Elsevier, Amsterdam.

Ducasse, O. 1983. Etude de populations du genre *Protoargilloecia* (Ostracodes) dans les faciès bathyaux du Paléogène Acquitain: Deuxième test éffectué en domaine profond. Comparaison avec le genre *Cytherella. Geobios* 16, 273–282.

Ellis, B.F. & Messina, A.R. 1952 to date. *Catalogue of Ostracoda*. Special Publications. American Museum of Natural History. (Over 23 volumes in loose-leaf form.)

Forester, R.M. 1983. Relationship of two lacustrine ostracode species to solute composition and salinity: implications for paleohydrochemistry. *Geology* 11, 435–438.

Griffiths, H.I. & Holmes, J.A. 2000. *Non-marine Ostracods and Quaternary Palaeoenvironments*. Technical Guide 8. Quaternary Reseach Association, London.

van Harten, D. 1992. Hydrothermal vent Ostracoda and faunal association in the deep sea. *Deep Sea Research* 39, 1067–1070.

van Harten, D. 2000. Variable noding in *Cyprideis torosa* (Ostracoda, Crustacea): an overview, experimental results and a model from Catastrophe Theory. *Hydrobiologia* 419, 131–139.

Hartmann, G. 1963. Zur Morphologie und Ökologie rezenter Ostracoden und deren Bedeutung bei der Unterscheidung mariner und nichtmariner Sedimente. *Forschritte in der Geologie von Rheinland und Westfalen* 10, 67–80.

Heip, C. 1976. The spatial pattern of *Cyprideis torosa* (Jones, 1850) (Crustacea: Ostracoda). *Journal of the Marine Biological Association UK* 56, 179–189.

Holmes, A. & Chivas, A.R. (eds) 2002. *The Ostracoda: applications in Quaternary research*. AGU Geophysical Mongraph, No. 103.

Horne, D.J. 1995. A revised ostracod biostratigraphy for the Purbeck-Wealden of England. *Cretaceous Research* 16, 639–663.

Horne, D.J. & Martens, K. 2000 (eds). Evolutionary biology and ecology of Ostracoda. Developments in hydrobiology. *Proceeding. Theme 3 of the 13th International Symposium on Ostracoda, Chatham, 1997*. Kluwer Academic Publications, (reprinted from *Hydrobiologica* 419, 1–197).

Hsu, K.J. & Weissert, H.J. 1985. *South Atlantic Palaeoceanography*. Cambridge University Press, Cambridge.

Jarvis, I., Carson, G.A., Cooper, M.K.E., Hart, M.B., Leary, P.N., Tocher, B.A., Horne, D.J. & Rosenfeld, A. 1988. Microfossil assemblages and the Cenomanian-Turonian (Late Cretaceous) Oceanic Anoxic Event. *Cretaceous Research* 9, 3–103.

Keen, M.C. 1977. Ostracod assemblages and the depositional environments of the Headon, Osborne and Bembridge Beds (upper Eocene) of the Hampshire Basin. *Palaeontology* 20, 405–446.

Kempf, E. 1980. Index and bibliography of non-marine Ostracoda. I. Index A, Supplement 1. *Geologischen Instituts an der Universität Zuköln Sanderveroeffentlichungen* 35, 1–188.

Kempf, E. 1986. Index and bibliography of marine Ostracoda. I. Index A. *Geologischen Instituts an der Universität Zuköln Sanderveroeffentlichungen* 50, 1–762.

Kempf, E. 1995. Index and bibliography of marine Ostracoda. I. Index A, Supplement 1. *Geologischen Instituts an der Universität Zuköln Sanderveroeffentlichungen* 100, 1–239.

Kempf, E. 1997. Index and bibliography of non-marine Ostracoda. I. Index A. *Geologischen Instituts an der Universität Zuköln Sanderveroeffentlichungen* 109, 1–142.

Kesling, R.V. 1951. Terminology of ostracode carapaces. *Contributions. Museum of Paleontology, University of Michigan* 9, 93–171.

Knox, L.W. & Gordon, E.A. 1999. Ostracodes as indicators of brackish water environments in the Catskill Magnafacies (Devonian) of New York State. *Palaeogeography, Palaeoclimatology, Palaeoecology* 148, 9–22.

Krömmelbein, K. 1979. African Cretaceous Ostracods and their relations to surrounding continents. *Proceedings of the 37th Annual Biology Colloquium (1976): historical biogeography, plate tectonics and the changing environment* Oregon State University, pp. 305–310.

Krutak, P.R. 1972. Some relationships between grain size of substrate and carapace size in modern brackish-water Ostracoda. *Micropalaeontology* 18, 153–159.

Lethiers, F. & Whatley, R.C. 1994. The use of Ostracoda to reconstruct the oxygen levels of Upper Palaeozoic Oceans. *Marine Micropalaeontology* 24, 57–69.

Majoran, S. & Dingle, R.V. 2001. Palaeoceanographical changes recorded by Cenozoic deep sea ostracod assemblages from the South Atlantic and the Southern Ocean (ODP Sites 1087 and 1088). *Lethaia* 34, 63–83.

Majoran, S. & Widmark, J.G.V. 1998. Response of deep sea ostracod assemblages to Late Cretaceous palaeoceanographical changes: ODP Site 689 in the Southern Ocean. *Cretaceous Research* 19, 843–872.

Majoran, S., Widmark, J.G.V. & Kucera, M. 1997. Palaeoecological preferences and geographical distribution of Late Maastrichtian deep sea ostracods in the South Atlantic. *Lethaia* 30, 53–64.

Martens, K. 1985. Salinity tolerance of *Mytilocypris henricae* (Chapman) (Crustacea, Ostracoda). *Hydrobiologica* 124, 81–83.

McKenzie, K.G. & Jones, P.J. (eds) 1993. *Ostracoda in the Earth and Life Sciences*. Balkema, Rotterdam, Brookfield.

Meisch, C. 2000. *Freshwater Ostracoda of Western Central Europe, Süsswasserfauna von Mitteleuropa*, 8/3. Spektrum Akad. Verlag, Heidelberg.

Moguilevsky, A. & Whatley, R. (eds) 1996. *Microfossils and Oceanic Environments*. University of Wales/Aberystwyth Press, Aberystwyth.

Moore, R.C. 1961. Ostracoda. In: Moore, R.C. (ed.) *Treatise on Invertebrate Paleontology. Part Q. Arthropoda 3: Crustacea*. Geological Society of America and University of Kansas Press, Lawrence, Kansas.

van Morkhoven, F. 1962–1963. *Post-Palaeozoic Ostracoda. Their morphology, taxonomy and economic use*, vol. 1 (1962, general); vol. 2 (1963, generic descriptions). Elsevier, Amsterdam.

Müller, K.J. 1979. Phosphatocopine ostracodes with preserved appendages from the Upper Cambrian of Sweden. *Lethaia* 12, 1–27.

Neale, J.W. (ed.) 1969. *The Morphology and Ecology of Recent Ostracoda*. Oliver & Boyd, Edinburgh.

Oertli, H.J. (ed.) 1971. *Colloque sur la paléoecologie des Ostracods*. Bulletin du Centre de Recherches Pau-SNPA, Pau, France.

Pokorny, V. 1958. *Grundzüge der Zoologischen Mikropaläontologie*. Berlin, VEB Deutscher Verlag der Wissenschaften.

Pollard, J.E. 1966. A non-marine ostracod fauna from the coal measures of Durham and Northumberland. *Palaeontology* 9, 667–697.

Schallreuter, R.E.L. & Siveter, D. 1985. Ostracods across the Iapetus Ocean. *Palaeontology* 28, 577–598.

Siveter, D.J. 1980. British Silurian Beyrichiacea (Ostracoda). *Monograph. Palaeontographical Society* 133 (issued as part of volume 133 for 1979), 76pp.

Siveter, D.J. & Willams, M. 1997. Cambrian Bradoriid and phosphatocopid arthropods of North America. *Special Papers in Palaeontology* 57, 1–69.

Siveter, D.J., Williams, M., Peel, J.S. & Siveter, D.J. 1996. Bradoriids (Arthropoda) from the Early Cambrian of North Greenland. *Transactions of the Royal Society of Edinburgh: Earth Sciences* 86, 113–121.

Smith, R.J. 2000. Morphology and ontogeny of Cretaceous ostracods with preserved appendages from Brazil. *Palaeontology* 43, 63–98.

Vannier, J., Shang, Qi, W. & Cohen, M. 2001. Leperditicopid arthropods (Ordovician-Late Devonian): functional morphology and ecological range. *Journal of Paleontology* 75, 75–95.

Wakefield, M.I. 1994. Middle Jurassic (Bathonian) ostracoda from the Inner Hebrides, Scotland. *Monograph of the Palaeonographical Society, London No. 596* 148, 1–89.

Wakefield, M.I. 1995. Ostracoda and palaeosalinity fluctuations in the Middle Jurassic Lealt Shale Formation, Inner Hebrides, Scotland. *Palaeontology* 38, 583–619.

Waloszek, D. 1999. On the Cambrian diversity of Crustacea. In: Schram, F.R. & von Vaupel Kein, J.C. (eds) *Crustaceans and the Biodiversity Crisis, Proceedings, Fourth International Crustacean Congress, Amsterdam, 1998* 1, 3–27.

Whatley, R.C. 1983. The application of Ostracoda to palaeoenvironmental analyses. In: Maddocks, R.F. (ed.) *Applications of Ostracoda*. University of Houston Geosciences, Houston, Texas, pp. 51–57.

Whatley, R.C. & Ayress, M.A. 1988. Pandemic and endemic distribution patterns in Quaternary deep sea Ostracoda. In: Hanai, T., Ikeya, N. & Ishizaki, K. (eds) *Evolutionary Biology of Ostracoda, its Fundamentals and Applications. Proceedings of the Ninth International Symposium on Ostracoda, Shizuoka, Japan*. Elsevier, Amsterdam. (Also published as *Developments in Palaeontology and Stratigraphy* 11, 739–755.)

Whatley, R.C. & Maybury, C. (eds) 1990. *Ostracoda and Global Events*. Chapman & Hall, London.

Whatley, R.C., Pyne, R.S. & Wilkinson, I.P. 2003. Ostracoda and palaeo-oxygen levels, with particular reference to the Upper Cretaceous of East Anglia. *Palaeogeography, Palaeoclimatology, Palaeoecology* 194, 355–386.

Williams, M. & Siveter, D.J. 1998. British Cambrian and Tremadoc Bradoriid and Phosphatocopid arthropods. *Monograph. Palaeontolographical Society* 152, 49pp.

Williams, M., Floyd, J.D., Salas, M.J., Siveter, D.J., Stone, P. & Vannier, J.M.C. 2003. Patterns of ostracod migration for the 'North Atlantic' region during the Ordovician. *Palaeogeography, Palaeoclimatology, Palaeoecology* 195, 193–228.

CHAPTER 21

Conodonts

Conodonts (or **euconodonts**) were a group of primitive jawless vertebrates, which ranged from the Upper Cambrian to the uppermost Triassic. They were the first vertebrates to produce a mineralized skeleton and are primarily known from scattered elements of their feeding apparatus. Individual elements are commonly 0.25–2 mm in size and are composed of calcium phosphate (calcium carbonate fluorapatite). Complete feeding apparatuses with 15 or more elements can be preserved in favourable conditions of low turbulence and rapid burial. Fossil evidence of the soft parts of these animals is extremely rare.

Conodonts have become the premier microfossils for dating Palaeozoic shallow marine carbonates and have been widely used in palaeoecological and biogeographical studies. Conodonts have become a central part of the ongoing debate on the origin of the vertebrate skeleton. Conodont colour alteration (CAI) has been applied in the interpretation of basin histories, regional metamorphic studies and in the search for hydrocarbons and minerals.

The study of conodonts was greatly advanced in 1983 by the discovery of complete conodont animals in the Carboniferous Granton Shrimp Bed near Edinburgh (Briggs *et al.* 1983). This lagerstätten has yielded 10 animal specimens attributable to at least two species. The excellent preservation of the material has provided detailed information on the anatomy of conodont animals and has indicated a chordate affinity for the group (Aldridge *et al.* 1986, 1993). A further animal of Lower Silurian age (Mikulic *et al.* 1985; Smith *et al.* 1987) and giant conodont animals of Upper Ordovician age (Aldridge & Theron 1993; Gabbott *et al.* 1995) have since been discovered. These animals plus ultrastructural studies of the elements (Sansom *et al.* 1992) have placed the conodonts firmly within the Chordata.

Soft anatomy

The Granton conodont animals (Fig. 21.1a) are small (c. 40 mm), laterally compressed and eel-like. Two of the Granton specimens preserve details of the head region distinguished by two lobe-shaped structures bilaterally disposed about the mid-line and have been interpreted as **sclerotic cartilages** that surrounded the eyes. Behind these, two small discs have been interpreted as the **optic capsules**. Indistinct transverse traces behind the head may be the remains of **branchial structures**. The head also bears the feeding apparatus (Fig. 21.1b) that allows for the orientation of the elements (see below). The main structures preserved in the trunk are the **notochord, chevron-shaped muscle blocks** and **caudal fin rays**.

Similar features have been found in specimens of the Upper Ordovician animals from the Soom Shale in South Africa (Fig. 21.2). More than 100 partial animals have been found from this deposit, with exquisite cellular level preservation. They differ from the Granton animals in the architecture of the apparatus, the size of the elements and the overall size of the animal, which could have reached 1 m in length. A single Soom specimen preserves the sclerotic capsules, **extrinsic eye musculature** and the trunk muscles that show details of **rod-like muscle fibres, myofibrils** and possibly **sarcomeres** (Gabbott *et al.* 1995).

A morphologically distinct conodont animal, *Panderodus unicostatus*, has been discovered in Lower Silurian strata (Mikulic *et al.* 1985). This animal is

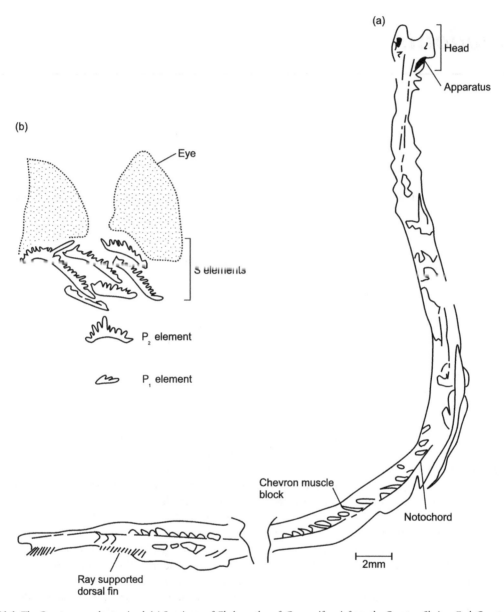

Fig. 21.1 The Granton conodont animal. (a) Specimen of *Clydagnathus* cf. *C. cavusiformis* from the Granton Shrimp Bed, Granton Sandstone (Dinantian), Edinburgh, Scotland showing the general anatomy. Scale bar = 2 mm. (b) Enlarged view of the counterpart of the apparatus in the Granton specimen, ×35. ((a) From Briggs *et al.* 1983 with permission.)

Chapter 21: Conodonts

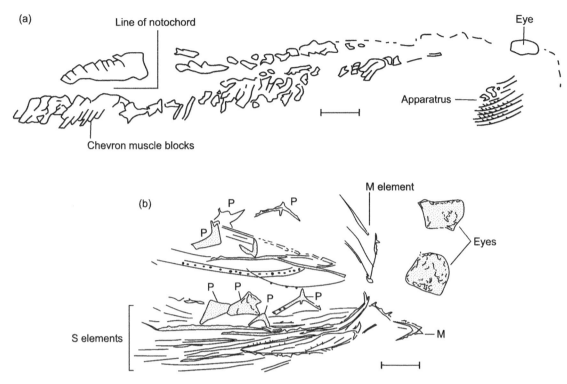

Fig. 21.2 The Soom conodont animal. (a) Specimen of *Promissum pulchrum* from the Soom Shale (Upper Ordovician), South Africa showing general anatomy. Scale bar = 10 mm. (b) Enlarged sketch of the *Promissum* apparatus. Scale bar = 2 mm. ((a) Traced from a photograph kindly provided by Professor R.J. Aldridge; (b) redrawn from Aldridge & Theron 1993, figure 2 with permission.)

poorly preserved, but appears to be dorso-ventrally compressed; importantly, the head contains well-preserved coniform conodont elements (Fig. 21.3).

Conodont elements

The resistant nature of the tooth-like conodont elements means they are the only commonly preserved parts of the animal. Most pre-Carboniferous conodont elements are constructed of two parts, the **crown** and the **basal body** (Fig. 21.4). The basal body occupies an opening, the **basal cavity**, in the crown; in many specimens the basal body is absent or lost and it is rare in post-Devonian conodont elements. The crown commonly comprises **hyaline**, lamella tissue with growth lines and an internal opaque tissue, the 'white matter', normally seen in the cores of the serrated **denticles** and the **cusp**, the often larger denticle above the tip of the basal cavity. The basal body is more variable and can preserve lamellar or spherulitic structure and may or may not contain tubules. Electron microscopy of polished and etched elements has led to the suggestion that the lamellar crown tissue is homologous with **enamel** (Fig. 21.4) and the white matter may also be a form of enamel, unique to conodonts. Tissues of the basal body bear comparison with **globular calcified cartilage** and a variety of **dentine** types (Fig. 21.4). However, these interpretations have not been universally accepted (e.g. Forey & Janvier 1993). Histochemical tests have been used to determine the nature of these tissues (Kemp & Nicoll 1996). The results of these are at odds with the structural interpretations and have yet to be verified, as does the claim of preserved DNA in Ordovician and Devonian conodonts (*op. cit.*).

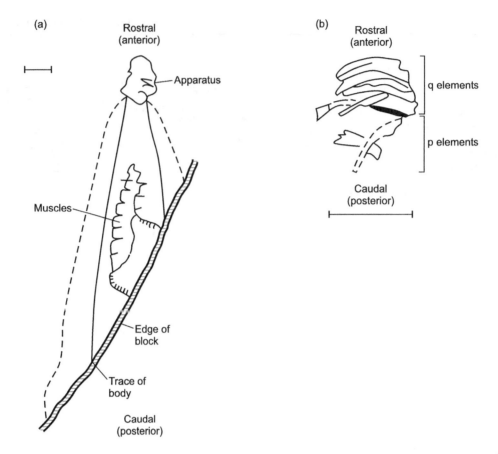

Fig. 21.3 The Waukesha conodont animal. Scale bar = 2 mm. (a) Partially preserved specimen of *Panderodus unicostatus* from the Brandon Bridge Formation (Silurian) Waukesha, Wisconsin. Scale bar = 1 mm. (b) Enlarged sketch of the apparatus found in the Waukesha animal. Scale bar = 1 mm. ((a) Redrawn from Smith *et al.* 1987.)

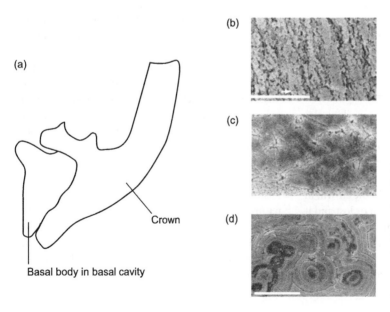

Fig. 21.4 Structure and histology of a conodont element. (a) Line drawing of a longitudinal section through *Cordylodus* illustrating the basal body and crown. (b) Crystallites of the crown in *Cordylodus* orientated transversely to the growth lamellae (arrow). Scale bar = 5 μm. (c) Detail of white matter in *Panderodus unicostatus* showing well-developed lacunae and interconnected and radiating canaliculi, image about 60 μm across. (d) SEM photomicrograph of the basal body of *Cordylodus* showing the spherulitic texture of the tissue, which has been interpreted as globular calcified cartilage or dentine. Scale bar = 20 μm. (Photomicrographs from Sansom *et al.* 1992, figures 1 and 2 (with permission, copyright MacMillan Magazines Ltd).)

Apparatus architecture

The three-dimensional architecture of relatively few conodont apparatuses is known. The apparatus of the ozarkodinids has become the stereotypical apparatus for all non-coniform species, though it is highly derived in evolutionary terms. Careful modelling of the elements and compaction studies of **natural assemblages**, complete apparatuses preserved on bedding planes, has enabled the elucidation of the three-dimensional architecture of this apparatus (e.g. Aldridge *et al.* 1987; Purnell & Donoghue 1998; Fig. 21.5).

Elements fall into at least two morphologically and functionally distinct domains. In non-coniform taxa, apparatuses can be described as comprising a rostral domain of paired **S elements** (Sb, Sc and Sd; plus a single **Sa** element on the midline), associated with a pair of dorso-lateral **M elements**, and a caudal domain comprising up to four pairs of **P elements** (the **Pa, Pb, Pc** and **Pd** elements). At rest the long axis of the S and M elements lay subparallel to the midline whilst that of the P elements was orientated dorso-ventrally (Fig. 21.5a,b).

In the majority of species, the location within each domain is interpreted from the shape category of the element and cannot be substantiated by bedding plane assemblages, these locations are labelled Pa, Pb etc., and this makes the comparison and description of

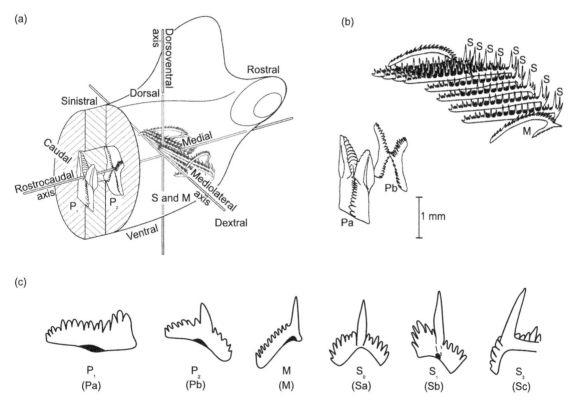

Fig. 21.5 Conodont apparatus, orientation and nomenclature. (a) Biological orientation in conodonts as applied in the head and apparatus of an ozarkodinid conodont. (b) Complete ozarkodinid conodont apparatus Scale bar = 1 mm. (c) Elements of *Ozarkodina confluens*. Locational terminology P_1–S_3 as advocated by Purnell *et al.* (2000) for locations confirmed by natural assemblages. Pa–Sc are not confirmed by natural assemblages but inferred from morphological similarity. The P_1–S_3 nomenclature represents a 'dental formula' comparable to that used for mammals. ((a) Reproduced from Purnell *et al.* 2000, figure 1 (with permission copyright, The Paleontological Society); (b) redrawn after Aldridge *et al.* 1987 and Purnell 1993b.)

homologous elements in different species unjustifiable. To address this problem Purnell et al. (2000) introduced new terms for orientation in conodonts and their elements (Fig. 21.5c) in which locations were known from animal associations and natural assemblages. Element locations were defined according to the relationships between elements with reference to the principal axes of the body confirmed in bedding plane assemblages. In the standard ozarkodinid apparatus it takes the form of letters with numeric subscripts (e.g. P_1, P_2, S_0–S_4, **M**). The S locations are numbered outwards from the central S_0. Though this may appear confusing to the general reader, it is important to retain the distinction between biological species in which homologous elements are known and can be compared in evolutionary studies and species in which homology can only be inferred.

The reconstruction and description of coniform apparatuses has lagged behind that of more complex conodonts and no real consensus exists as to the nomenclature to be applied to the shape and location of these elements in apparatuses. Sansom et al. (1994) proposed a scheme for panderodontid conodonts based on diagenetically fused clusters of elements and the natural assemblage preserved in the *Panderodus* animal (Fig. 21.6). The apparatus can be divided into a rostral domain containing **q** (costate) elements (including **qa**, **qg** and **qt** elements) and a caudal domain of **p** (acostate) elements (including **pf** and **pt**). Both p and q elements were paired and at rest lay across the midline of the animal. The single, symmetrical **ae** element is thought to have been located on the midline, probably in a rostro-dorsal position within the oropharyngeal cavity (Fig. 21.6). The location within each domain is interpreted from the shape category of the element, for example the qa location in the apparatus is occupied by an arc-shaped shaped (arcuatiform) element.

Function

Conodont elements are now accepted as the components of oropharyngeal feeding apparatus. Two functional paradigms have been proposed: that conodonts were **microphagous** suspension feeders in which the S and M elements supported a soft, ciliated sieve structure to filter food particles, whereas the P elements crushed these particles between tissue-covered surfaces. The alternative is that conodonts were **macrophagous**, with the S and M elements actively grasping the food that was then sliced and crushed by the P elements; in this model the elements are not considered to have been tissue covered when the animal was feeding. Functional modelling (Aldridge et al. 1987; Purnell 1993a; Purnell & Donoghue 1998), growth studies (Purnell 1994) and the discovery of microwear facets (Purnell 1995) on the denticles of some elements support a grasping and processing function for the apparatus. Similarly, three-dimensional modelling of the *Panderodus* apparatus (Sansom et al. 1994) indicates that the rostral domain could be everted to fulfil a grasping function whilst the caudal domain was retained in the pharynx and processed the food particles.

The question of homology between coniform and non-coniform conodonts is unresolved. Natural assemblages of few conodonts are known, particularly from the Lower Ordovician, and the apparent differences in orientation of elements from the rostral domains of coniform and non-coniform groups make demonstration of homologies difficult. However, if conodonts are a monophyletic clade it is likely that the number of elements in fully developed apparatuses was fairly constant and that only a small number of apparatus plans existed. This would have been in much the same way as only a few basic dentitions, relating to the feeding style, occur in mammals.

Growth

Conodonts grew by the polycyclic, appositional addition of layers of crown enamel, so that the inner lamellae are the oldest. This is unusual in that in other vertebrates, the eruption of the permanent teeth destroys the enamel organ and enamel layers cannot be overgrown. Armstrong & Smith (2001) defined the lamellae within the crown as comprising minor and major increments and concluded minor increments are equivalent to the cross-striations in hominoid enamel. In *Protopanderodus varicostatus* minor increments typically have a minimum thickness of ~1 μm,

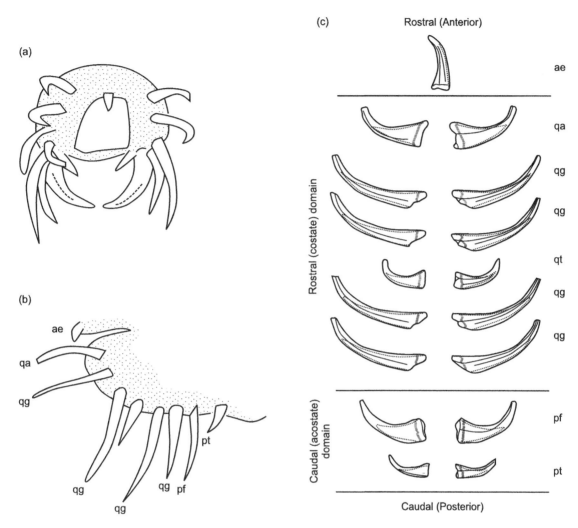

Fig. 21.6 Diagramatic representation of the *Panderodus unicostatus* apparatus. (a) Rostral view in grasping mode. (b) Lateral view in grasping mode. (c) Locational terminology, elements are approximately in proportion and approximately ×40. Left hand column of elements is viewed from the unfurrowed side, right hand column from the furrowed side, dashed lines indicate the outline of the basal cavity. ((c) Redrawn after Sansom *et al.* 1994.)

and by analogy with other vertebrates, were likely to be deposited in a day. However, some minor increments in this species were up to 7 μm thick, probably representing growth episodes, lasting up to a week. Major increments, which are marked by major growth discontinuities, incorporate about 1 month's growth. Intervening periods of function have unknown durations and it is therefore impossible to use these data for assessing the age of conodont animals.

Shape and orientation

Though conodont elements vary tremendously in shape a number of recurrent shape categories have been recognized based on the number and disposition of primary processes (Figs 21.7–21.9). Process direction is defined by the position on the cusp from which they arise. The orientation of conodont elements in the majority of taxa can now be related to their orientation

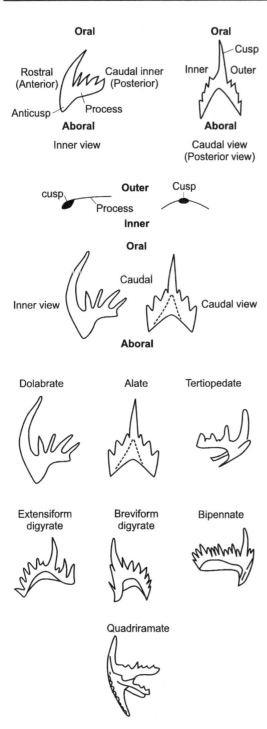

Fig. 21.7 Orientation and shape categories of M and S elements. (Redrawn from Sweet 1988.)

in the animal. The **oral** surface of the P elements is that which occludes and the **aboral** surface has the basal cavity or groove.

Dolabrate elements only have a caudal process and are commonly pick shaped. **Alate** elements are bilaterally symmetrical and have a caudal process and two lateral processes. In **modified alate** elements the caudal process is reduced to a slight inflation of the basal cavity margin. **Bipennate** elements have a caudal and rostral process; the rostral is usually shorter and commonly curves or is deflected inwards. **Digyrate** elements are broadly a similar shape to alate elements but they are asymmetrical; the caudal process is only rarely developed and the lateral processes are usually unequally developed and variably twisted in opposite directions. **Extensiform** digyrate elements have long lateral processes, curving towards their distal ends. **Breviform** digyrate elements have short lateral processes that curve from the base of the cusp. Digyrate elements can occupy P and S locations and the true orientations are not known from natural assemblages. It is therefore assumed a priori that where a P element is digyrate the longer of the two processes is ventral and when in an S location it is caudal. **Tertiopedate** elements have a caudal process and lateral processes that are asymmetrically disposed about the cusp. **Quadriramate** elements have rostral, caudal and two lateral processes. **Multiramate** is reserved for elements with more than four processes, although none are currently known.

Elements with a dorsal process are again divided into categories based on the number of primary processes (Fig. 21.8). **Segminate** elements possess a dorsal process that can bear one or more rows of nodes or ridges. **Carminate** and **angulate** elements have dorsal and ventral processes. A carminate element has an essentially straight aboral margin, whereas in angulate elements this is arched. **Pastinate** elements have three primary processes, dorsal, ventral and a rostral or caudal process. The processes may be adenticulate represented only by a conspicuous flange. **Stellate** elements have four primary processes that may bifurcate to form secondary processes.

These elements can also be **scaphate** (laterally expanded and entirely excavated on the aboral surface) or **planate** with the basal cavity reduced to a narrow

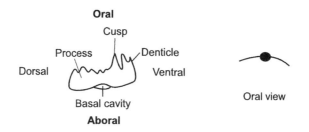

Fig. 21.8 Shape categories of platform (P) elements. Abbreviations: bp, basal pit; bg, basal groove; bc, basal cavity. (Redrawn from Sweet 1988.)

groove or pit by a zone of **recessive basal margin**. Elements can thus be described using a variety of combinations of terms, for example pastiniscaphate, pastiniplanate, etc. (Fig. 21.8).

Coniform elements (Fig. 21.9) comprise a more or less expanded base that encloses the **basal cavity**; the **cusp** is solid and tapers to a point. Some members of the Panderodontidae have an indented **furrow** along the entire length of the cusp that is thought to have been for muscle or ligament attachment. In **non-geniculate** elements there is a smooth transition from base to cusp (Fig. 21.9a), whereas in **geniculate** elements the concave edge of the base joins the cusp at an acute angle (Fig. 21.9b). **Rastrate** coniform elements develop denticles along the concave edge of the cusp (Fig. 21.9c).

Shape categories have been defined for elements attributed to members of the family Panderodontidae and are probably more widely applicable among coniform taxa. These categories are recognized on the basis of cusp curvature and cross-sectional symmetry. All *Panderodus* elements are non-geniculate. **Falciform** elements (pf in Fig. 21.6) are laterally compressed with an oval cross-section and have a short cusp; both the concave and convex edges are drawn into low **keels**. **Tortiform** (pt in Fig. 21.6) elements are spatulate and the cusp is twisted (relative to the base) away from the furrowed face, their convex margins are drawn out into sharp edges. **Graciliform** elements (qg in Fig. 21.6) are slender and have a low **costa** (a narrow ridge) running up each face; in general they have a keyhole-shaped cross-section. Both asymmetrical and symmetrical

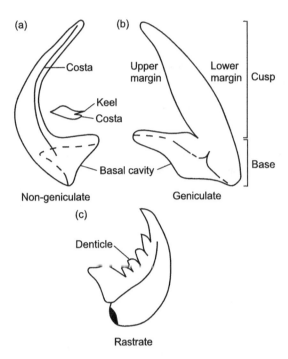

Fig. 21.9 Morphological terminology applied to coniform elements. (a) Non-geniculate. (b) Geniculate. (c) Rastrate. (Redrawn from Sweet 1988.)

high- and low-based forms can be found. **Arcuatiform** elements (qa in Fig. 21.6) have a costa running up one face; the cusp is variously twisted towards the unfurrowed face; rarely the concave edge may be serrate. **Truncatiform** (qt in Fig. 21.6) elements are short, and the unfurrowed face is drawn into a slight edge along the convex margin. The cusp is elongate, recurved and varies in the degree of twisting from species to species. **Aequaliform** elements (ae in Fig. 21.6) are truly symmetrical with a furrow developed on both faces.

Classification

Conodonts were first illustrated in 1856 when Pander described them as the remains of an unknown group of Palaeozoic fish. Hinde (1879) also considered conodonts to be fish teeth and one of his specimens from the Devonian of New York State preserved a cluster of conodont elements which he interpreted as the apparatus of a single species. Despite this discovery subsequent work utilized **form taxonomy**, describing each element as a separate species.

Complete natural assemblages and the availability of increasingly large collections led to the realization that individual conodont elements formed part of a much more complex multi-element apparatus. **Multi-element taxonomy** in which the whole apparatus is reconstructed from discrete elements and classified as a single species was first applied from the early 1960s. Walliser (1964) and Sweet & Bergström (1969) were instrumental in the development of this new and more biological taxonomy and by 1981 the conodont 'Treatise' had largely gone over to the multi-element system of classifying conodonts.

Since 1970 a number of classification schemes have been proposed to take into account the multi-element nature of conodont apparatuses. The most comprehensive is that proposed by Clark (with others in the *Treatise on Invertebrate Palaeontology*) as modified by Sweet (1988) and Aldridge & Smith (in Benton 1993, pp. 563–573). The scheme is at best considered provisional and has many limitations, not least that: the apparatuses of many taxa are incompletely known, that almost all have not been proven in natural assemblages and that the scheme is not based upon cladistic or other classificatory methods. Indeed, the Conodonta in this scheme is a grade of organization acquired independently in two coniform ancestral lineages that made their first appearance in the Upper Cambrian. The *Teridontus* lineage is thought to have been ancestral to all familiar conodont taxa whilst the *Proconodontus* lineage, in comparison, was impoverished (Sweet & Donoghue 2001). Sweet (1988) included the taxa in the *Teriodontus* lineage in the class Conodonti, which included five orders and 34 families, only some of these are monophyletic (include the ancestor and all its descendants). The five orders include the Protopanderodontida, Panderodontida, Prioniodontida, Prioniodinida and Ozarkodinida. Box 21.1 presents a familial level classification; space precludes illustrations of the full apparatus for eponymous genera, but further illustrations of additional elements can be found in Sweet (1988). For coniform taxa with no known natural assemblages the caudal and rostral domains are inferred from the *Panderodus* apparatus plan.

> **Box 21.1 Familial level classification of conodonts**
>
> **Order PROCONODONTIDA**
>
> Protoconodontidae (U. Camb.)
> Cordylodontidae (U. Camb.-L. Ord.)
> Fryxellodontidae (U. Camb.-L.Ord.)
>
> *Tentatively assigned*
>
> Ansellidae (Llanv.-Ash.)
> Belodellidae (Arg.-Fam.)
>
> **Order PROTOPANDERODONTIDA**
>
> Acanthodontidae (Trem.-Llanv.)
> Clavohamulidae (U. Camb.-Llanv.)
> Cornudontidae (Trem.-Crd.)
> Dapsilodontidae (Llan.-Lock.)
> Drepanoistodontidae (Trem.-Ash., Lock.)
> Oneotodontidae (Mer.-Arg.)
> Protopanderodontidae (Mer./Trem.-Ash./Llan.)
> Strachanognathidae (Arg.-Ash.)
>
> **Order PANDERODONTIDA**
>
> Panderodontide (Arg?/Llanv.-Giv.)
>
> **Order PRIONIODONTIDA**
>
> Balognathidae (Arg.-Ash.)
> Pygodontidae (Ord.)
> Cyrtoniodontidae (Arg.-Ash.)
> Distomodontidae (Sil.-Dev.)
> Icriodellidae (Ord.-Sil.)
> Icriodontidae (Lly?/Lud.-Fam.)
> Multioistodontidae (Arg.-Crd.)
> Oistodontidae (Trem.-Arg.)
> Periodontidae (Arg.-Ash.)
> Plectodinidae (Llanv.-Ash.)
> Prioniodontidae (Trem.-Ash.)
> Polyplacognathidae (Llanv.-Crd.)
> Pterospathodontidae (Lly.-Wen.)
> Rhipidognathidae (Arg.-Ash.,?Lud.)
>
> **Order PRIONIODINIDA**
>
> Bactrognathidae (Tou.-Vis./Spk?)
> Chirognathidae (Arg.-Crd.)
> Ellisoniidae (Bsh.-Trissic)
> Gondolellidae (Bsh.-Rht.)
> Prioniodinidae (Llan./Crd.-Gze.)
>
> **Order OZARKODINIDA**
>
> Anchignathodontidae (Tou.-Die.)
> Cavusgnathidae (Fam.-Sak./Art?)
> Elictognathidae (Fam.-Tou.)
> Gnathodontidae (Fam.-Spk.)
> Idiognathodontidae (Bsh.-Art.)
> Kockelellidae (Ash.-Lud.)
> Mestognathidae (Tou.-Spk.)
> Palmatolepidae (Giv.-Fam.)
> Polygnathidae (Pra.-Vis.)
> Sweetognathidae (Vis.-Gri.)

Order Proconodontida

This order contains the oldest known euconodonts with an apparatus comprising multiple pairs of deeply excavated, smooth cones. These are subsymmetrical or oval in cross-section and develop keels on the convex and/or concave margins. Families can be distinguished on the degree of differentiation of the apparatus. *Protoconodontus* (U. Camb., Fig. 21.10a) has an apparatus comprising a single morphotype of non-geniculate coniform elements with relatively large, deeply excavated basal cavities and a cusp with a subsymmetrical, oval cross-section and keels on the concave and convex margins. *Cordylodus* (U. Camb.-L. Ord., Fig. 21.10b) has an apparatus comprising two, perhaps up to six types of dolabrate elements. Elements were constructed of enamel, 'white matter' and a basal body, interpreted as being composed of either globular calcified cartilage (Sansom *et al.* 1992) or dentine. *Fryxellodontus* (U. Camb.-L. Ord., Fig. 21.10c) has an apparatus tentatively divided into two domains, containing four element morphotypes. All the elements are non-geniculate, smooth and basically coniform.

The Proconodontida is also likely to include families currently assigned within the order Belodellida (Sweet & Donoghue 2001). These include conodonts with elements that are non-geniculate, coniform, thin-walled and smooth. Elements of the caudal domain develop lateral costae and keels, and in *Ansella* (Llanv.-Ash., Fig. 21.10d) small serrations are developed along the concave edge of the q elements. The degree of element differentiation within the rostral and caudal domains appears to vary through time. *Ansella* contains five distinct element types whereas *Belodella* has only three. *Belodella* (Aren.-Fam., Fig. 21.10e) typically includes laterally compressed erect elements; the pf is falciform,

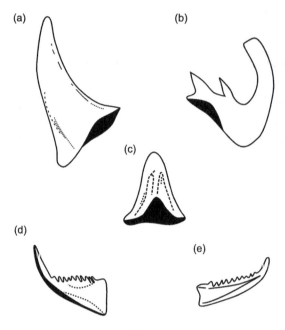

Fig. 21.10 Diagrams of the characteristic elements of members of the order Proconodontida, magnifications approximately ×20. (a) *Protoconodontus*. (b) *Cordylodus*. (c) *Fryxellodontus*. (d) *Ansella*. (e) *Belodella*. (Redrawn from figures in Sweet 1988.)

qa arcuatiform and ae bicostate. *Walliserodus* (Ord.-Sil., Fig. 21.16a–f) contains erect, deeply excavated elements with variable numbers of pronounced costae, the ae element is highly characteristic at species level.

Order Protopanderodontida

This order contains the majority of conodonts that have apparatuses composed entirely of coniform elements in which the basal cavity is short relative to the cusp. Smith (1991) illustrated a diagenetically fused cluster of *Parapanderodus* (Fig. 21.11i) which contains in order a single large qa element, a pair of qg elements, a single compressed and twisted qt element and a further pair of qg elements similar to those found in *Drepanodus* (Ord., Fig. 21.6g–k). Homology is inferred by comparison with the rostral domain of *Panderodus*. Families can be separated by the degree of morphological differentiation of the elements in both the rostral and caudal domains.

The Protopanderodontidae (e.g. *Protopanderodus* Mer./Tre.-Ash./Lly., Fig. 21.11a) includes genera of **hyaline** (lacking white matter), non-geniculate coniform elements with fine longitudinal striations.

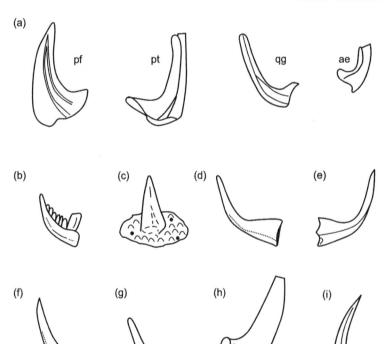

Fig. 21.11 Diagrams of the characteristic elements of members of the order Protopanderodontida, each genus is represented by a P element. (a) The apparatus plan of *Protopanderodus*, magnifications approximately ×20. This apparatus contains multiple pairs of qg elements. (b) *Belodina*. (c) *Clavohamulus*. (d) *Cornuodus*. (e) *Dapsilodus*. (f) *Drepanoistodus*. (g) *Oneotodus*. (h) *Strachanognathus*. (i) *Parapanderodus*. ((a), (h) Redrawn from figures in Armstrong 2000; (b)–(g), (i) from Sweet 1988.)

Belodina (Trem.-Llanv., Fig. 21.11b) has a panderodontid apparatus with a geniculate qa and rastrate qg, qt, pf and ae elements, but appears to lack a pt element. The Clavohamulidae contains a number of conodont genera that appear to have apparatuses constructed of a single element morphotype. In *Clavohamulus* (U. Camb.-Llanv., Fig. 21.11c) the basal part of the element is spread laterally to form a hemispherical mound ornamented with nodes. In other genera the base may be granulose or spinose. *Cornuodus* (Trem.-Crd., Fig. 21.11d) comprises non-geniculate elements and may have had an apparatus similar to *Panderodus*. *Drepanoistodus* (Drepanoistodontidae, Trem.-Ash., Fig. 21.11f) includes coniform elements with short flared bases and has a three-domain apparatus, interpreted as including recurved, costate/keeled, erect q elements, more or less symmetrical p elements and an ae element. The Oneotodontidae (e.g. *Oneotodus* Mer.-Arg., Fig. 21.11g) includes conodonts with one or two domain apparatuses of non-geniculate, commonly hyaline coniform elements with finely striate surfaces. The more advanced members of the family have q elements with longitudinal costae or ridges. *Strachanognathus* (Strachanognathidae, M.-U. Ord., Fig. 21.11h) appears to have a two- or three-domain apparatus with non-geniculate coniform elements with a single striate denticle. The Dapsilodontidae (Llanv.-Lock., e.g. *Dapsilodus*, Fig. 21.11e) contains taxa with laterally compressed non-geniculate coniform elements, including pf, ae and multiple pairs of qg elements (Armstrong 1990). This is atypical for a member of this order.

Order Panderodontida

The family Panderodontidae is only represented by *Panderodus* (Figs 21.6, 21.16l–p), which is excluded from the order Protopanderodontida on the basis of containing laterally furrowed elements. In other respects the apparatus plans are likely to be identical. Sansom *et al.* (1994) proposed that all coniform apparatuses which exhibit differentiation into a rostral domain (qa-qg), a caudal domain (pf-pt) and a symmetrical ae component should be classified within the order Panderodontida. Panderodontid conodonts were some of the most abundant in a wide range of marine habitats and are often the only conodonts present. The apparent lack of facies restriction suggests they were predominantly nektonic or pelagic animals.

Order Prioniodontida

This order arose in the Tremadoc and radiated to dominate many Ordovician faunas. By mid-Silurian times the order had declined, becoming extinct at the end of the Devonian. Members of this order are united in the possession of pastinate (or their platform equivalents) P elements, where an S_0 element is known this is alate. The apparatus plan of at least some prionoidontids (e.g. *Amorphognathus* Figs 21.12a, 21.15n–u) has been reconstructed based upon over 100 natural assemblages of *Promissum pulchrum*. *Promissum* contains 19 elements: two pairs of each of P_1, P_2, P_3 and P_4 elements, a single S_0 element and eight other S elements and a pair of rostro-lateral M elements. Unlike the apparatus of the Granton conodont animal (an ozarkodinid conodont) a dorsal row of P_1–P_3 are horizontally aligned as opposed pairs, with the P_4 elements arranged slightly below the P_2 elements. A pair of M elements is located to the rostero-dorsal of the P elements. A ventral domain contained S elements that form an oblique array below the P elements (Fig. 21.26).

Other prioniodontid species have been reconstructed from discrete element collections or partial natural assemblages and do not conform exactly to the *Promissum* plan. It is uncertain if this is due to biological difference. *Phragmodus* (Ord., Fig. 21.12b) has only two pastinate P elements, a dolabrate M and alate, tertiopedate and bipennate S elements, and is more likely not a member of this order. The elements of the rostral domain more closely resemble that found in prioniodinid conodonts. *Pygodus* (Llanv.-Crd., Fig. 21.12c) has an apparatus plan with three pairs of P elements and lacked an Sa element. Species of this genus are useful in the biostratigraphy of the Middle Ordovician and are most common in outer-shelf and deep-water biofacies. *Distomodus* (Sil. Fig. 21.12d) has a stelliscaphate Pa, pastinate Pb, modified tertiopedate M, and bipennate S elements. *Icriodella* (M. Ord.-Lly., Fig. 21.12e) has a pastinate Pa, pyramidal pastinate Pb, alate Sa and pyramidal tertiopedate

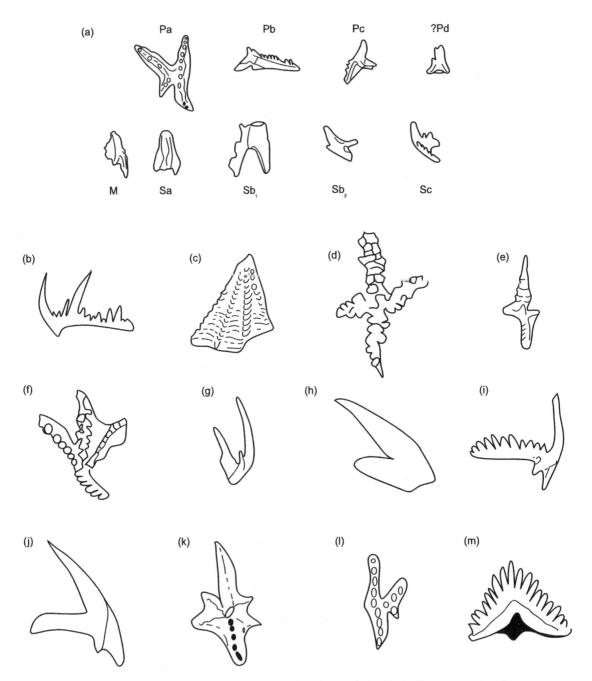

Fig. 21.12 Diagrams of the characteristic elements of members of the order Prioniodontida. (a) The apparatus plan of *Amorphognathus*, magnifications approximately ×20. The Sb_1 and Sb_2 elements are likely to be homologous with the S_1 and S_2 elements in *Promissum pulchrum* (Fig. 21.2). (b)–(l) Constituent genera represented by a P element, (m) is an Sa element: (b) *Phragmodus*; (c) *Pygodus*; (d) *Distomodus*; (e) *Icriodella*; (f) *Pedavis*; (g) *Multioistodus*; (h) *Oistodus*; (i) *Plectodina*; (j) *Prioniodus*; (k) *Polyplacognathus*; (l) *Pterospathodus*; (m) *Rhipidognathus*. ((a) Redrawn from figures in Armstrong 1990; (b)–(m) redrawn from figures in Sweet 1988.)

S elements. The Icriodontidae (Lly?/Lud.-Fam., e.g. *Pedavis*, Fig. 21.12f) includes conodonts with segminiscaphate Pa, pastinate Pb and compressed pastinate or coniform M element, and if there are elements in other locations these are variably coniform or weakly denticulate. The Multioistodontidae (Arg.-Crd., e.g. *Multioistodus*, Fig. 21.12g) includes apparatuses with elements that are essentially denticulated coniforms in which the Pa and Pb are similar pastinate elements, together with geniculate M elements and variously alate, tertiopedate and bipennate S elements. *Plectodina* (Plectodinidae, Llanv.-Ash., Fig. 21.12i) has an apparatus with a pastinate Pa, angulate Pb, a dolabrate or bipennate M, alate, digyrate, dolabrate and bipennate S elements. In *Prioniodus* (Prioniodontidae, Fig. 21.12j) the P locations are apparently filled by identical pastinate elements. The angulate M element has a markedly reclined cusp and in some taxa a caudal process which is significantly shorter than the rostral. The S elements can be alate, tertiopedate, bipennate and quadrate. Taxa assigned to the Polyplacognathidae (Llanv.-Car.) appear to have no known S elements. In *Polyplacognathus* (Fig. 21.12k) the dorsal domain contains a pair of stelliplanate and two pairs of pastiniplanate elements. *Pterospathodus* (Pterospathodontidae, Lly.-Wen., Fig. 21.12l) has three pairs of P elements (Pa-Pc) and lacks a truly symmetrical Sa element (Männik & Aldridge 1989). In *Rhipidognathus* (Rhipidognathidae, Arg.-Ash., ?Lud., Fig. 21.12m) a carminate Pa, angulate Pb, a modifed alate M, alate Sa, breviform digyrate Sb and a bipennate Sc complete the apparatus. Species of this genus are some of the only conodonts to inhabit hypersaline environments.

Whilst the family Oistodontidae has been placed in this order it includes essentially coniform genera. *Oistodus* (Trem.-Arg., Fig. 21.12h), for example, includes species with variably costate and laterally compressed, geniculate coniform P and M elements; the Pb and non-geniculate Sb elements may be pastinate, and elements in the rostral domain are non-geniculate.

Order Prioniodinida

The oldest known representatives of this order are taxa assigned to the Chirognathidae, the youngest to the Gondolellidae. This is a monophyletic order thought to have been derived from the Prioniodontida. Prioniodinids have digyrate elements in the caudal domain. In this order ecological specialism seems to have modified or reduced the apparatus in some taxa. The basic plan is exemplified by *Periodon* (Periodontidae, Ord., Figs 21.13a, 21.15g-m) which has an apparatus with extensiform digyrate Pa, digyrate Pb and a dolabrate M with denticles along the concave edge of the cusp. The rostral domain contains at least a breviform digyrate Sa, two pairs of Sb and a pair of Sc elements.

Clusters and natural assemblages are known from a single *Hibbardella* specimen from the Devonian Gogo Formation of western Australia (Nicoll 1977), an incomplete *Idioprioniodus* from the lower Namurian of Germany (Purnell & von Bitter 1996), *Neogondolella* from the Middle Triassic of Switzerland (Orchard & Rieber 1996) and a *Kladognathus* assemblage from the Mississippian of the USA (Purnell 1993b). Purnell (1993b) interpreted the apparatuses of *Hibbardella* and *Kladagnathus* to have the same basic plan and compared their architectures to that of the ozarkodinids.

The Chirognathidae (Arg.-Crd.) are the archetypal prioniodinid conodonts. *Erraticodon* (Fig. 21.13b) is the oldest chirognathid and has digyrate Pa and breviform digyrate Pb elements. S elements are essentially alate (Sa), tertiopedate (Sb) and bipennate (Sc). Families and genera are distinguished on the morphology of the P elements. Members of the Ellisoniidae (Bsh.-Tr., e.g. *Ellisonia*, Fig. 21.13c) have an apparatus very similar to that of *Idioprioniodus* (Prioniodinidae) comprising digyrate Pa, Pb and M elements. *Oulodus* (Prioniodinidae Llanv./Crd.-Gze, Fig. 21.13d) has extensiform digyrate P and Sb elements and a dolabrate M element. If correctly assigned to this order the Bactrognathidae (Tou.-Vis./Spk?) and the Gondolellidae (Bsh.-Rht.) are the only taxa in the order to possess platform elements in the Pa position (e.g. *Bactrognathus*, Fig. 21.13e). The Gondolellidae includes conodonts with an apparatus of a segminate Pa, breviform digyrate Pb and M elements, alate Sa, extensiform digyrate Sb and bipennate Sc elements. The S elements all develop a large number of slender denticles. This family includes the only abundant British

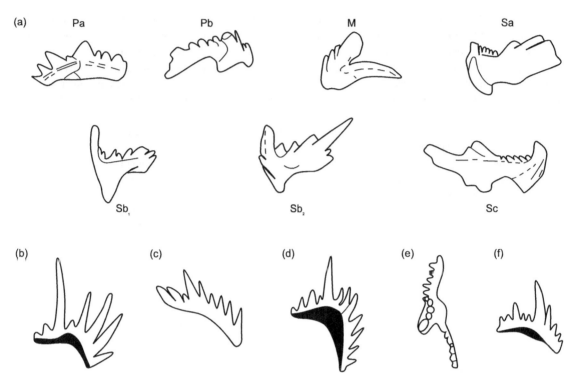

Fig. 21.13 Diagrammatic representation of the characteristic elements of members of the order Prioniodinida. (a) The apparatus plan of *Periodon*, magnifications approximately ×50. (b)–(f) Constituent genera represented by a P element: (b) *Erraticodon*; (c) *Ellisonia*; (d) *Oulodus*; (e) *Bactrognathus*; (f) *Merrillina*. Magnifications approximately ×30. ((a) Redrawn after illustrations in Armstrong *et al.* 1996; (b)–(f) redrawn after illustrations in Sweet 1988.)

Permian species *Merrillina divergens* (Fig. 21.13f). The origin of the Gondolelleidae is obscure. Carboniferous gondolellids characterize deep-water biofacies. It has been suggested that the acquisition of a segminate Pa element accompanied a reduction of element differentiation or reduced mineralization of elements in the rostral domain.

Order Ozarkodinida

The ozarkodinid apparatus comprises 15 elements (Fig. 21.5). At rest the P elements were orientated vertically with their long axes dorso-ventral and denticles opposed across the midline. The M and S elements were orientated with their long axes slightly oblique to the long axis of the apparatus, forming a V-shape opening rostrally. The S elements were inclined downwards to the rostral (Aldridge *et al.* 1987; Purnell & Donoghue 1998). The major family, the Spathognathodontidae, contains probably the best known conodont genus *Ozarkodina* (Fig. 21.5c) and is well represented by clusters and natural assemblages (Purnell & Donoghue 1998). The members of this family exhibit considerable morphological variability and are generally long ranging. *Ozarkodina* has a carminate P_1 and angulate P_2 element. Other families are largely distinguished on the morphology of the Pa element. The remaining elements include a single alate S_0, angulate P_2, digyrate, bipennate or dolabrate M, and two pairs of bipennate $S_{1/2}$ and digyrate $S_{3/4}$ elements (e.g. *Ozarkodina confluens*, Sil., Fig. 21.15a–f).

The Anchignathodontidae (Tou.-Die.) includes conodonts with a segminate Pa element. The family is exemplified by a succession of Carboniferous to earliest Triassic species assigned to *Hindeodus* (Fig. 21.14a). Constituent taxa of the Cavusgnathidae (Fam.-Sak./

Chapter 21: Conodonts 265

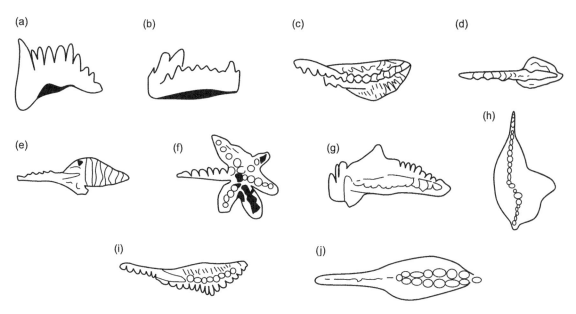

Fig. 21.14 Diagrammatic representation of the characteristic P elements of members of the order Ozarkodinida. (c)–(f), (h)–(j) Are oral views. The apparatus plan is illustrated in Fig. 21.3 (a) *Hindeodus*. (b) *Cavusgnathus*. (c) *Siphonodella*. (d) *Gnathodus*. (e) *Idiognathodus*. (f) *Kockelella*. (g) *Mestognathus*. (h) *Palmatolepis*. (i) *Polygnathus*. (j) *Sweetgnathus*. Magnifications approximately ×50. (Redrawn after figures in Sweet 1988.)

Art? e.g. *Cavusgnathus*, Fig. 21.14b) have a carminiscaphate Pa in which the ventral process is short and the dorsal process expanded to form a platform with rows of nodes or ridges along the margins. The Granton conodont animal, a species of *Clydagnathus*, is placed within this family. *Siphonodella* (Elictognathidae, Fam.-Tou., Fig. 21.14c) has a carminiplanate Pa. A complete elictognathid apparatus reconstruction has yet to be published. The Gnathodontidae (Fam.-Spk.) form an important and diverse stock of Devonian to Carboniferous conodonts that occupied a wide range of habitats. The oldest members of this family *Bispathodus* and *Gnathodus* (Fig. 21.14d) are known from natural assemblages. In *Gnathodus* the platform of the P_1 element is markedly asymmetrical. Idiognathodontid conodonts (Bsh.-Art.) are characterized by a carminiscaphate P_1 element with an inner surface that typically bears three longitudinal rows of nodes or denticles; the central row (the **carina**) is a continuation of the ventral free blade (e.g. *Idiognathodus*, Fig. 21.14e). *Kockelella* (Kockelellidae, Ash.-Lud., Fig. 21.14f) bears stelliscaphate Pa and angulate Pb elements. Through

evolution the Pa element becomes broader and develops processes. *Mestognathus* (Mestognathidae, Tou.-Spk., Fig. 21.14g) has a carminiplanate Pa with a high **blade** and V-shaped platform with ridged margins. The aboral surface has a grooved median keel that encloses a small basal pit. Species of *Mestognathus* are typically found in nearshore facies in the Carboniferous. In *Palmatolepis* (Palmatolepidae, Giv.-Fam., Fig. 21.14h) the carminiplanate Pa element has a nodose oral surface. The Pb is a bowed anguliplanate element with a high blade-like ventral process and a dorsal process that is superficially platform-like with a nodose rim. The processes of the S elements have needle-like denticles separated at intervals by much larger denticles. *Polygnathus* (Polygnathidae, Pra.-Vis., Fig. 21.14i) has a carminiscaphate Pa, angulate, dolabrate (or bipennate) Pb, alate Sa, extensiform digyrate Sb and bipennate Sc element. This is one of the few polygnathid apparatus reconstructions published. The Sweetognathidae (Vis.-Gri.) comprise a major group of Carboniferous and Permian conodonts. In *Sweetgnathus* (Fig. 21.14j) the Pa element

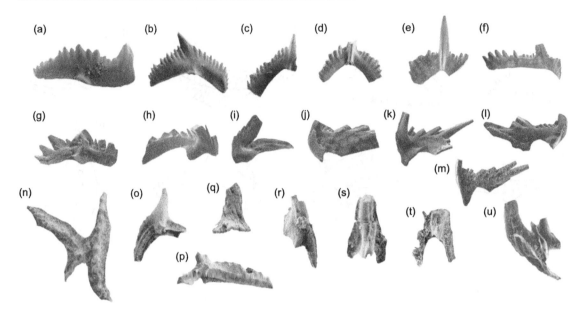

Fig. 21.15 SEM photomicrographs of the elements of selected conodont species. (a)–(f) *Ozarkodina confluens* (Ozarkodinida), Silurian: (a) sinistral P_1 element, caudal view, ×20; (b) sinistral P_2 element, caudal view, ×19; (c) sinistral M element, caudal view, ×19; (d) S_0 element, caudal view, ×17; (e) sinistral $S_{1/2}$ element, caudal view, ×17; (f) sinistral $S_{3/4}$ element, caudal view, ×18. (g)–(m) *Periodon aculeatus* (Prioniodinida), Lln-Ash. In the absence of any known bedding plane assemblages or natural clusters of this species' non-biological terminology is retained: (g) sinistral Pa element, posterior view, ×47; (h) sinistral Pb element, posterior view, ×47; (i) dextral M element, inner lateral view, ×35; (j) Sa element, lateral view, ×42; (k) dextral Sb element, posterior view, ×45; (l) sinistral Sc element, inner lateral view, ×39; (m) sinistral Sb element, inner lateral view, ×39. (n)–(u) *Amorphognathus ordovicicus* (Prioniodontida), Upper Ordovician. Biological orientation based upon homology with *Promissum pulchrum*: (n) Pa element, oral view, ×37; (o) Pc element, rostral view, ×46; (p) Pb element, rostral view, ×40; (p) ?Pd element, caudal view, ×46; (r) dextral M element, lateral view, ×24; (s) Sa element, caudal view, ×43; (t) Sb element, caudal view, ×27; (u) dextral Sc element, lateral view, ×55. ((b), (d) From Aldridge 1975 (with permission, copyright, Geological Society of London); (g)–(m) specimens previously illustrated by Armstrong 1997, plate 2 (reproduced with permission, copyright The Palaeontological Association); (n)–(u) specimens previously illustrated by Armstrong et al. 1996, figures 6.1–6.11 (reproduced with permission, copyright Yorkshire Geological Society).)

is carminiscaphate and appears to be **homeomorphic** (morphologically similar due to evolutionary convergence) with a number of Devonian and Carboniferous genera. The oral surface of the oval platform has two median rows of denticles or nodes.

Conodont affinities

A number of weakly phosphatized elements bearing a superficial resemblance to conodonts, the **protoconodonts** and **paraconodonts**, are known from the Cambrian and Ordovician (Fig. 21.17). These have been lumped together in the order Protoconodontida by some authors. They have a different internal structure and mode of growth to the true conodonts or euconodonts. Bengtson (1976) suggested paraconodonts may have evolved from protocondonts, but nobody has been able to substantiate this evolutionary relationship. Protoconodonts may represent the grasping spines of chaetognaths ('arrow worms' Szaniawski 1982).

Despite all the new evidence from the conodont animals there is still a continuing debate over the affinities of the conodonts. The presence of a notochord and chevron muscle blocks is limited to cephalochordates and craniates. Only the craniates have caudal fin rays and only the vertebrates possess eyes with extrinsic musculature and secrete calcium phosphatic skeletal elements. The presence of homologues of enamel

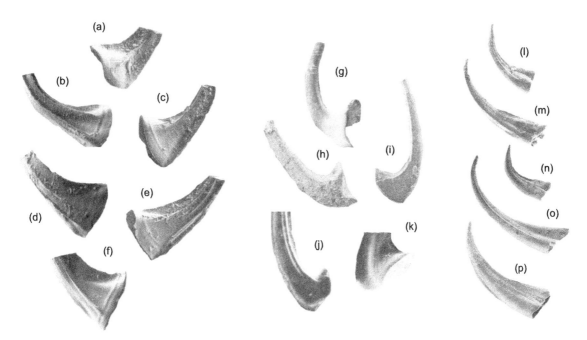

Fig. 21.16 SEM photomicrographs of selected coniform conodont species. In the absence of any orientation evidence for all but *Panderodus unicostatus* and its close relatives, it is premature to apply a biological terminology. In *P. unicostatus* the approximate orientation relative to the rostro-caudal and medio-lateral axes is known but dorsal and ventral cannot be determined. Features are therefore orientated in the traditional manner relative to the cusp and curvature. (a)–(f) *Walliserodus curvatus* (?Proconodontida), Silurian, the tips of the elements are missing: (a) ?qt element, inner lateral view, ×63; (b) ae element, lateral view, ×60; (c) q element, inner lateral view, ×52; (d) q element, outer lateral view, ×60; (e) q element, inner lateral view, ×60; (f) p element, outer lateral view, ×49. (g)–(k) *Drepanodus arcuatus* (Protopanderodontida), Ordovician, previously illustrated by Armstrong (2000, plate 3): (g) pt element, inner lateral view, ×36; (h) pf element, outer lateral view, ×33; (i) qt element, outer lateral view, ×33; (j) qg element, outer lateral view, ×37; (k) ?ae element, inner lateral view, ×40. (l)–(p) *Panderodus acostatus* (Panderodontida), Silurian: (l) ae element, inner lateral view, ×48; (m) qa element, inner lateral view, ×50; (n) qt element, inner lateral view, ×37; (o) qg element, inner lateral view, ×42; (p) pf element, lateral view, ×46. ((a)–(f) Previously illustrated by Armstrong 1990, plate 21, figures 6–15 (reproduced with permission, copyright, Geological Survey of Denmark and Greenland); (l)–(p) previously illustrated by Sansom *et al.* 1994, text figure 2 (reproduced with permission, copyright The Palaeontological Association).)

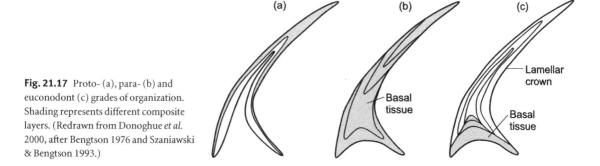

Fig. 21.17 Proto- (a), para- (b) and euconodont (c) grades of organization. Shading represents different composite layers. (Redrawn from Donoghue *et al.* 2000, after Bengtson 1976 and Szaniawski & Bengtson 1993.)

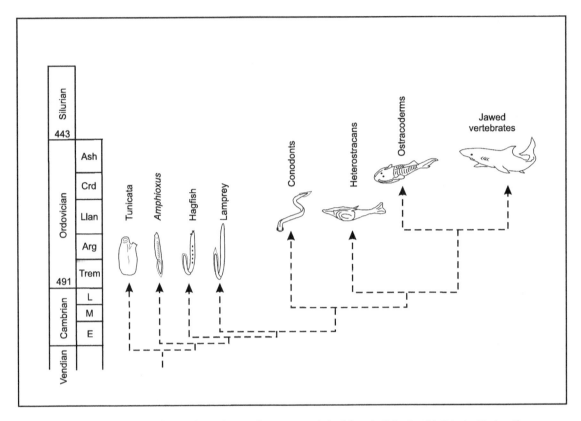

Fig. 21.18 Phylogeny of early vertebrates placing the conodonts as more derived than the living hagfish (Myxinoidea) and lampreys (Petromyzontida). (From Donoghue *et al.* 2000.)

and dentine in conodont elements also supports a vertebrate affinity.

Cladistic analysis of primitive vertebrates including the conodonts indicates conodonts are best considered stem gnathostomes though they lacked jaws (Fig. 21.18). Conodonts were the first members of this group to develop a phosphatic skeleton. If conodont elements functioned as teeth then the first parts of the vertebrate skeleton to evolve were the teeth and not bony scales, contrary to earlier hypotheses. The evolution of teeth would have served to increase the feeding efficiency of the animal, initiating what is a consistent feature of later vertebrate evolution. If it can be demonstrated that conodont elements and the teeth of jawed vertebrates (which appeared 100 million years later) are homologous then the whole of early vertebrate evolution needs to be re-evaluated (Smith & Hall 1993).

Mode of life palaeoecology and palaeobiogeography

Because conodonts are extinct, functional morphology, faunal associations and facies distribution have been used to infer their mode of life and palaeoecology. Conodonts were exclusively marine occurring in a wide range of habitats from hypersaline to bathyal, even abyssal. Conodont elements are also found in bedded chert successions considered to have been deposited beneath the calcite compensation depth (CCD), but it is likely these were nektonic or pelagic animals. Occasionally up to 20,000 elements per kilogram have been recovered from shallow marine tropical and subtropical limestone samples, suggesting shoals of conodont animals were dominant members in the communities. The Granton animals are found in a shallow, enhanced salinity, quiet water environment, susceptible to periodic dysoxia. *Promissum pulchrum* specimens occupied a periglacial marine environment. These finds appear atypical as most conodont species are found in open marine environments. Diversity was highest in equatorial latitudes.

The majority of conodont animals show some degree of facies dependence which indicates they lived close to the sea floor. Both the Granton and Soom animals possess characters that indicate conodont animals were active nektobenthonic predators or scavengers. However, the majority of coniform taxa are found across a much wider range of facies and this suggests they may have been nektonic or pelagic. The dorsal–ventral flattening of the body of the *Panderodus* animal may have been an adaptation to this mode of life. Some species (e.g. *Scaliognathus* and *Bispathodus* in the Carboniferous) are found in deep-water black shales with no associated benthos and were likely to have been nektonic, if not pelagic.

Functional morphological studies of the feeding apparatus indicate conodont animals were macrophagous, feeding on living or recently dead prey. In the absence of a jaw it is unlikely conodont animals could bite and they probably pulled chunks from the prey, much in the same way as the modern hagfish.

Recent work on the distribution of Middle and Upper Ordovician conodonts supports a differentiation of nektobenthonic continental shelf taxa including mainly prioniodinids and nektonic/pelagic taxa which included largely protopanderodontids and prioniodontids (Armstrong & Owen 2002). The latter apparently show depth stratification and or adaptation to specific water masses, a feature also suggested for some Carboniferous prioniodinid conodonts (e.g. Sandberg & Gutschick 1979).

The distribution patterns of Silurian conodont species (e.g. Aldridge 1976; Aldridge & Mabillard 1983) show no simple correlation between inshore to offshore shelly benthos and conodont biofacies. The primary ecological controlling factors of these conodonts are unclear. Some evidence exists for ecophenotypic variation along ecological gradients. Purnell (1992) demonstrated changes in position of the blade of the Pa element in *Taphrognathus varians*, in response to increasing environmental restriction.

Near-shore, high-energy facies tend to contain large robust conodonts with large basal cavities and large platform-bearing Pa elements, whilst quiet offshore facies have more delicate elements. Offshore facies also appear to be dominated by coniform species or non-coniform taxa bearing long, slender denticles.

The fact that conodonts show marked provincialism at various times in their history suggests they were sensitive to temperature. During the Ordovician there were distinct conodont faunas in high and low latitudes, the North Atlantic and American Midcontinent Provinces respectively, though this simple differentiation may have been more complex (see Armstrong & Owen 2002 for a review). The declining degree of endemism between these provinces has been used to plot the closure of the Iapetus Ocean (Armstrong & Owen 2002). The Late Ordovician glaciation led to the virtual elimination of the North Atlantic Province and stocks that survived into the Silurian were cosmopolitan with an apparent low to mid latitude distribution, comprising mainly prioniodinids, panderodontids and rare prioniodontids.

The breakdown of provinciality in the Late Ordovician was perhaps the result of the return of an equitable climate, end Ordovician mass extinction or the amalgamation of palaeocontinental areas. Devonian conodonts were restricted to the tropics and show a degree of endemism between different epeiric seas (Klapper & Johnson 1980). Carboniferous and Permian conodonts show little provincialism. By the Triassic a Tethyan Province had developed on the eastern side of the Pangaea supercontinent and a Mushelkalk Province in central Europe, though this provincialism declined in the Late Triassic prior to the extinction of the group.

Evolutionary history

Euconodonts (order Proconodontida) first appeared in the Late Cambrian and the vast majority at this time bore apparently simple apparatuses of coniform elements. The Proconodontida flourished briefly and became extinct by the end of the Tremadoc. Meanwhile, other orders became established and conodonts reached their peak diversity by the Early Ordovician with appearance of the Protopanderodontida, Panderodontida and Prioniodontida. This major radiation is mirrored in a large number of invertebrate groups and correlates with a eustatic rise in sea level and the opening of new shelf niches (Smith *et al.* 2002). The end Ordovician mass extinction heralded the onset of a general decline which continued for the remainder of their history. By the mid-Silurian faunas were largely dominated by species with ozarkodinid apparatus plans and panderodontids. Coniform conodonts became extinct at the end of the Devonian.

The Prioniodinida and Ozarkodinida arose in the mid-Ordovician and came to dominate the later Palaeozoic. Following a general decline through the Permian the Prioniodinida diversified during the Triassic and were the final surviving order of conodonts becoming extinct at the very end of the Triassic Period. The Ozarkodinida radiated in the Silurian and Middle to Late Devonian, declined through the Early Carboniferous and finally became extinct in the Late Triassic mass extinction. The fact that conodont extinction rates were highest in the Norian and not the Rhaetic suggests their ultimate extinction may have been the cumulative result of several factors and not a single catastrophic event (Sweet 1988).

Applications

Conodonts have become the premier group for the global biostratigraphy of Palaeozoic shallow marine environments with regional and international biozonations developed using traditional and quantitative methodologies. Marine strata of Cambrian to Triassic age are divided into approximately 150 conodont biozones and many of the Palaeozoic stage and period boundaries are defined on the first appearance of conodont species. The application of the group to palaeoecological investigations and biogeography has expanded since the discovery of the whole animals. A useful source of case studies can be found in Clark (1984). Epstein *et al.* (1977) pioneered the use of conodonts as geothermometers and depth of burial indicators; an example of the application of this technique can be found in Armstrong *et al.* (1994). Conodont element geochemistry is increasingly being used in studies of palaeoclimate and palaeoceanography (e.g. Wright *et al.* 1984; Armstrong *et al.* 2001).

Further reading

Sweet (1988) provides a useful, if somewhat personal, introduction to conodonts. Aspects of their palaeobiology can be found in Aldridge (1987) and Aldridge *et al.* (1993). A number of case studies illustrating the applications of conodonts can be found in Austin (1987). A useful summary essay on the affinities of conodonts can be found in Purnell *et al.* (1995) and more detailed treatments in Donoghue *et al.* (2000) and Sweet & Donoghue (2001).

REFERENCES

Aldridge, R.J. 1975. The stratigraphic distribution of conodonts in the British Silurian. *Journal of the Geological Society, London* 131, 607–618.

Aldridge, R.J. 1976. Comparison of macrofossil communities and conodont distribution in the British Silurian. In: Barnes, C.R. (ed.) *Conodont Paleoecology. Special Paper. Geological Association Canada* 15, 91–104.

Aldridge, R.J. (ed.) 1987. *Palaeobiology of Conodonts.* E. Horwood for the British Micropalaeontological Society, Chichester.

Aldridge, R.J. & Mabillard, J.E.M. 1983. Local variations in the distriubtion of Silurian conodonts an example from the *amorphognathoides* interval of the Welsh Basin. In: Neale, J.W. & Brasier, M.D. (eds) *Microfossils from Recent and Fossil Shelf Seas*. Ellis Horwood, British Micropalaeontological Association, pp. 10–16.

Aldridge, R.J. & Theron, J.N. 1993. Conodonts with preserved soft tissue from a new Ordovician Konservat-Lagerstätte. *Journal of Micropalaeontology* 12, 113–119.

Aldridge, R.J., Briggs, D.E.G., Clarkson, E.N.K. & Smith, M.P. 1986. The affinities of conodonts-new evidence from the Carboniferous of Edinburgh, Scotland. *Lethaia* 19, 279–291.

Aldridge, R.J., Smith, M.P., Norby, R.D. & Briggs, D.E.G. 1987. The architecture and function of Carboniferous polygnathacean conodont apparatuses. In: Aldridge, R.J. (ed.) *Palaeobiology of Conodonts*. Ellis Horwood, Chichester, pp. 63–75.

Aldridge, R.J., Briggs, D.E.G., Smith, M.P., Clarkson, E.N.K. & Clark, N.D.L. 1993. The anatomy of conodonts.

Philosophical Transactions Royal Society, London B340, 405–421.

Armstrong, H.A. 1990. Conodonts from the Lower Silurian of the north Greenland carbonate platform. *Bulletin. Grønland Geologiske Undersøgelse* 159, 151pp.

Armstrong, H.A. 1997. Conodonts from the Shinnel Formation, Tweeddale Member (middle Ordovician), Southern Uplands, Scotland. *Palaeontology* 40, 763–799.

Armstrong, H.A. 2000. Conodont micropalaeontology of mid-Ordovician aged limestone clasts from LORS conglomerates, Lanark and Strathmore basins, Midland Valley, Scotland. *Journal of Micropalaeontology* 19, 45–59.

Armstrong, H.A., Johnson, E.W. & Scott, R. 1996. Conodont biostratigraphy of the attenuated Dent Group (Upper Ordovician) at Hartley Ground, Broughton in Furness. *Proceedings of the Yorkshire Geological Society* 51, 9–23.

Armstrong, H.A. & Owen, A.W. 2002. Euconodont diversity changes in a cooling and closing Iapetus Ocean. In: Crame, J.A. & Owen, A.W. (eds) *Palaeobiolgeography and Biodiversity Change: a comparison of the Ordovician and Mesozoic-Cenozoic radiations. Geological Society, Special Publications* 194, 85–98.

Armstrong, H.A. & Smith, C.J. 2001. Growth patterns in euconodont crown enamel: implications for life history and mode of life reconstruction in the earliest vertebrates. *Proceedings. Royal Society, Series B* 268, 815–820.

Armstrong, H.A., Smith, M.P., Tull, S. & Aldridge, R.J. 1994. Conodont colour alteration as a guide to the geothermal history of the North Greenland carbonate platform. *Geological Magazine*, 131, 219–230.

Armstrong, H.A., Pearson, D.G. & Greselin, M. 2001. Thermal effects on rare earth element and strontium isotope chemistry in single conodont elements. *Geochimica Cosmochimica Acta* 65, 435–441.

Austin, R. (ed.) 1987. *Conodonts: Investigative techniques and applications.* Ellis Horwood, Chichester.

Bengtson, S. 1976. The structure of some Middle Cambrian conodonts, and the early evolution of conodont structure and function. *Lethaia* 9, 185–206.

Benton, M.J. 1993. *The Fossil Record 2*. Chapman & Hall, London.

Briggs, D.E.G., Clarkson, E.N.K. & Aldridge, R.J. 1983. The conodont animal. *Lethaia* 16, 1–14.

Clark, D.L. (ed.) 1984. Conodont biofacies and provincialism. *Memoir. Geological Society of America* 196, 1–340.

Donoghue, P.C.J., Forey, P.L. & Aldridge, R.J. 2000. Conodont affinity and chordate phylogeny. *Biological Reviews* 75, 191–251.

Epstein, A.G., Epstein, J.B. & Harris, L.D. 1977. Conodont Color Alteration an index to organic metamorphism. *Geological Survey Professional Paper* 995, 27pp.

Forey, P.L. & Janvier, P. 1993. Agnathans and the origin of jawed vertebrates. *Nature* 361, 129–134.

Gabbott, S.E., Aldridge, R.J. & Theron, J.N. 1995. A giant conodont with preserved muscle tissue from the Upper Ordovician of South Africa. *Nature* 374, 800–803.

Hinde, G.J. 1879. On conodonts from the Chazy and Cincinnati group of the Cambro-Silurian, and from the Hamilton and Genesee Shale divisions of the Devonian in Canada and the United States. *Quarterly Journal of the Geological Society, London* 35, 351–369.

Kemp, A. & Nicoll, R.S. 1996. Histology and histochemistry of conodont elements. *Modern Geology* 20, 287–302.

Klapper, G. & Johnson, J.G. 1980. Endemism and dispersal of Devonian conodonts. *Journal of Paleontology* 54, 400–455.

Männik, P. & Aldridge, R.J. 1989. Evolution, taxonomy and relationships of the Silurian conodont *Pterospathodus*. *Palaeontology* 32, 893–906.

Mikulic, D.G., Briggs, D.E.G. & Kluessendorf, J. 1985. A Silurian soft-bodied biota. *Science* 228, 715–717.

Nicoll, R.S. 1977. Conodont apparatuses in an Upper Devonian palaeoniscoid fish from the Canning Basin, Western Australia. *Bureau of Mineral Resources Journal of Australian Geology and Geophysics* 2, 217–228.

Orchard, M.J. & Rieber, H. 1996. Multielement clothing for *Neogondolella* (Conodonta, Triassic). Sixth North American Paleontological Convention Program with Abstracts. *Paleontological Society Special Publication* 8, 297.

Purnell, M.A. 1992. Conodonts of the Lower Border Group and equivalent strata (Lower Carboniferous) in northern Cumbria and the Scottish Borders, U.K. *Royal Ontario Museum Life Sciences Contributions, No.* 156.

Purnell, M.A. 1993a. Feeding mechanisms in conodonts and the function of the earliest vertebrate hard tissues. *Geology* 21, 375–377.

Purnell, M.A. 1993b. The *Kladognathus* apparatus (Conodonta, Carboniferous): homologies with ozarkodinids and the prioniodinid Bauplan. *Journal of Paleontology* 67, 875–882.

Purnell, M.A. 1994. Skeletal ontogeny and feeding mechanisms in conodonts. *Lethaia* 27, 129–138.

Purnell, M.A. 1995. Microwear on conodont elements and macrophagy in the first vertebrates. *Nature* 374, 798–800.

Purnell, M.A. & von Bitter, P.H. 1996. Bedding-plane assemblages of *Idioprioniodus*, element locations, and the bauplan of prioniodinid conodonts. In: Dzik, J. (ed.) *Sixth*

European Conodont Symposium, Abstracts, Instytut Paleobiologii PAN, Warszawa, 48.

Purnell, M.A. & Donoghue, P.C.J. 1998. Architecture and functional morphology of the skeletal apparatus of ozarkodinid conodonts. *Palaeobiology* 41, 57–102.

Purnell, M.A., Aldridge, R.J., Donoghue, P.C.J. & Gabbott, S.E. 1995. Conodonts and the first vertebrates. *Endeavour* 19, 20–27.

Purnell, M.A., Donoghue, P.C.J. & Aldridge, R.J. 2000. Orientation and anatomical notation in conodonts. *Journal of Paleontology* 74, 113–122.

Sandberg, C.A. & Gutschick, R.C. 1979. Guide to conodont biostratigraphy of Upper Devonian and Mississippian rocks along the Wasatch Front and Cordilleran hingeline. *Brigham Young University Geology Studies* 26, 107–133.

Sansom, I.J., Armstrong, H.A. & Smith, M.P. 1994. The apparatus architecture of *Panderodus* and its implications for coniform classification. *Palaeontology* 37, 781–799.

Sansom, I.J., Smith, M.P., Armstrong, H.A. & Smith, M.M. 1992. Presence of the earliest vertebrate hard tissues in conodonts. *Science* 256, 1308–1311.

Smith, M.P. 1991. Early Ordovician conodonts of East and North East Greenland. *Meddr Grønland* 26, 81pp.

Smith, M.M. & Hall, B.K. 1993. A developmental model for evolution of the vertebrate exoskeleton and teeth: the role of cranial and trunk neural crest. *Evolutionary Biology* 27, 387–448.

Smith, M.P., Briggs, D.E.G., & Aldridge, R.J. 1987. A conodont animal from the lower Silurian of Wisconsin, U.S.A., and the apparatus architecture of panderodontid conodonts. In: Aldridge, R.J. (ed.) *Palaeobiology of conodonts*. Ellis Horwood, Chichester, pp. 91–104.

Smith, M.P., Donoghue, P.J. & Sanson, I.J. 2002. The spatial and temporal diversification of Early Palaeozoic vertebrates. In: Crame, J.A. & Owen, A.W. (eds) *Palaeobiogeography and Biodiversity Change: the Ordovician and Mesozoic-Cenozoic Radiations. Geological Society Special Publication No.* 194, 69–85.

Sweet, W.C. 1988. *The Conodonta: morphology, taxonomy, paleoecology and evolutionary history of a long extinct animal phylum*. Oxford Monographs in Geology and Geophysics 10. Oxford University Press, New York.

Sweet, W.C. & Bergström, S.M. 1969. The generic concept in conodont taxonomy. *Proceedings. North American Paleontological Convention* 1, 29–42.

Sweet, W.C. & Donoghue, P.J. 2001. Conodonts: past, present and future. *Journal of Paleontology* 75, 1174–1184.

Szaniawski, H. 1982. Chaetognath grasping spines recognized among Cambrian protoconodonts. *Journal of Paleontology* 56, 806–810.

Szaniawski, H. & Bengtson, S. 1993. Origin of euconodont elements. *Journal of Paleontology* 67, 640–654.

Ulrich, E.O. & Bassler, R.S. 1926. A classification of the toothlike fossils, conodonts, with descriptions of American Devonian and Mississippian species. *Proceedings.U.S. National Museum* 68, Art. 12.

Walliser, O.H. 1964. Conodonten des Silurs. *Abhandlungen hessisches Landesamt für Bodenforschung, Wiesbaden* 41, 1–106.

Wright, J. Seymour, R.S. & Shaw, R.F. 1984. REE and Nd isotopes in conodont apatite: variations with geological age and depositional environment. *Special Paper. Geological Society of America* 196, 325–340.

Appendix – Extraction methods

Before microfossils can be properly examined they must, of course, be extracted from the rocks, prepared and then mounted. Each palaeontologist tends to have favourite methods for these procedures, some of which may be rather elaborate and tailored to particular needs or geared to laboratories with skilled technical assistance. There are, none the less, many simple, safe and inexpensive methods that can be used for more 'reconnaissance' investigations. To prepare your own material at all stages also has some advantages: it allows for greater flexibility and it increases the pleasure of discovery. Only these reconnaissance techniques will be dealt with here. For more detailed methods readers are referred to references in the chapters in this book.

Sample collection

At surface outcrops either spot or channel sampling may be employed. Spot sampling consists of taking samples at predetermined stratigraphical levels. Channel samples are more continuous collections through longer stratigraphical intervals (say up to 3 m), which tend to blur the detailed story but avoid the risk of being totally barren. Subsurface outcrops may be sampled with various coring devices; simple manual ones such as the Dutch auger and the Hiller corer are suitable only for softer sediments. Retrieved cores can then be spot or channel sampled. More usually, though, the less expensive method of studying chippings brought up by subsurface drilling is followed. In studying drill chips care must be taken to avoid contamination from microfossils in the drilling mud and to take account of the phenomenon of 'caving' in which younger fragments have fallen downwards and become mixed with those of greater age. In this case the youngest rather than the oldest stratigraphical occurrences are more reliable for zoning. When collecting from outcrops take care that the rock is not weathered and uncontaminated by recent vegetation or by hammers, chisels, trowels and the like. The sample bag should be absolutely clean inside. The amount of sample placed in the bag will depend, in part, on how much you can carry and store; it is always best to collect enough for resampling without having to return to the outcrop. For general reconnaissance studies about 500 g should be enough. Labelling of samples is a very personal business. It is advisable, however, to note the sample number, rock unit, horizon above or below a known geological datum, locality (as accurately as possible), date and your name or initials. Some or all of this information can be placed on the outside of the sample bag and should also be placed on a card within the bag before sealing. At the same time the relevant data should be entered into the field note book.

Sample preparation

The sample must next be disaggregated to release the microfossils. Ideally this requires a small laboratory with an adequate sink (preferably with a sediment trap), hotplates, oven and, if possible, a fume cupboard. Other than the chemicals noted below, the laboratory will also need some heat-resistant bowls (e.g. stainless steel), evaporating dishes, petri dishes, pestle and mortar, glass measuring cylinders, glass beakers, filter funnels and filter papers, retort stands and clamps, flat-bottomed glass tubes, plastic buckets with lids and sets of sedimentological sieves, especially 1-mm, 250-μm, 74-μm and 63-μm mesh sizes. More specialist nylon sieves are required for palynological preparation with 20-μm and 10-μm meshes. The water taps should be fitted with moderate lengths of rubber tubing to facilitate wet sieving. Many of the following sample processes require the rock to be broken into fragments. For small samples a pestle and mortar are sufficient, but larger and harder samples will require a rock splitter or a rock crusher. If these are not available, place the sample inside several clean polyethylene bags on a firm surface and strike with a hammer.

WARNING! Treat all chemicals and equipment with sober respect, particularly stong acids and alkalis, and be aware of associated health hazards particularly with organic solvents and heavy liquids which are carcinogenic.

A Pulverization method

This simple and speedy technique may be used to extract coccoliths and organic-walled microfossils (e.g. spores, pollen, acritarchs) from well-indurated rocks such as chalks and mudstones. It can also be used to liberate radiolarians and foraminifera with varying success.
1 Place 5–20 g of fresh sample in a mortar. Add a few drops of distilled water and crush by pounding (not grinding) with a pestle until the largest fragments do not exceed 2 mm in diameter.
2 Flush the sample into a jar or bottle with a jet of distilled water.
3 If the sample is very argillaceous, the clay can be dispersed by placing the container in an ultrasonic bath and letting it shake for 2 minutes to 2 hours, according to results. Note, however, that this can destroy some of the more delicate microfossil structures.
4 Wash and concentrate as in methods G to L.

B Scrubbing-brush method

This is another easy method useful for extracting calcareous and siliceous microfossils from partially indurated limestones (e.g. chalks and marly limestones), sandstones and shales.

1 Fill a clean bucket with water.
2 Take a fresh piece of the rock and scrub it under the water with a hard bristle toothbrush or a scrubbing brush. The action should be as gentle as possible to obtain a residue without damaging the microfossils.
3 For larger calcareous and siliceous microfossils, strain the cloudy water through a 74-μm sieve, flush out the residue into an evaporating dish, decant off the surplus water in the dish and dry the residue at a low temperature. For organic-walled microfossils, strain the water through a 20-μm sieve and flush out the remaining residue into a glass bottle. Coccoliths can be obtained from the water flushed through these sieves if they are allowed to settle out for an hour or so.
4 Concentrate the microfossils using methods H, J, K, L or M.

C Solvent method

Partially indurated argillaceous rocks (excepting black and dark grey shales), marls and soft limestones can be disaggregated by this method.
1 Break the fresh rock into fragments about 1–10 mm in diameter. The harder rocks will need smaller fragments with a greater surface area.
2 Dry at a low temperature in an oven, remove and allow to cool.
3 Pour on petroleum ether, petrol or turpentine substitute and allow to stand in a fume cupboard until the rock is saturated (usually from 30 minutes to 8 hours). Handle with care.
4 Pour off the excess solvent. This can be collected for further use by straining through a filter paper.
5 Pour on some hot water to cover the sample and allow to stand until the rock shows no signs of breaking down further (usually from 5 to 30 minutes).
6 If the disaggregation is only partial and further residue is required, repeat stages 2–5, or follow by method D.
7 Wash and concentrate as in methods G to M.

D Na_2CO_3 method

This washing-soda method is cheap, safe and effective with partially indurated argillaceous rocks, marls and soft limestones. However, it is not effective with black or dark grey shales, mudstones, chalks and porcelaneous limestones.

1 Break the fresh rock into fragments about 1–10 mm in diameter. The harder rocks will require a smaller size of fragment.
2 Place the rock fragments in a stable heat-resistant bowl or beaker, cover with water and add one or two large spoonfuls of Na_2CO_3.
3 Set the liquid to boil and allow to simmer until the rock shows no further signs of breaking down. It is best to keep the water level topped up while boiling proceeds.
4 Wash and concentrate as in methods G to M.

E NaClO method

Ordinary domestic bleach (sodium hypochlorite) is a useful agent for disaggregating indurated carbonaceous black shales, mudstones, clays and coals for study of their organic-walled microfossils and conodonts. As it also bleaches dark organic tissues, making them clearer for microscopy, it can be used in conjunction with other methods. Disaggregation is relatively slow compared with the foregoing techniques, and a fume cupboard is advisable to reduce the smell of chlorine.
1 Break the fresh rock into fragments from 1 to 10 mm in diameter and place in an evaporating dish, bowl or beaker.
2 Cover with a 15–20% solution of NaClO in water and place a cover over it to prevent contamination from the air.
3 Leave until a sufficient quantity of the rock has broken down (usually one day to several weeks). Top up the solution if evaporation occurs.
4 Decant the supernatant liquid over a filter-lined funnel and cover the remaining residue with distilled water. Flush the filtrate with distilled water until no salt crystals are left. Remove the filter paper and flush this residue back into the evaporating dish.
5 Wash and concentrate the residue as in methods G to M.

F Acid digestion and maceration

Non-calcareous microfossils can be released from calcareous rocks by treatment with 10–15% acetic acid or commercial grade formic acid. Although the former is more expensive, it has the advantage of being quick and suitable for both pure and argillaceous limestones and dolomites. The residues obtained in this way may contain conodonts, radiolarians, diatoms, organic-walled microfossils and archaeocopid ostracods. Silicified or phosphatized microfossils may also be liberated from limestones in this way.
1 Break the rock into fragments of about 10–30 mm in diameter.
2 Place about 500 g in a 2-gallon (10-litre) plastic bucket and cover with about 1 litre of 10% acetic acid (handle with care!). Adding 25–50g of $CaCO_3$ can act as a buffer against the vigour of the reaction and increase recovery and quality of preservation, particularly of phosphatic microfossils. Top this up to the 2-gallon (10-litre) mark with hot (c. 80°C) water. Smaller samples can be treated in smaller vessels with lesser quantities but the acid should again be diluted to about 10–15% and equal in volume more than two times the volume of the rock to be dissolved. Cover with a lid and place in a safe, well-ventilated spot or in a fume cupboard.
3 After 6–24 hours most of the limestone should have dissolved and any effervescence ceased. Any remaining rock fragments can be retreated if necessary.
4 Wash the sample as in method G. If microfossils smaller than 44 µm are to be studied, allow the sample to settle out through the spent acid and decant off the clear supernatant liquid. Filter the fine residue through a filter-lined funnel and then flush the filtrate into a beaker for further examination.
5 Wash and concentrate as in methods G to M.

Organic-walled microfossils can also be extracted from rocks by digesting the rock matrix in concentrated mineral acids; hydrochloric acid (HCl) for calcareous rocks and hydrofluoric acid (HF) for argillaceous rocks and fine sandstones.

Warning: these chemicals should only be using by professional technicians in a properly equipped laboratory.

1 50–100 g of pea-sized fragments are placed in a plastic pot and dampened with distilled water.
2 Sufficient HCl is added to initially cover the sample and then sufficient to remove all carbonate fraction.
3 Once the reaction has ceased the spent acid is decanted and the sample neutralized with distilled water.
4 40% or 53% hydrofluoric acid is then added to digest the non-calcareous fraction. Samples are left in HF until an organic sludge is left in the beaker, perhaps up to a week depending upon age and degree of induration.
5 The sample is neutralized and sieved through a 10-µm nylon sieve to remove fine debris.

6 The sieved residue can be further concentrated and mounted as in method H.

G Washing and wet sieving

Once disaggregated, many of the samples will now consist of muds (or silty and sandy muds). These clay minerals can obscure the microfossils and can be removed by washing over 74-, 63- and 44-μm mesh sieves; this procedure will also serve to concentrate some of the larger microfossils. Washing out the clay minerals will cause the loss of coccoliths and the smallest spores, pollen, acritarchs and diatoms. The finest sieve size should therefore be chosen with care. For general purposes the following procedure may be followed.

1 Put aside about 20 cm^3 of wet sample in water if it is intended to study the smaller spores, pollen, microplankton and coccoliths (see method H).
2 Wash the bulk of the sample gradually through a clean, fine-mesh sieve with a gentle jet of water from the tap (for diatoms, spores and pollen, use distilled water). A 44-μm sieve will retain many small diatoms and organic-walled microfossils whilst 74- and 63-μm sieves will retain most of the smaller radiolarians, foraminifera and conodonts and a 250-μm sieve the ostracods. If there are many shells or clasts greater than 2 mm in diameter, use a 1-mm top sieve to retrieve the coarser material. If the clays are difficult to disperse from the residue, it may be necessary to boil the sample with Na_2CO_3 (method D), but it is quicker to place the sample in a beaker of water within an ultrasonic cleaner device for several minutes.
3 Flush the residue into an evaporating dish with a jet of distilled water. For coccoliths, diatoms, radiolarians, silicoflagellates and organic-walled microfossils, place this residue in a clean bottle with distilled water. For foraminifera, radiolarians, silicoflagellates, tintinnids, conodonts and ostracods, decant off the supernatant liquid and set residue to dry at a low temperature in an oven.
4 Sort and concentrate as in methods H to L.

Sorting and concentration

The discovery and analysis of microfossils is greatly speeded by sorting them into size classes and separating them according to their specific gravities. A variety of methods can be employed here.

H Decanting and smear slides

Smear slides provide a rapid method for producing slides of coccoliths and diatoms.

1 A small amount of the disaggregated sample is placed in distilled water and a drop of cellosize added to act as a dispersant.
2 The cover slip is left to dry on a warm hotplate.
3 To make permanent mounts allow the slide and residue to dry at a low temperature away from possible sources of contamination. Place a drop of mounting medium (e.g. Canada Balsam) on a clean cover slip and drop this over the residue. Allow to dry before examining with transmitted light.

Pre-concentration of the microfossils can often improve the quality of the slides, reducing the amount of background debris. Organic-walled microfossils (e.g. arcritarchs, dinoflagellates, spores and pollen) can be separated by careful swirling in a large-diameter watch glass, rather like panning for gold. Decanting is almost as quick as centrifugation and can be done with a minimum of facilities. It is especially suitable for sorting and concentrating the coccoliths and organic-walled microfossils.

1 Place six clean, flat-bottomed glass tubes in a row on a stable surface. Before each place six clean glass slides and have ready glass cover slips, distilled water, a pippette and a watch or a clock with a second hand.
2 Take the bottle of sample 'fines' (which have been pulverized, scrubbed or otherwise disaggregated and are either washed or unwashed, this could include the organic residues from acid maceration) in water, swirl them around gently to place the fine material in suspension and then decant into tube 1.
3 Allow to settle for 30 seconds and then carefully decant the supernatant liquid into tube 2.
4 Allow to settle for 60 seconds and then decant the supernatant liquid into tube 3.
5 Allow to settle for 2 minutes and then decant the supernatant liquid into tube 4.
6 Allow to settle for 5 minutes and then decant the supernatant liquid into tube 5.
7 Allow to settle for 10 minutes and then decant the supernatant liquid into tube 6.

8 Allow to settle for 20 minutes and then decant the supernatant liquid back into the original bottle.

9 As the decanting proceeds, spare moments can be used to make temporary mounts on the glass slides. With a pipette draw up a little of the residue from each tube, drop some on to the glass slide and cover with a cover slip.

10 Prepare a permanent mount using the method described above.

I Dry sieving

Calcareous, siliceous and phosphatic microfossils can vary greatly in size so it is useful to sieve the dried residue into various fractions using, for example, 1-mm to 63-μm sieves. Each fraction should then be placed in a bag or bottle with a label bearing relevant data and examined separately (see method O).

Separation by heavy liquids

Microfossils in washed residues can be concentrated by treatment with a variety of heavy liquids. Care should be taken to avoid both breathing and touching these liquids as they are toxic. A fume cupboard and careful preparation of the equipment are therefore necessary.

J Carbon tetrachloride (CCl_4; specific gravity (SG) 1.58)

This can be used to concentrate buoyant foraminifera, radiolarians and diatoms. Unfortunately it does not concentrate thick-shelled, infilled or fragmentary microfossils or ostracods.

1 Place the washed and dried sample residue in a beaker.

2 Add two or three times this volume of CCl_4 and stir vigorously with a clean glass rod or a disposable tooth pick. The lighter microfossils mentioned above will float to the surface if present. (Handle with care!)

3 Pour this 'float' into a filter-lined funnel arranged over a collecting vessel.

4 Stir the sample again (adding more CCl_4 if necessary) and pour off the float, repeating the process until none is left in the beaker.

5 Allow the filter paper and residue to dry until odourless. Filter off the remaining sediment and allow this to dry until this is also odourless. Return the filtered CCl_4 to the bottle.

6 Put the float and remaining sediment in separate containers for subsequent microscopic study (method O).

K Carbon tetrachloride (second method)

To concentrate organic-walled microfossils, the following variant of the above method may be employed.

1 Filter the water from the sample through a filter-lined funnel or through a fine-meshed sieve.

2 Remove the $CaCO_3$ from the residue with formic or acetic acid (see method F).

3 Rinse the sample with distilled water through a filter-lined funnel or a fine-meshed sieve.

4 Flush the filtrate off the filter paper or sieve into a beaker using a fine jet of acetone (Care!). Decant the acetone plus water into a filter-lined funnel and repeat until all the water has been removed. Flush the filtrate back into the beaker again with as little acetone as possible and let this evaporate off in a fume cupboard.

5 Pour some CCl_4 into the beaker and stir gently until the residue is well dispersed. Cover the beaker and allow it to stand for at least 2 hours. Handle with care!

6 Decant off the float (bearing the organic residue) into a filter-lined funnel over a beaker. When the CCl_4 has filtered through, flush the light residue with a little jet of acetone into a bottle of distilled water. Allow the acetone to evaporate off.

7 Prepare temporary and permanent mounts as in method H.

L Bromoform ($CHBr_3$; SG 2.8–2.9) and tetrabromoethane ($C_2H_2Br_4$; SG 2.96)

Handle these liquids with care and only use in a fume cupboard. These heavy liquids can be used to concentrate conodonts as well as calcareous and siliceous microfossils; conodonts separate in the heavy fraction and calcareous and siliceous microfossils in the lighter fraction.

1 Prepare a retort stand with brackets to hold a lower filter funnel and a higher one a short distance above the other with a 250- or 500-ml separating funnel.

Place a clean beaker below the lower funnel, which should be lined with a fast, strong quality (grade 4) filter paper.

2 Pour about 75 ml of the heavy liquid in to the separating funnel, add the sample and top up with heavy liquid to about 150 ml. Stir with a glass rod.

3 Allow the sample to separate out (which may take up to 2 hours), stirring occasionally.

4 Open the tap and allow the heavy residue (with the conodonts) to drain into the filter-lined funnel beneath. The heavy liquid that filters into the beaker below can now be put aside in a bottle for further use. Wash the residue with acetone to remove any residual heavy liquid. The washings should be stored for recycling (see below).

5 Open the tap and allow the light fraction to drain into a separate filter paper (with foraminifera, ostracods, radiolarians and organic-walled microfossils). Retrieve the heavy liquid and wash with acetone as before.

6 Allow the filter papers to dry out in the fume cupboard until odourless and then place them in a warm oven for further drying.

7 To retrieve the heavy liquids from the acetone, place the washings in a side-arm flask. Water can be circulated through the washings through the bung, pierced with a short length of pyrex tube. Allow the water to wash through the flask for at least 2 hours. The specific gravity of the heavy liquids can be checked by placing a calcite rhomb in the flask. This will again float freely on the liquid when all the acetone has been removed.

M Electromagnetic separation

If available this apparatus can be used to concentrate various kinds of microfossils, especially conodonts. Some hints on its use are given by Dow (in Kummel & Raup 1965, pp. 263–267).

N Stained acetate peels

Well-indurated limestones are generally difficult to disaggregate so that their microfossils (e.g. larger foraminifera, calcareous algae, radiolarians) are often studied in petrographic thin sections. A quicker method that destroys less material is the acetate peel technique. This takes a detailed impression of an acid-etched surface, but although it gives a clear indication of the gross morphology of a microfossil, the ultrastructure is usually better seen in proper thin sections.

1 Cut the limestone vertically into slabs with a rock saw, each slab about 10 mm thick.

2 Polish the faces with successively finer grades of corundum powder and water on a glass plate until the surface is quite smooth.

3 Rinse the limestone slabs with distilled water and allow to dry. Do not touch these polished faces.

4 Prepare the following solutions: (a) 0.2 g of Alizarin Red dissolved in 100 ml of 1.5% HCl; (b) 2.0 g of potassium ferricyanide dissolved in 100 ml of 1.5% HCl. Store these two separately, and handle with care.

5 Mix the above together in the ratio $a : b = 3 : 2$ just before use. Pour into a stable receptacle wide enough to allow the submersion of one polished face of limestone.

6 Immerse one face of the limestone in the solution, agitating it slightly whilst holding between finger and thumb. Etching may take from 15 to 60 seconds according to the age and induration of the limestone. At this stage calcite and aragonite should stain pink to red, ferroan calcite stain royal blue or mauve and dolomite remain unstained. Ferroan dolomite will stain a pale to deep turquoise blue.

7 Wash the stained and etched slab gently in a dish of distilled water and allow to dry. Drying can be hastened by flushing with acetone. Take care!

8 The following stage will require a well-ventilated place, preferably a fume cupboard. Place the etched and stained limestone slab face uppermost on a sheet of rubber foam underlain by a tray to collect residual liquid. Make sure this face is horizontal. Also have ready some pre-cut strips of clean and transparent cellulose acetate sheeting. This can be bought in rolls of varying thicknesses but it should be both flexible and strong enough not to tear during handling.

9 Squirt a thin layer of acetone over the whole of the etched face; the first coating usually evaporates quickly and it is better to wait until the surface is nearly dry again before squirting on another layer of acetone. Handle this with care.

10 Take a pre-cut acetate strip in both hands and quickly align it with the polished slab. Bring the longest

edge of the strip and the slab into contact and, with moderate speed, bring the strip down to rest over the polished and etched face, avoiding the development of bubbles by pushing a little wall of acetone in front.

11 Allow to dry for at least 3 minutes, during which time the peel must not be touched. To remove the peel, take one free corner of the strip and peel back with a firm, even pressure. Trim off any surplus acetate sheet from around the peel immediately after pulling (to prevent wrinkling). Place the peel between paper towel and press for about 30 minutes under a pile of books (to prevent curling).

12 Several peels can be taken from the same prepared surface without re-etching. Serial sections can be prepared by grinding the limestone face down a little further between each peel. In this case, re-etching and staining will be necessary every time.

13 The peels can be labelled by writing directly on the 'etched' side with Indian ink or a biro. Store them in labelled envelopes and examine with transmitted light between glass slides taped together.

O Picking and mounting dried samples

Dried residues are best scanned on a picking tray. This should be flat with a black surface (for calcareous and siliceous microfossils) or white surface (for dark phosphatic microfossils) divided into 1-cm squares, each one preferably numbered. It should also be easy to clean to reduce the risk of contamination. A small portion of the sample should be gently tapped from its container and scattered lightly and evenly over the tray. The grains and microfossils can be manoeuvred with the aid of a good quality 00 sable hair paint brush and removed to a Franke slide for storage and examination. To do this pick up the microfossil with the fine point of a water-moistened brush and dislodge it by stroking the brush gently on the mounting surface of the slide. Adhesion of microfossils is improved by brushing the slide's surface beforehand with a weak solution of Gum Tragacanth to which a drop of Clove Oil has been added (to reduce fungal growth). Franke slides, which are the most popular means of storing dried microfossils, can be purchased commercially.

REFERENCE

Kummel, B. & Raup, D. (eds) 1965. *Handbook of Paleontological Techniques*. W.H. Freeman, San Francisco.

Systematic Index

Abies 111, (Fig. *13.11*)
Acantharia 188, 190, 192, 197
Acanthocircus 192, (Fig. *16.1*)
Acanthocythereis 241, (Fig. *20.13*)
Acanthodiacrodium 71, 74, (Fig. *9.1*)
Acanthometra 192, (Fig. *16.1*)
Acanthomorphitae 74, (Fig. *9.2*)
Acavatitriletes (Box *13.1*)
Acer (Fig. *13.11*)
Achnanthes (Fig. *17.1*)
Acritarcha 73, 88
Actinocyclus (Fig. *17.5*)
Actinomma 192, (Fig. *16.1*)
Actinopoda 192
Actinoptychus (Fig. *17.2*)
Aechmina 236, (Fig. *20.10*)
Ahumellerellaceae (Box *14.1*)
Alabamina (Fig. *15.10*)
Albaillella 192, (Fig. *16.1*)
Albaillellaria 192
Aldanella 53, (Fig. *7.3*)
Aletes (Box *13.3*)
Allogromia 160, (Fig. *7.6*), (Fig. *15.9*)
Allogromiina 151, 157, 160, (Fig. *15.32*)
Alnus (Fig. *13.11*)
Alveolinacea 169
Ambitisporites 116, (Fig. *13.12*)
Ammobaculites 160, (Fig. *15.14*)
Ammodiscus 151, 160, (Fig. *15.4*), (Fig. *15.31*)
Ammonia 153, 175, (Fig. *15.9*), (Fig. *15.27*), (Fig. *15.31*)
Ammonidium 73
Ammovertella 160, (Fig. *15.4*)
Amoebae (Fig. *6.4*)

Amorphognathus 261, (Fig. *21.12*), (Fig. *21.15*)
Amphistegina 157, 174, (Fig. *15.10*), (Fig. *15.25*)
Anabarites 53, (Fig. *7.3*)
Anabarochilina 232, (Fig. *20.7*)
Anacystis (Fig. *8.3*)
Anchignathodontidae 264
Ancyrochitina 97, (Fig. *11.1*)
Ancyrospora 116, (Fig. *13.12*)
Anemia (Fig. *13.6*)
Animalia 5–7, (Fig. *1.2*), (Fig. *6.4*)
Ansella 259, 260 (Fig. *21.10*)
Aparchites 238
Appendicisporites 129, (Fig. *13.13*)
Aquifex (Fig. *6.4*)
Aquilapollenites 111, (Fig. *13.11*)
Araphidineae 201
Archaea 42, 45
Archaebacteria 5, 42, (Fig. *1.2*), (Fig. *6.4*)
Archaeocopida 221, 230, 231, (Fig. *20.14*)
Archaeomonas 214, (Fig. *18.5*)
Archaeosphaeroides (Fig. *8.3*)
Archaeospira 51, (Fig. *7.3*)
Archaeotrichion (Fig. *6.6*)
Archaias 155, 169, (Fig. *15.21*)
Archamoebae 6
Archeoentactiniidae 192
Archezoa 5, 6, (Fig. *1.2*)
Areoliera 91
Argilloecia 229, 233, (Fig. *20.9*)
Arkangelskiellaceae (Box *14.1*)
Arpylorus 90
Articulina 169, (Fig. *15.20*)
Aschemonella 164, (Fig. *15.4*)
Asteraceae 121

Asterigerina 174, (Fig. *15.25*)
Asterolampra (Fig. *17.5*)
Astromphalus (Fig. *17.5*)
Astrorhiza 164, (Fig. *15.4*)
Athalamida 178
Azonomonoletes (Table *13.3*)
Azpeitia (Fig. *17.5*)

Bachmannocea (Fig. *18.2*)
Bachmannocena 212
Bacillariophyta 214
Bacteria (Fig. *1.2*)
Bactrognathidae 263
Bactrognathus 263, (Fig. *21.13*)
Bairdia 234, (Fig. *20.1*), (Fig. *20.9*), (Fig. *20.15*)
Bairdiocopina 234
Bairdiocypris (Fig. *20.9*)
Baltisphaeridium 71, 74, 76
Baragwathania 113
Barrandeina 117
Bathropyramis 192, (Fig. *16.1*)
Bathysiphon 153, 164, 179, (Fig. *15.4*), (Fig. *15.10*)
Beggiatoa 61
Beggiatoales 61
Belodella 259, 260, (Fig. *21.10*)
Belodellida 259
Belodina 260, (Fig. *21.11*)
Belonaspis 196, (Fig. *16.1*)
Betula (Fig. *13.11*), (Fig. *13.15*)
Beyrichia 236, (Fig. *20.10*)
Bigenerina 164, (Fig. *15.14*)
Bilateratia (Fig. *1.2*)
Bilidinea 89
Biliphyta 7, (Fig. *1.2*)
Biraphidineae 201
Biscutaceae (Box *14.1*)

280

Bispathodus 265, 269
Bolivina 151, 153, 173, (Fig. *15.10*), (Fig. *15.24*), (Fig. *15.31*)
Braarudosphaera 137, (Fig. *14.2*), (Fig. *14.3*)
Braarudosphaeraceae 132, (Box *14.1*)
Bradleya 241, (Fig. *20.15*)
Bradoriida 230, 232
Brightwellia (Fig. *17.5*)
Bryophyta 104
Bulimina 155, 173, (Fig. *15.24*)
Buliminacea 173, (Fig. *15.24*), (Fig. *15.31*)
Buliminella 174, (Fig. *15.24*)
Bythoceratina 230, 236, (Fig. *20.6*), (Fig. *20.9*)

Calamites 117
Calamospora 117
Calcarina 157, 175, (Fig. *15.27*)
Calcidiscus (Fig. *14.5*)
Calciosoleniaceae (Box *14.1*)
Calluna (Fig. *13.15*)
Calpionella 217, (Fig. *19.2*)
Calpionellidae 217
Calpionelloidea 217
Calyculaceae (Box *14.1*)
Calyptosphaeraceae (Box *14.1*)
Campylacantha 192, (Fig. *16.1*)
Candona (Fig. *20.15*)
Cannabis 122
Cannopilus 211, (Fig. *18.1*)
Carbonita 233, (Fig. *20.9*)
Carpinus (Fig. *13.11*)
Carterina 168, 183, (Fig. *15.19*), (Fig. *15.31*), (Fig. *15.32*)
Carterinina 168, (Fig. *15.19*), (Fig. *15.31*)
Caryophyllaceae (Fig. *13.15*)
Cassidulina 153, 176, (Fig. *15.10*), (Fig. *15.29*)
Cassidulinacea 176, (Fig. *15.29*)
Caulobacter (Fig. *8.2*)
Caulobacteraceae 60
Cavatomonoletes (Box *13.3*)
Cavusgnathidae 264
Cavusgnathus 265, (Fig. *21.14*)
Cedrus (Fig. *13.15*)
Celtia (Fig. *20.13*)
Centrales 200, 202, 204

Ceratium 86, (Fig. *10.7*)
Ceratobulimina 170, (Fig. *15.22*)
Ceratolithaceae (Box *14.1*)
Cercozoa 177
Cestodiscus (Fig. *17.5*)
Chaetoceras 205, (Fig. *17.4*)
Challengerianum 192, (Fig. *16.1*)
Charophyta 7
Cheirolepidiaceae 119
Chenopodiaceae (Fig. *13.15*)
Chiastozygaceae (Box *14.1*)
Chirognathidae 263
Chitinozoa 7, 96-9
Chlamybacteriales 60
Chlamydia 43
Chlorarachina (Fig. *1.2*)
Chlorophyta 7
Chordata 249
Chromista 5, 6, 135-6, 204, (Fig. *1.2*)
Chromobiota 135
Chroococcales (Fig. *8.3*)
Chrysophyceae 211
Chrysophyta 92, 129, 136, 204
Chuaria 73
Cibicides 153, 174, 181, (Fig. *15.9*), (Fig. *15.25*), (Fig. *15.31*)
Cibicidoides 27
Ciliata 217
Ciliophora 215, 217
Circinatisphaera 73
Cladarocythere (Fig. *20.15*)
Cladocopina 239
Classopollis. See *Corollina*
Clavatipollenites 120, (Fig. *13.12*), (Fig. *13.13*)
Clavohamulidae 261
Clavohamulus 261, (Fig. *21.11*)
Cloudina 51
Clydagnathus (Fig. *21.1*)
Cnidaria 54
Coccolithaceae (Box *14.1*)
Coccolithophyceae 136
Coccolithus 131, (Fig. *14.5*)
Collosphaeridae 188
Colomiellidae 217
Conchoecia 228
Conochitinidae 97, (Fig. *11.3*)
Conodonta 258
Conodonti 258
Contusotruncana 8

Cooksonia 113
Corbicula (Fig. *20.16*)
Corbisema 210-12, (Fig. *18.1*)
Cordylodus 259, (Fig. *21.4*), (Fig. *21.10*)
Cornua 212, (Fig. *18.2*)
Cornuodus 261, (Fig. *21.1*)
Corollina 119, (Fig. *13.12*)
Corticata 6, (Fig. *1.2*)
Corylus (Fig. *13.11*)
Coscinodiscophyceae 202
Coscinodiscus 202, 205, (Fig. *17.2*), (Fig. *17.4*), (Fig. *17.5*)
Coskinolina 164, (Fig. *15.15*)
Craspedobolbina (Fig. *20.13*)
Craspedodiscus (Fig. *17.5*)
Crepidolithaceae (Box *14.1*)
Crinopolles 119
Crustacea 219, 230, 232
Crytochites 51
Ctenophora 54
Cupressaceae 111
Cupressus 111
Cyamocytheridea (Fig. *20.16*)
Cyclammina 165, (Fig. *15.10*), (Fig. *15.13*)
Cyclococcolithina 136, (Fig. *14.2*)
Cyclogyra 168, (Fig. *15.20*)
Cyclolina 165, 181, (Fig. *15.15*)
Cyclopsinella 165, (Fig. *15.15*)
Cymatiogalea 73, 74, 76, (Fig. *9.1*)
Cymatiosphaera 51, 74, 76, (Fig. *9.1*)
Cypridea 233, (Fig. *20.9*)
Cyprideis 229, 230, 234, (Fig. *20.6*), (Fig. *20.9*)
Cypridina 228, 239, (Fig. *20.6*), (Fig. *20.11*)
Cypridinoidea 238
Cypridoidea 228
Cypridopsis 226, (Fig. *20.1*), (Fig. *20.15*)
Cyprinotus (Fig. *20.13*)
Cypris (Fig. *20.6*), (Fig. *20.9*)
Cyrtocapsa 192, (Fig. *16.1*)
Cystites (Table *13.3*)
Cystosporites 116, (Fig. *13.12*)
Cytheracea (Fig. *20.6*)
Cytherella 229, 233, (Fig. *20.8*), (Fig. *20.15*)

Cytherelloidea (Fig. *20.13*)
Cytherocopina 233
Cytheroidea 233
Cytheromorpha (Fig. *20.15*)
Cytheropteron (Fig. *20.6*)
Cytherura 236, (Fig. *20.9*)

Dapsilodontidae 260
Dapsilodus 261, (Fig. *21.11*)
Darwinula 228, 233, (Fig. *20.9*)
Darwinulinacea 228, (Fig. *20.6*)
Darwinulocopina 233
Deflandrea 89–91, (Fig. *10.3*)
Deinococcus 43
Densosporites 117, (Fig. *13.12*), (Fig. *13.13*)
Dentalina (Fig. *15.31*)
Denticulopsis 207, (Fig. *17.5*)
Desmochitina 97, 99, (Fig. *11.1*)
Desmochitinidae (Fig. *11.3*)
Desulfovibrio 64
Deunffia 74, 76, 80
Diacromorphitae 74, (Fig. *9.2*)
Diatomea 204
Dictyastrum 192, (Fig. *16.1*)
Dictyocha 211, 212, (Fig. *17.4*), (Fig. *18.1*), (Fig. *18.4*)
Dictyochaceae 211
Dictyozoa 89
Dicyclina 165, (Fig. *15.15*)
Dicyclinidea 165
Diexallophasis 71–72, (Fig. *9.3*)
Dinoflagellata 89
Dinogymodinium (Fig. *10.6*)
Dinokaryota 88
Dinophysida 89, (Fig. *10.8*)
Dinozoa 88
Discoaster 137, 139, (Fig. *14.2*)
Discoasteraceae (Box *14.1*)
Discocyclina 175, (Fig. *15.26*)
Discorbacea 174, (Fig. *15.25*)
Discorbis 153, 174, (Fig. *15.9*), (Fig. *15.25*)
Discosphaera 135, (Fig. *14.3*), (Fig. *14.5*)
Distephanus 209, 212, 213, (Fig. *17.4*), (Fig. *18.1*), (Fig. *18.4*)
Distomodus 261, (Fig. *21.12*)
Dodecaactinella (Fig. *7.3*)
Domasia 74, 76, 80

Drepanodus 260, (Fig. *21.16*)
Drepanoistodontidae 261
Drepanoistodus 261, (Fig. *21.11*)
Duostomina 171, (Fig. *15.22*)
Duostominacea 228

Earlandinita 165, (Fig. *15.16*)
Ebria 92, (Fig. *10.10*)
Ebridians 80, 92, (Fig. *10.10*)
Ebriophyceae 92
Eiffelia (Fig. *7.3*)
Eiffellithaceae (Box *14.1*)
Elictognathidae 265
Ellisonia 263, (Fig. *21.13*)
Elphidium 151, 175, (Fig. *15.27*), (Fig. *15.31*)
Embryophyta 6
Emiliania 131, (Fig. *11.3*), (Fig. *14.5*)
Endothyra 167, (Fig. *15.17*)
Endothyracea 167, (Fig. *15.16*), (Fig. *15.17*)
Entactinosphaera 192, (Fig. *16.1*)
Entocythere 228
Entomoconchidae 239
Entomoconchus 239, (Fig. *20.11*)
Entomozoacea 238
Entophysalis (Fig. *8.3*)
Eoastrion 61
Eobacterium 61
Eosphaera (Fig. *7.3*)
Eotetrahedrion (Fig. *7.3*)
Epactridion (Fig. *7.3*)
Ephedra 111, 119, (Fig. *13.12*), (Fig. *13.15*)
Epistominella 155, 181
Eridochoncha 232, (Fig. *20.7*)
Eridostracoda 230, 232
Erraticodon 263, (Fig. *21.13*)
Estiastra 74, 75
Ethmodiscus 204, 207
Eubacteria 5, 42, 43, (Fig. *1.2*), (Fig. *6.4*)
Eubacteriales 61
Eucapsis (Fig. *8.3*)
Euchromista 204, (Fig. *1.2*)
Eucommiidites 119, (Fig. *13.12*)
Eukaryota 42, (Fig. *6.4*)
Eunice (Fig. *12.1*)
Eunicida 101
Eurychilina 238, (Fig. *20.10*)

Fasciculithaceae (Box *14.1*)
Fasciolites 169, (Fig. *15.20*)
Favella 217
Florinites 117, 119, (Fig. *13.12*)
Fohsella 8
Fontbotia 27, 31, 155, (Fig. *4.3*)
Foraminiferida 159
Fragilaria 199, (Fig. *17.1*), (Fig. *17.4*)
Frondicularia 170, (Fig. *15.22*)
Fryxellodontus 259, (Fig. *21.10*)
Fungi 5, 7, (Fig. *1.2*)
Fusulina 165, (Fig. *7.6*), (Fig. *15.17*), (Fig. *15.32*)
Fusulinacea 167, 179, (Fig. *15.17*)
Fusulinella (Fig. *15.19*)
Fusulinina 145, 147, 165, 177, 178 (Fig. *15.16*), (Fig. *15.17*)

Galba (Fig. *20.16*)
Gallionella 60
Gephyrocapsa (Fig. *14.5*)
Giardia (Fig. *6.4*)
Gigantocypris 228
Ginkgo 111
Girvanella 66
Gladius (Fig. *17.5*)
Glaucophyta 7
Glenobotrydion (Fig. *7.2*)
Globigerina 142, 157, 172, 189, 226, 233, (Fig. *15.23*), (Fig. *15.32*)
Globigerinacea 173
Globigerinella 179
Globigerinidae 180
Globigerinina 146, 172, (Fig. *15.23*), (Fig. *15.31*)
Globigerinoides 14, 31, 157, 179, 182, (Fig. *4.3*), (Fig. *14.8*), (Fig. *15.9*), (Fig. *15.31*)
Globorotalia 8, 10, 157, 173, (Fig. *15.9*), (Fig. *15.23*)
Globorotalidae 180
Globorotalinacea 173
Globotruncana 173, (Fig. *15.23*)
Globotruncanacea 173
Globotruncanidae 181
Gnathodontidae 265
Gnathodus 265, (Fig. *21.14*)
Gnetum 111
Gondolellidae 263
Goniolithaceae (Box *14.1*)

Goniomonas (Fig. *1.2*)
Gonyaulacysta 89, 90, (Fig. *10.9*)
Gonyaulax 89, (Fig. *10.3*)
Gramineae (Fig. *13.15*)
Grypania 48, (Fig. *7.3*)
Gunflintia (Fig. *8.4*)
Guttulina 170, (Fig. *15.22*)
Gymnodiniales 75
Gymnodinium 87, 89, (Fig. *10.6*)
Gymnodinoidia 89, (Fig. *10.8*)
Gymnomyxa 6, (Fig. *1.2*)

Halobacterium (Fig. *6.4*)
Halocypridina 239
Halocypris 226, 239, (Fig. *20.11*)
Haplocytherida (Fig. *20.15*)
Haplofragmoides (Fig. *15.10*)
Haptophyta 135, 136
Hastigerinella 157, 171, (Fig. *15.23*)
Hastigerinoides 171, (Fig. *15.23*)
Healdia 233, (Fig. *20.8*)
Helianthemum (Fig. *13.15*)
Helicosphaeraceae (Box *14.1*)
Heliolithaceae (Box *14.1*)
Helicopontosphaera 135, (Fig. *14.2*)
Helicosphaera 131, (Fig. *14.5*)
Heliobacterium 43
Heliozoa 188, 198
Helminthoidichnites (Fig. *7.3*)
Hemiaulus 205, (Fig. *17.5*)
Hemicytherura 229
Hemidiscus (Fig. *17.5*)
Hemisphaerammina (Fig. *15.4*)
Henryhowella 241
Herkomorphitae 73, (Fig. *9.2*)
Hermesinum (Fig. *10.10*)
Hesslandona (Fig. *7.3*)
Heterohelicacea 171
Heterohelix 171, (Fig. *15.23*)
Hexangulaconularia 53, (Fig. *7.3*)
Hibbardella 263
Hilates 109, (Box *13.3*)
Hindeodus 264, (Fig. *21.14*)
Hoeglundina 171, (Fig. *15.22*)
Hollinella 236, (Fig. *20.10*)
Hormosina 164, (Fig. *15.10*), (Fig. *15.13*)
Huroniospora (Fig. *8.3*)
Hyphomicrobiales 61

Hystrichosphaera 89
Hystrichosphaeridium 89, (Fig. *10.3*)

Icriodella 261, (Fig. *21.12*)
Idiocythere (Fig. *20.15*)
Idiognathodus 265, (Fig. *21.14*)
Idioprioniodus 263–8
Illinites 119, (Fig. *13.12*)
Ilyocypris (Fig. *20.13*)
Involutina 168, (Fig. *15.19*), (Fig. *15.31*)
Involutinina 168, (Fig. *15.32*)
Islandiella 174, (Fig. *15.24*)
Isoetales 106

Juniperus 111

Kakabekia 61
Kidstonella (Fig. *8.3*)
Kirbya (Fig. *20.10*)
Kladognathus 263
Kloedenella 233, (Fig. *20.8*)
Kockelella 265, (Fig. *21.14*)
Kockelellidae 265
Krithe 228, 229, 236, (Fig. *20.9*)
Kunmingella 232

Laevigatosporites 120
Lagena 153, 170, (Fig. *15.4*), (Fig. *15.22*)
Lagenicula 117
Lagenida (Fig. *7.6*)
Lagenina 148, 170, 179, (Fig. *15.22*), (Fig. *15.31*), (Fig. *15.32*)
Lagenochitina 97, (Fig. *11.1*)
Lagenochitinidae 100, (Fig. *11.3*)
Laminatitriletes (Table *13.2*)
Leiocopida 238, 239
Leiofusa 73, 74, (Fig. *9.3*)
Leiosphaeridia 103
Leiosphaeridium 74, 101, (Fig. *9.3*)
Lenticulina 153, 228, (Fig. *15.22*)
Leperditia (Hermannina) 232
Leperditia 232, (Fig. *20.7*), (Fig. *20.13*)
Leperditicopida 230, 237, (Fig. *20.14*)
Lepidocyclina 157, 175, (Fig. *15.26*)
Lepidodendrales 111

Lepidodendron 162, (Fig. *13.2*), (Fig. *13.13*)
Lepidostrobus (Fig. *13.13*)
Leptocythere 229
Liliaceae (Fig. *13.15*)
Limnocythere 228, 234, (Fig. *20.6*), (Fig. *20.9*), (Fig. *20.13*)
Limnocytheridae 228
Linderina 174, (Fig. *15.25*)
Lingulodinium 122
Lithostromationaceae (Box *14.1*)
Lituolacea 228–9, (Fig. *15.14*), (Fig. *15.15*)
Loftusia 165, (Fig. *15.13*)
Loxoconcha 229
Loxostomum 177, (Fig. *15.29*)
Lueckisporites 119, (Fig. *13.12*)
Lycopodium (Fig. *13.6*)
Lycopodophyta 153
Lycospora 162, (Fig. *13.13*)
Lyramula 212, 213, (Fig. *18.2*)

Macrocypris 261, (Fig. *20.13*)
Mallomonas 214
Manawa 236
Marsileales 111
Martinssonia (Fig. *7.3*)
Medullosa 166
Melanocyrillium 118
Melania (Fig. *20.16*)
Melonis 177, 233, (Fig. *15.30*)
Melosira 201, (Fig. *17.2*)
Merrillina 263, (Fig. *21.13*)
Mesocena (Fig. *18.1*)
Mesocypris 226, (Fig. *20.6*)
Mestognathidae 265
Mestognathus 265, (Fig. *21.14*)
Metacopina 226, 232, 233, 234, (Fig. *20.8*), (Fig. *20.14*)
Metakaryota (Fig. *1.2*)
Metallogenium 89
Metamonada 11
Methanobacterium (Fig. *6.4*)
Micrhystridium 72, 74, 76, 107
Micromitra (Fig. *7.3*)
Microrhabdulaceae (Box *14.1*)
Microsporidia 11
Miliammellus 205, 227, (Fig. *15.22*), (Fig. *15.31*)
Miliammina 160, (Fig. *15.13*)

Miliolidae   212, (Fig. *15.10*)
Miliolina   152, 153, 179, 180, 204,
   226, 234, (Fig. *15.20*), (Fig. *15.21*),
   (Fig. *15.31*), (Fig. *15.32*)
Miliolinella   (Fig. *15.31*)
Mobergella   53
Moenocypris   243, (Fig. *20.15*)
Mongolodus   (Fig. *7.3*)
Monoraphidineae   201, 282
Multioistodontidae   263
Multioistodus   263, (Fig. *21.12*)
Myocopida   228
Myodocopa   219, 221
Myodocopida   221, 224, 232, 238,
   (Fig. *20.11*), (Fig. *20.14*)
Mytilocypris   228
Myxinoidea   (Fig. *21.18*)
Myxococcoides   (Fig. *8.3*)

Nannoceratopsida   89, (Fig. *10.8*)
Nannoceratopsis   126, (Fig. *10.6*)
Nannoconaceae   (Box *14.1*)
Nannoconus   217
Nassellaria   188–90, 192, 197
Naviculopsis   212, (Fig. *18.1*)
Negibacteria   (Fig. *1.2*)
Neocyprideis   (Fig. *20.1*), (Fig. *20.15*)
Neogloboquadrina   218, 239
Neogondolella   263
Neoschwagerina   168, 225,
   (Fig. *15.17*)
Neoveryhachium   75, 107
Netromorphitae   74, (Fig. *9.2*)
Nitzschia   (Fig. *17.5*)
Noctiluca   120, (Fig. *10.6*)
Nodella   (Fig. *20.10*)
Nodosaria   211, 228, (Fig. *15.22*)
Nodosariacea   234
Nodosinella   167, (Fig. *15.16*)
Nodospora   113
Nonion   233, (Fig. *15.30*)
Nonionacea   (Fig. *15.30*)
Nostoc   76
Nostocales   (Fig. *8.3*)
Nothofagidites   (Fig. *13.13*)
Nubeculariidae   153
Nubeculinella   226, (Fig. *15.20*)
Nummulites   157, 177, 180,
   (Fig. *15.28*)
Nutallides   155

Octoedryxium   76, 105
Oistodontidae   263
Oistodus   263, (Fig. *21.12*)
Oneotodontidae   261
Oneotodus   261, (Fig. *21.11*)
Ooidium   105
Oolithotus   (Fig. *14.5*)
Oomorphitae   105, (Fig. *9.2*)
Operculatifera   142, 149
Operculodinium   120, 128, (Fig. *10.7*)
Orbitoidacea   175, 230, (Fig. *15.26*)
Orbitolina   165, (Fig. *15.15*)
Orbitolinidae   165
Orbitolites   227, (Fig. *15.21*)
Orbulina   33, 157–8, 179, 229
   (Fig. *15.23*)
Ornithocercus   89, (Fig. *10.6*)
Ortonella   95
Osangularia   177, 233, (Fig. *15.30*)
Oscillatoria   99
Ostracoda   230–46, (Fig. *7.6*)
Oulodus   261, (Fig. *21.13*)
Ozarkodina   264, (Fig. *21.5*),
   (Fig. *21.15*)
Ozarkodinida   258, 263, 264–6,
   (Fig. *21.14*), (Fig. *21.15*)

Palaeocopida   221, 232–41, 239,
   (Fig. *20.10*)
Palaeospiculumidae   192
Palaeotextularia   167, (Fig. *15.16*)
Palmatolepidae   265
Palmatolepis   265, (Fig. *21.14*)
Panderodontidae   258, 260, 270,
   (Fig. *21.16*)
Panderodontidae   257, 258
Panderodus   249, 254, 257, 260,
   261, 269, (Fig. *21.3*), (Fig. *21.4*),
   (Fig. *21.6*), (Fig. *21.16*)
Paracypris   (Fig. *20.9*)
Paradoxostoma   228, 236, (Fig. *20.6*),
   (Fig. *20.9*)
Paradoxostomatidae   233
Parafusulina   (Fig. *15.19*)
Parakrithe   241
Parapanderodus   260, (Fig. *21.11*)
Paraparchites   238, (Fig. *20.11*)
Parathuramminacea   164, (Fig. *15.16*)
Patellifera   188
Patellina   168, 224, (Fig. *15.19*)

Pavonina   230, (Fig. *15.24*)
Pedavis   263, (Fig. *21.12*)
Pedicythere   241
Peneropolis   157, (Fig. *15.21*)
Pennales   100, 201
Pennyella   241
Peridinea   120, 126
Peridiniales   89
Peridinium   89, 122, (Fig. *10.3*)
Peridinoidia   89, (Fig. *10.8*)
Perinotriletes   (Table *13.2*)
Periodon   263, (Fig. *21.13*),
   (Fig. *21.15*)
Periodontidae   263
Petrianna   (Fig. *20.13*)
Petromyzontida   (Fig. *21.18*)
Phaeodaria   188–9, 192, 265–6
Phosphatocopida   232
Phragmodus   261, (Fig. *21.12*)
Phycodes   53
Phytomastigophora   210
Picea   111, (Fig. *13.11*), (Fig. *13.15*)
Pinnularia   100, (Fig. *17.1*)
Pinus   111, (Fig. *13.11*), (Fig. *13.15*)
Pityosporites   119, (Fig. *13.12*)
Planispirillina   168, (Fig. *15.31*)
Planobina   (Fig. *20.16*)
Planorbulina   230, (Fig. *15.25*)
Plantae   (Fig. *1.2*)   *see plants* in
   General index
Plantago   (Fig. *13.15*)
Platycopina   226, 233–4, (Fig. *20.8*),
   (Fig. *20.14*)
Plectodina   263, (Fig. *21.12*)
Plectodinidae   263
Pleurophrys   160, 221, (Fig. *15.4*)
Pleurostomella   177, 232, (Fig. *15.29*)
Poaceae   120, (Fig. *13.15*)
Podocarpus   111, (Fig. *13.11*)
Podocopa   219, 224, 230
Podocopida   228, 232–41, 234,
   (Fig. *20.8*)
Podocopina   233–4, (Fig. *20.9*),
   (Fig. *20.14*)
Podocyrtis   192, (Fig. *16.1*)
Podorhabdaceae   (Box *14.1*)
Polycope   228, 239, (Fig. *20.11*),
   (Fig. *20.13*)
Polycyclolithaceae   (Box *14.1*)
Polycystina   188–9

Systematic Index 285

Polycystinea 192
Polygnathidae 265
Polygnathus 265, (Fig. *21.14*)
Polygonomorphitae 105, (Fig. *9.2*)
Polykrikos 120, (Fig. *10.6*)
Polymorphina 228, (Fig. *15.22*)
Polyodryxium (Fig. *9.1*)
Polyplacognathidae 263
Polyplacognathus 263, (Fig. *21.12*)
Pontosphaeraceae (Box *14.1*)
Pontosphaerea 184
Porostrobus 117, 164, (Fig. *13.13*)
Poseidonamicus 230
Posibacteria (Fig. *1.2*)
Potamocypris (Fig. *20.13*)
Potonieisporites 119, (Fig. *13.12*)
Prediscosphaera 188, (Fig. *14.2*)
Prediscosphaeraceae (Box *14.1*)
Primaevifilum (Fig. *6.6*)
Primitiopsis (Fig. *20.10*)
Prinsiaceae (Box *14.1*)
Prioniodinida 258, 263, 270, (Fig. *21.15*)
Prioniodinidae 263
Prioniodontida 258, 261, 263, 270, (Fig. *21.15*)
Prioniodontidae 263
Prioniodus 263, (Fig. *21.12*)
Prismatomorphitae 105, (Fig. *9.2*)
Proconodontida 259–60, 270, (Fig. *21.16*)
Proconodontus 258
Profusulinella 225, (Fig. *15.17*)
Prokaryota 67
Promissum 261, 269, (Fig. *21.2*), (Fig. *21.12*), (Fig. *21.15*)
Propontocypris (Fig. *20.13*)
Prorocentroidia 124
Prorocentrum 124, (Fig. *10.6*)
Prosomatifera 100, (Fig. *11.1*), (Fig. *11.2*)
Proteobacteria 43
Protocentroidia (Fig. *10.8*)
Protoconodontida 266
Protoconodontus 259, (Fig. *21.10*)
Protohaploxypinus 119, (Fig. *13.12*)
Protohertzina 53, 79
Protohyenia 113
Protopanderodontida 258, 260–1, (Fig. *21.16*)

Protopanderodontidae 260
Protopanderodus 254, 260, (Fig. *21.11*)
Protoperidinium 128, (Fig. *10.7*), (Fig. *10.9*)
Protozoa 5–11, 159, 192, 203, (Fig. *1.2*)
Pseudoemiliania 188, (Fig. *14.2*)
Pseudoeunotia (Fig. *17.5*)
Pseudomonadales 89
Pseusosaccititriletes (Fig. *13.2*)
Psilophyta 153
Pteridophyta 153
Pteromorphitae 103, (Fig. *9.2*)
Pterophyta 153
Pterospathodontidae 263
Pterospathodus 263, (Fig. *21.12*)
Pterospermella 103, (Fig. *9.3*)
Pullenatina 18
Pulvinosphaeridium 105, 107
Puncioidea 89
Pygodus 261, (Fig. *21.12*)
Pyxilla (Fig. *17.5*)

Quercus (Fig. *13.11*)
Quinqueloculina 151, 152, 224, (Fig. *15.9*), (Fig. *15.20*)

Radiata (Fig. *1.2*)
Radiolaria 105, 126, 188, 265–74
Radiozoa 188, 192
Rectobolivina 173, 230, (Fig. *15.24*)
Renalcis 93
Reophax 164, (Fig. *15.9*), (Fig. *15.13*)
Reticulatisporites (Fig. *13.13*)
Reticulomyxa 179, 234
Reticulosa 159
Rhabdammina 164, (Fig. *15.10*)
Rhabdosphaera 137, (Fig. *14.2*), (Fig. *14.5*)
Rhabdosphaeraceae (Box *14.1*)
Rhaetogonyaulax 127
Rhagodiscaeae (Box *14.1*)
Rhipidognathidae 263
Rhipidognathus 263, (Fig. *21.12*)
Rhizammina 164, (Fig. *15.4*)
Rhizopoda 141, 159
Rhizoselenia (Fig. *17.4*)
Rhodophyta 12

Richteria (Fig. *20.11*)
Rivularia (Fig. *8.4*)
Robertina 228, (Fig. *15.22*)
Robertinida (Fig. *15.31*)
Robertinina 205, 228, (Fig. *15.22*), (Fig. *15.31*), (Fig. *15.32*)
Robertinoides (Fig. *15.31*)
Rocella (Fig. *17.5*)
Rossiella (Fig. *17.5*)
Rotaliacea 230–1, (Fig. *15.27*), (Fig. *15.31*)
Rotaliina 180, 205, 208, 229, 234, (Fig. *7.6*), (Fig. *15.23*), (Fig. *15.24*), (Fig. *15.25*), (Fig. *15.26*), (Fig. *15.27*), (Fig. *15.28*), (Fig. *15.29*), (Fig. *15.32*)
Rotaliporacea 229

Saccaminopsis 165, (Fig. *15.16*)
Saccammina 164, (Fig. *15.31*)
Saipanetta 256
Salpingella 217, (Fig. *19.2*)
Salpingellina 217, (Fig. *19.2*)
Salvinales 111
Sarcodina 159
Scaliognathus 269
Schizosphaerellaceae (Box *14.1*)
Schulzospora 119, (Fig. *13.12*)
Schwagerina 168, 225, (Fig. *15.17*)
Scyphosphaera 131, 184, (Fig. *14.3*)
Scytonema (Fig. *8.4*)
Selaginellales 111
Serpula (Fig. *20.16*)
Shepheardella 160, 221
Sigillaria 162
Silicoloculinida (Fig. *15.31*)
Silicoloculinina 227, (Fig. *15.22*), (Fig. *15.31*)
Siphonina 174, 230, (Fig. *15.25*)
Siphonodella 264, (Fig. *21.14*)
Siphotextularia (Fig. *15.31*)
Skeletonema 105
Sollasitaceae (Box *14.1*)
Soritacea 233
Sorosphaera 164, (Fig. *15.12*)
Spathognathodontidae 264
Sphaerochitinidae 103
Sphaeromorphitae 103, (Fig. *9.2*)
Sphaerotilus 89
Sphenolithaceae (Box *14.1*)
Sphenophyta 155

Spiniferites 89, 120, 126, 128, (Fig. *10.3*), (Fig. *10.9*)
Spirillina 225, (Fig. *15.19*), (Fig. *15.31*)
Spirillinina 205, 225, (Fig. *15.19*), (Fig. *15.31*), (Fig. *15.32*)
Spirochetes 43
Spiroclypeus 157, 232, (Fig. *15.28*)
Spirocyclina 165, (Fig. *15.13*)
Spirotrichea 217
Spirotrichida 217
Sporangiostrobus 162, (Fig. *13.13*)
Spumellaria 188, 192–5, 265–6,
Stephanolithaceae (Box *14.1*)
Stephanopyxis (Fig. *17.5*)
Strachanognathidae 261
Strachanognathus 261, (Fig. *21.11*)
Striatopodocarpites (Fig. *13.13*)
Suessia 126, (Fig. *10.7*), (Fig. *10.8*)
Sulfolobus (Fig. *6.4*)
Sweetognathidae 265
Sweetognathus 265, (Fig. *21.14*)
Synechocystis (Fig. *8.3*)
Synedra (Fig. *17.5*)
Syracospaeraceae (Box *14.1*)
Synura 214

Tanuchitinidae 100
Taphrognathus 269
Tasmanites 71, 76, 77, (Fig. *9.3*)
Taxaceae 111
Taxodiaceae 111
Technitella 164
Tectatodinium 84
Teridontus 258
Terrestricytheroidea 226

Tetrahedrales 107, (Fig. *13.12*)
Tetrataxis 167, (Fig. *15.16*)
Textularia 164, (Fig. *15.14*), (Fig. *15.32*)
Textulariina 145, 147, 153, 160, 179, 180, (Fig. *7.6*), (Fig. *15.14*), (Fig. *15.15*), (Fig. *15.31*)
Thalassicola 192, (Fig. *16.1*)
Thalassionema (Fig. *17.4*), (Fig. *17.5*)
Thalassiosira 202, 205, (Fig. *17.2*), (Fig. *17.4*), (Fig. *17.5*)
Thalassiothrix (Fig. *17.5*)
Thalassocythere 241
Thaumatocypridoidea 239
Thaumatocypris (Fig. *20.11*)
Theoduxus (Fig. *20.16*)
Thoracosphaeraceae (Box *14.1*)
Tintinnina 215, 216, 217
Tintinnopsella 217, (Fig. *19.2*)
Tintinnopsis 217, (Fig. *19.1*), (Fig. *19.2*)
Tolypammina 164
Tracheophyta 104
Treptichnus 53
Tretomphalus 174, (Fig. *15.25*)
Triceratum (Fig. *17.5*)
Tricolpites 120, (Fig. *13.12*)
Triletes (Box *13.1*)
Triloculina 169, (Fig. *15.9*)
Trinacria (Fig. *17.5*)
Trochammina 165, (Fig. *15.10*), (Fig. *15.14*)
Troqetrorhabdulaceae (Box *14.1*)
Tsuga 111, (Fig. *13.11*)
Tuberculatisporites (Fig. *13.13*)

Tunisphaeridium 71
Tytthocorys 217, (Fig. *19.2*)

Umbellosphaera (Fig. *14.5*)
Usbekistania 164, (Fig. *15.4*)
Uvigerina 27, 31, 155, (Fig. *4.3*), (Fig. *15.10*)

Vallacerta 212–20, (Fig. *18.1*)
Vargula 225, (Fig. *20.13*)
Variramus 212, (Fig. *18.2*)
Velatachitina 97, (Fig. *11.1*)
Verneuilina 165, (Fig. *15.14*)
Veryhachium 73, 76, 80
Vestrogothia 231, (Fig. *20.7*)
Virgulinella 177, (Fig. *15.29*)
Viriplantae 7, (Fig. *1.2*)
Visbysphaera 71, 74, (Fig. *9.1*)
Vittatina 119, (Fig. *13.12*)

Walliserodus 260, (Fig. *21.16*)
Wetzeliella 91, (Fig. *10.7*), (Fig. *10.9*)
Wilsonites 119, (Fig. *13.12*)
Wollea (Fig. *8.4*)
Woodhousia 111, (Fig. *13.11*)

Xanioprion (Fig. *12.1*)
Xestoleris 229

Yunnanodus (Fig. *7.3*)

Zonomonoletes (Box *13.3*)
Zosterophyllum 113
Zygacantha 196, (Fig. *16.1*)
Zygodiscaceae (Box *14.1*)
Zygodiscus 137, (Fig. *14.2*)

# General Index

abdomen 192
aboral surface 256, 265
abyssal plains 147, 153, 155, 159, 188, 204, 228, 229, 268
accommodation space 21
acetate peels 278
acetic acid 275, 277
acetone 277–8
acritarch alteration index 77
acritarchs 3, 7, 23, 35, 48, 71–7, 89, 92, 96, 99, 101, 274, 276
   apical 73
   binary fission 75
   central cavity 71
   circinate suture 73
   crests 71, 73–5, 77
   cryptosuture 72
   double wall 71
   epityche 73–4
   equator 73
   equatorial flange 74. *See also* ala
   excystment structures 71–2
   flanges 75
   fusiform 74
   herkomorphs 75–6
   lateral rupture 72, 74
   median split 72
   munitium 73
   netromorphs 75–6
   operculum 73
   ornament 72
   plates 74
   polygonomorphs 75–6
   processes 71, 73–4
   pylome 73
   sculpture 74–5
   sphaeromorphs 49, 75–6
   spines 75, 77
   tabulation 75
   trabeculum 71
   vesicle 71–4
aerobic bacteria 57
agglutinated tests 57, 145, 160, 178
aggradational 22, 23
ala 74. *See also* equatorial flange of spores and pollen
Albian 91, 120, 182, 217
alcohol 123
algal blooms 64
Alizarin Red 278
allopatric speciation 10
alpha index 155
alternation of generations 80, 104
alveolinids 157
ammonia 64. *See* ammonium ions
ammonium ions 64. *See* ammonia
anaerobic bacteria 57, 63
anaerobic zone 88
anagenesis 8
anemophily 109
angiosperm 104, 109, 111, 119–20
anoxygenic bacteria 39
Antarctic Bottom Water 155, 182
Apex Chert 46
apomorphic 13
apparent oxygen utilization (AOU) index 30
Aptian 91, 120, 217
aragonite 45, 131, 142, 147, 168, 170, 278
archaeocyathan sponges 53
Archean 39, 43–6, 66
Arenig 76, 99
argillaceous rocks 92, 99, 246, 274, 275
asexual reproduction 4, 51, 99, 144
assemblage biozone 18
Atdabanian Stage 53
athalassic 228
Atlantic Ocean 86–7, 134, 137, 180
autotrophy 4, 88

bacteria 5, 39, 59–62, 63–6, 67, 142, 153, 159, 189, 215, 226
   binary fission 62, 63
   budding 89
   clotted microtexture 66
   flagellum 59, 88
   heterocyst cells 62
   hormogonia 62
   processes 59
   sheathed 88
   stalked 88
   trichome 62–3
bacterial mats 43, 45, 59. *See also* biofilms
banded iron formations (BIFs) 44, 61
bauxites 65
Belt Supergroup 76
benthic foraminifera 10, 22, 27, 28, 30–2, 57, 142, 144, 145, 148, 153, 156, 157, 160, 177, 181, 183
Berriasian 89, 216
binomial system 65
biofilms 59, 64. *See also* bacterial mats
biomarkers 43, 46, 49, 75, 89
biomineralization 53, 57, 131, 203
biostratigraphy 5, 16, 18, 21
biozone 16–18
bisaccate pollen 21, 117, 119
Bitter Springs Chert 50, 75
Black Sea 87, 123
black shales 96, 191, 269, 274
bog iron ores 60

General Index

Botomian Stage   53
brackish water   28, 132, 153–4, 160, 175, 198, 201, 216, 218, 228, 229, 233, 236
bradoriids   53
bromoform   277
brood pouches   225, 236. *See also* crumina
brown algae   5, 76, 136
bryophytes   104, 109
budding   61, 99

Caedax   140, 214
calcareous microfossils   197, 200, 207, 275
calcareous nannoplankton   9, 16, 30, 57, 129, 129–40, 241
  cross bar   137
  elliptical ring   137
  flagella   129, 137
  haptonema   129
  lateral bars   211
  lattice   131
  plates   130
  radial elements   137
  radial plates   137
  rays   129, 137
  rods   130
  sculpture   131
  shields   131, 137
  spiral flange   137
  stellate calcareous nannofossils   129
calcification   10, 66, 129, 224
calcite   25, 30, 35, 57, 66, 129, 131, 135, 142, 145, 147, 148, 157, 159, 168, 176, 216, 223, 278
calcite compensation depth (CCD)   159
calcium carbonate fluorapatite   249
calcium phosphate   35, 86, 249
calpionellids   57, 215, 217
  agglutinated   215
  hyaline   218
Cambrian explosion   48, 51, 55, 76
Cambrian   48, 55, 57, 75, 76, 89, 142, 160, 179, 180, 188, 195, 217, 219, 232, 249, 266, 270
Canada Balsam   123, 140, 208, 214, 246, 276
Caradoc   99

carapace   57
carbon   30, 32, 39, 44, 46, 54, 200
carbon cycle   46, 57
carbon dioxide   39, 208
carbon isotopes   25, 30, 43
carbon tetrachloride   277
carbonaceous chondrites   39
carbonate ooze   131
Carboniferous   57, 61, 76, 96, 99, 117, 119, 120, 122, 157, 165, 192, 217, 229, 239, 249, 251, 264, 265, 269
carotenoids   210
catagenesis   35
caving   18, 273
celestine   188
cell membrane   4, 39, 129
cellosize   276
Cenomanian   120, 101
Cenozoic   16, 18, 76–7, 86, 89, 120, 137–40, 142, 159, 181, 183, 188, 191, 197, 205, 207, 211, 212, 217, 232, 240, 243
centric diatoms   200, 202, 204
cephalochordates   268
cephalothorax   219
chalks   129, 135, 140, 159, 184, 246, 274, 275
channel samples   273
charophytes   21
chemoautotrophy   88
chemolithoautrophy   88
chemolithotrophic bacteria   39
chemostratigraphy   48
cherts   46, 65, 191, 198, 204
chitin   35, 96, 216, 219
chitinozoa   96, 99
  aboral end   96, 97
  aboral pole   96
  annulations   99
  aperture   96
  body chamber   96
  chamber   97
  copula   96, 97
  flanges   97
  lips   97
  neck   96, 97
  operculum   96, 97
  oral end   96
  oral pole   96
  ornament   97

  prosome   97
  sleeve   96, 97
  spines   96, 97
chlorinity   228
chlorophylls   7, 210
chlorophytes   157
chloroplasts   4, 75, 82, 199, 204
chromosomes   61, 75, 82, 144
chronostratigraphy   16
chrysophytes   200
cilia   4, 6, 215
cladistics   13
cladogenesis   8–10
cladogram   12–13
class   5
clines   10
club-mosses   106
coal   22, 35, 65, 117, 120, 233
coccolithophores   129, 131–2, 137–8
coccolithophorids   10
coccoliths   23, 57, 129–41, 145, 159, 215, 217, 274, 276
coccosphere   129, 135
colpi   111. *See also* spores and pollen, monocolpate; spores and pollen, tricolpate
commensal   228
composite standard reference section (CSRS)   19
concurrent range zone   18
Coniacian   91
conifers   104, 117, 119
conodont animals   249, 255, 266, 268
conodont colour alteration (CAI)   35, 249
conodonts   21, 35, 241, 249–72, 275, 276, 277–8
  ae element   254, 260
  aequaliform   258
  alate   256
  angulate   256
  arcuatiform   258
  basal body   251, 259
  basal cavity   251, 256, 257, 260
  basal groove   256
  bipennate   256
  blade   265, 269
  branchial structures   249
  breviform digyrate   256
  carina   265

carminate 256
caudal domain 253, 254, 260, 261, 263
caudal fin rays 249
chevron muscle blocks 249, 266
coniform elements 257, 260, 261, 269
costae 260
cross striations 254
crown 251, 254
cusp 251, 256, 258, 260, 261, 263
denticles 251, 254, 257, 263, 265, 266, 269
digyrate 256
dolabrate 256
dorsal 254, 256, 261, 263, 269
dorsal process 256, 265
extensiform digyrate 256
extrinsic eye musculature 249
falciform 257
furrow 257
geniculate 257, 259, 261, 263
graciliform 257
head 249
hyaline 251, 260
keels 257, 374
lamellae 251
lateral 249, 256
ligament 369
longitudinal striations 260
M elements 253, 254, 261, 263, 264
major increments 255
minor increments 254
multi-element taxonomy 258
multiramate 256
muscle fibres 249
myofibrils 249
natural assemblages 253, 256, 258, 261, 263, 265
non-geniculate 257, 260, 263
optic capsules 249
oral surface 256, 265
p (acostate) elements 254
P elements 253, 254, 256, 261, 263, 264
pastinate 256
primary processes 256
q (costate) elements 254
quadriramate 256
rastrate 257

recessive basal margin 257
rostral domain 253, 254, 259, 260, 263, 264
S elements 253, 261, 263, 264
sarcomeres 249
scaphate 257
sclerotic capsules 249
sclerotic cartilages 249
segminate 256
spherulitic structure 251
spines 266
stellate 256
tertiopedate 256
tortiform 257
truncatiform 258
trunk 245
white matter 251, 259, 261
continental slope 22, 87, 181, 189
convergence 13, 169, 229, 266
copepod crustaceans 135, 191
coprolites 65, 122
cordaitaleans 117, 119
cordaites 117
craniates 266
Cretaceous 8–10, 16, 23, 28, 46, 76, 89, 111, 119, 120, 137, 139, 142, 165, 174, 181, 182, 196, 204, 207, 210–14, 216, 217
crumina 225, 233. *See also* brood pouches
cryptarchs 46, 66
cryptic speciation 12
cryptolamellar calcite walls 174, 177
cryptospores 109
cyanobacteria 5, 12, 39–43, 45, 59, 61, 66
  akinetes 62, 63
  coccoid cyanobacteria 75
  endolithic cyanobacteria 66
  oxygenic cyanobacteria 39
  skeletal envelope 66
cycadophytes 111
cycads 111, 119
cyclothems 120
cytoplasm 4, 85, 111, 142, 145, 147, 152, 183, 188, 189, 196, 200, 211, 213

daughter cells 4, 62, 129, 144, 149
deep-sea ooze 129

deltas 23, 28, 113, 117
dendritic 153
denitrifying bacteria 59, 64
dentine 251, 259, 268
depth of burial 35, 270
Devensian 121
Devonian 57, 75, 76, 96, 99, 101, 102, 104, 116, 119, 120, 122, 179, 195, 239, 251, 258, 261, 263, 265, 270
diagenesis 22, 28, 35, 146, 198
diatomites 33, 203, 207, 208, 211, 214
diatoms 5, 6, 16, 40, 92, 137, 142, 144, 153, 157, 159, 189, 191, 197, 198, 200, 208, 215, 217, 226, 275, 276, 277
  binary fission 200
  central nodule 201
  central vacuole 200
  discoidal 202
  epivalve 202
  flagella 200
  frustule 198, 200, 204
  girdle 200, 202
  hypovalve 200, 202
  imperforate 200
  naviculoid valves 201
  plates 200
  polar nodules 201
  pseudoraphe 201, 202
  punctae 200, 202
  radial punctae 202
  raphe 201, 202
  sieve membranes 200
  spines 200
  statospore 202
  valves 200, 201, 204
  valve view 201
diderms 39
dimorphism 224, 226, 231, 236, 238
dinoflagellates 3, 5, 6, 10, 35, 71, 75, 80–92, 144, 157, 204, 215, 217, 276
  antapical 84
  antapical horns 82, 86, 89
  antapical series 82, 89
  apical archaeopyle 89
  apical series 82, 84, 89
  apical horn 82
  archaeopyle 85, 89
  armoured 80–2, 85, 89
  autocyst 82

dinoflagellates (cont'd)
  autophragm 82
  binary fission 84
  bioluminescent 80
  cavate 83, 84, 89, 91
  chorate cysts 82, 84
  cingular archaeopyle 89
  cingular plates 82, 89
  cingulum 82–4, 89
  crests 83, 89
  cysts 10, 23, 76–7, 80–92
  dinokaryon 80
  discoidal 82
  ectophragm 82
  epitheca 82
  eye spots 82
  flagella 80–2, 85, 88
  flanges 89
  furrow 82
  fusiform 82
  hypocyst 84
  hypotheca 82
  intercalary archaeopyle 89
  intercalary plates 82
  intergonal processes 84
  intertabular ornament 84
  motile stage 80
  operculum 84, 89
  ornament 83–4
  pellicle 80, 89
  pericoels 83–4
  peridinioid shape 89
  periphragm 83
  phragma 82
  plates 80, 89
  postcingular plates 82, 89
  precingular archaeopyle 89
  precingular plates 82, 84, 89
  processes 83–4, 87, 89
  proximate 82, 89–90
  proximate cysts 82, 90
  proximochorate cysts 82
  pusules 82
  sculpture 82, 88
  spines 82
  sulcal plates 82, 89
  tabulation 82, 88, 90
  unarmoured 80, 87, 89
dinosporin 82
dinosterane 89

dinoxanthin 80
diploid 75, 104, 144, 199
discoasters 129, 137
disruptive selection 12
distal germination 119
distilled water 123, 208, 274–8
dithecism 184
DNA 39, 75, 177, 251
dolomites 275
dominant species 155
double fertilization 109
drill chips 273
DSDP (Deep Sea Drilling Project) 27, 56, 182, 217
Duoshantuo Formation 53
dyads 109, 111, 113

ebridians 92–3
  actines 93
  bars 92
  hafts 93
  hoops 93
eccentricity 28
echinoderm ossicles 7
ecological gradient 8, 269
ecophenotypes 12, 180
Ediacara biota 53–4, 76
eggs 99, 225
empire 5
Emsian 116
enamel 251, 254, 259, 266
endemism 190, 230, 241, 269
endoplasmic reticulum 6, 82, 204
endospores 61–2
entomophily 109
Eocene 28, 32, 91, 137, 175, 177, 180, 183, 197, 205, 207, 212, 217, 229, 241, 243
epicuticle 223
epicystal 89
epoch 16
Equator 152, 158, 190
era 16
estuaries 28, 113, 207
euconodonts 249, 259, 266
euglenoids 5
eukaryotes 6, 48–50, 54, 57, 62, 80, 178
eustacy 21. See also relative sea level
eustatic cycles 18

eutaxa 113
eutrophic 153, 157
evolute 147, 170, 176
evolution 4, 8, 12, 13

faecal pellets 66, 135, 191, 204
faecal pumping 57
Famenian 119
family 5
fan 22. See also lowstand wedge
fermentation 64
ferns 104, 119, 123
ferric oxide cement 145
filter-feeding 226, 229
first appearance datum (FAD) 18
fish teeth 258
flagella 4
Flandrian 121
flexibacteria 61
Florida Bay 28
food vacuoles 144
foraminifera 3, 6, 8–10, 12, 21–2, 25–30, 57, 87, 89, 139, 142–86, 217, 226, 246, 274, 276, 277, 278
  agamont generation 144
  agglutinated 22, 57, 142, 145, 148, 153, 154, 160, 161, 178, 180, 182
  alar prolongations 176
  alveoli 167
  aperture 144, 145, 148, 149, 151, 157–8, 164, 168, 170, 177
  axopodia 6, 188, 189
  benthic foraminifera 10, 22, 25–8, 30–53, 57, 142, 144, 145, 148, 152, 153, 157, 158, 160, 177, 180, 181
  biconvex profile 175
  bilamellar 148, 170, 175, 177
  biloculine 226
  biserial 147, 162, 166, 171, 174
  branched alveoli 147
  branched chambers 149, 153, 164
  brevithalamous 148
  bullae 171
  bullate aperture 151
  chamber 144, 147–9, 160, 164, 168, 171, 174, 177
  chamber shape 147–9, 160
  chamberlets 146, 157, 165, 168, 169, 176
  chomata 168

clavate chambers 157
coiling directions 182
complex septate growth 147
concavo-convex 153
contained growth 147
continuous growth 147
costate 149
cuniculus 168
dentate aperture 149
diaphanotheca 167, 224
discoidal 149, 153, 157, 160, 166, 175
ectoplasm 144, 145, 147, 172
endoplasm 144, 147, 148, 149, 155, 172
epitheca 166
flabelliform 169, 173
flagella 144
float chamber 174
foramen 142, 149
fusiform shape 157, 160, 167, 169
glomospiral coiling 160
high trochospiral coil 168, 174
hyaline 146, 148, 152, 153, 156, 168, 173, 175, 176
hyaline perforate 146, 148
imperforate 146, 148, 153, 154, 168
inner lamella 148
interseptal buttresses 169
involute 149, 169, 170, 175, 176
keels 145, 149, 157
keriotheca 167
labiate aperture 149
labyrinthic wall 160
lamellae 151
lateral apertures 173
lateral layers 176
longithalamous 151
marginal zone 164
megalospheric 145, 176
microgranular 146, 160, 166, 168, 170
microspheric 145, 166, 176
MinLoc 151
monolamellar 148
monothalamous chamber 142
multilamellar 148
multilocular test 149, 160, 164, 170
multiple apertures 149
multithalamous chambers 142

mural pores 146, 147, 166
non-laminar wall structure 148
organic test 160
ornament 149, 160
outer lamella 148
pillars 157, 160, 164, 175, 176
plano-convex 153, 174
porcellaneous 145
primary aperture 149, 171, 174
pseudo-fibrous, wall structure 147
pseudopodia 142, 144, 148, 149, 153, 170, 240
pseudo-radial wall structure 147
radial hyaline calcite 175
radial pores 147
radial septulae 160, 164
radial zone 164
rate of chamber expansion 149
rate of translation 149
reticulate zone 165
retral processes 175
rods 175
rotaliid canal 147
rotaliid septa 175, 176
sculpture 145, 149
secondary apertures 149, 157
secondary chambers 174
septal filaments 176
septal flap 148
septate periodic growth 151
septulae 165, 168
septum 151
simple septate growth 151
spines 144, 145, 146, 148, 149, 157, 168, 172, 175
spiral side 149, 174
spiroloculine 169
streptospiral coiling 168
teeth 149, 174
terminal aperture 160, 168, 176
transverse septulae 168
triloculine 169
trimorphic 145
triserial 149, 174
trochospiral 149, 160, 168, 172, 175, 176
umbilical boss 149, 174
umbilical side 174
uniserial 149, 160, 164, 168, 173, 176
whorl 149, 168

form taxonomy 258
formic acid 39, 275
fossil assemblages 21, 84, 104, 191, 204, 217, 243
fossil fuels 30
fossil record 7, 9, 12, 45, 53–4, 55, 59, 64, 66, 75, 129, 159, 188, 196, 217, 218
fragmentation 113, 153
fragmentation of bacteria 61, 62
fucoxanthin 211
fungi 5, 39, 49, 65, 66
fuschin 123
fusulinids 157

gametogenesis 144
gametophyte 104, 109
gamont generation 144
gastropods 99, 145
genera 5, 12
ginkgophytes 111
glacial 10, 27, 28, 39, 62, 102, 120, 122, 138, 181, 207, 245
glacial erratics 102
glauconitic clays 216
globigerina ooze 142, 159, 226, 233
globular calcified cartilage 251, 259
glycerine 123, 246
gnathostomes 268
Golgi body (dichosomes) 11, 82, 131, 144
gram-positive, bacteria
 gram-positive (High G + C; Archaea) 43
 gram-positive (Low G + C) bacteria 43
graphical correlation 18–19, 22, 71
Gray's spore stain 123
Gulf Stream 92
Gunflint Chert 46, 61
gymnosperms 109, 111, 117, 119

hagfish 269
haploid 85, 104, 144
hardgrounds 22
heterococcoliths 130, 135
heterokonts 5
heteromorphs 225, 236
heterosporous 104, 117, 119
heterotrophy 4, 60

high magnesium calcite   146
highstand systems tract   21, 40
holococcoliths   130, 135
homology   82, 254, 260
homosporous   104
horsetails   117
Hox genes   55. *See also* regulatory genes
hydrochloric acid   123, 275
hydrogenosomes   11
hydrothermal hypothesis   39
hypersaline   62, 66, 153, 228, 236, 263, 268
hypersaline lakes   28
hyperthermophile bacteria   39
hystrichospheres   88

Iapetus Ocean   245, 269
ingroup   13
instars   225, 229
interglacial   27–8, 121, 138, 245
International Code of Botanical Nomenclature (ICBN)   88, 113
International Code of Zoological Nomenclature (ICZN)   88
interval biozones   18
iron and manganese ores   59, 65
iron pyrites   61
Isua Group   39, 43, 45

Jurassic   40, 75, 76, 90, 116, 119, 137, 165, 180, 181, 183, 216, 217, 228, 239, 243

kingdom   5
Kofoidian System   82
K-strategists   155

lacustrine environments   16, 113, 217, 239
lagoons   28, 62, 153, 155, 201, 229
last appearance datum (LAD)   18
lignites   123
limestones   39, 65, 75, 100, 159, 184, 198, 208, 217, 246, 274, 275, 278
limnic ostracods   229, 243
line of correlation (LOC)   19
lipids   43
lithostratigraphy   16
Llandovery   116

local range zone   18
low magnesium calcite   35, 131, 168, 171
lowstand systems tract   21, 22
lowstand wedge   22. *See also* fan
Ludlow   76, 116
lunar cycle   145
lycopsids   104, 117
lysocline   159

Maastrichtian   91, 182
macroevolution   8–9
macrofossils   3, 16, 28
macropalaeontology   3
macrophagous   254, 269
malachite green   123, 184
manganese nodules   65
mails   140, 198, 216, 274, 275
Mars   4, 39
marshes   152, 155, 199
mass extinctions   9
maximum flooding surface   21, 22, 40
Mediterranean Sea   28, 180
megasphaeromophs   49
megasporangium   109. *See also* ovule
megaspores   113, 117, 122
meiosis   5, 61, 75, 82, 104, 106, 144
Mesozoic   9, 16, 28, 57, 76–7, 80, 86, 89, 119, 120, 122, 131, 138, 139, 142, 146, 159, 181, 188, 191, 197, 212, 215, 216, 218, 232, 241, 245
Messian Salinity Crisis   183
metagenesis   35
metamorphism   28, 35
metanauphilus   225
metazoans   53–4, 57, 86
methanogenic bacteria   64
4α-methyl-13-ethylcholestane   75, 89
Mg/Ca   182, 229
micrite envelope   66
microbial carbonates   65
microevolution   8–9, 196
microfossil record   4
micromolluscs   53
micropalaeontology   3–4, 12, 181
microphagous   254
microspore   104, 116, 123
Milankovitch   28
Miocene   10, 16, 28, 33, 138, 175, 180, 183, 205, 207, 212, 217, 242

miospores   21, 113, 123
mitochondria   75, 61, 82, 144
mitosis   75, 61, 104
mixotrophic   4
molecular clocks   55
molluscs   91, 142
monads   109, 111, 113
monocrystalline wall   146
monoderms   39
monolete   106
monophyletic groups   13, 54, 254, 258, 263
monsoonal upwelling   159, 191
Monterey Event   33
morphon   113
mosaic evolution   89
mucilaginous sheath   59, 61
mudstones   75, 77, 96, 123, 140, 197, 274, 275
multinucleate   144, 145
multiple fission   145

nannoconids   129
Neanderthals   122
Nemakit-Daldynian Stage   53
nematode worms   145
Neogene   16, 91, 120, 142, 196, 211, 229
Neomuran Hypothesis   48, 75
Neoproterozoic   50, 54, 75
neritic   87, 91, 203, 216, 217, 229
nested heirarchy   12
nitrate   64, 86, 153, 203, 228
nitrifying bacteria   63
nitrogen-fixation   63
non-arboreal pollen (NAP)   121
North Atlantic Deep Water   155, 182
North Atlantic Ocean   86
North Sea   21–3, 90, 180
Northern Hemisphere glaciation   77
notochord   249, 266
nucleus   4, 5, 11, 75, 82, 109, 129, 144, 145, 188, 200, 211
numerical taxonomy   13. *See also* phenetics
nummulitic sands   157

obliquity   28
ocean anoxic events   92
ocean currents   76, 87, 182

ocean stratification 53, 182
ODP (Ocean Drilling Programme) 91, 182, 217
oil shales 65
Oligocene 28, 32, 91, 157, 177, 183, 197, 205, 211, 212, 217, 228
oligotophic 157
ontogeny 149, 225
opal revolution 197
opaline silica 146, 170, 188, 192, 200, 211
Oparin-Haldane hypothesis 39
optically radial calcite 170, 172
orbitolinids 157
orbuline trend 171
order 5
Ordovician 75–7, 99, 101, 109, 113, 138, 165, 180, 219, 232, 239, 245, 249, 251, 254, 261, 266, 269, 270
organelles 4, 5–11, 73, 80, 142
organic-walled microfossils 3, 66, 274, 275, 276, 277, 278
Orsten microbiota 53
osmosis 147
ostracods 3, 7, 16, 21, 219, 219–48, 275, 276, 277, 278
  adductor muscles 217, 236
  adductor muscle scars 232, 233, 236, 239
  adont hinge 224, 233, 238
  alae 224, 226, 230, 232, 236
  amphidont hinge 224, 229, 232
  antennae 226, 228, 232, 233, 238
  antennula 215, 221
  appendages 219, 221, 226, 228, 231, 233, 239
  bio-luminescent 225
  branched pore canals 229
  brancial plate 219, 226
  buccal cavity 215
  carapace 219–29, 230, 239, 242, 246
  cardinal angles 236, 238
  caudal rami 219
  central muscle-scar 221
  chitinous 219, 224, 225, 233, 236
  copulatory appendages 221
  dorsal 219, 221, 232, 233, 239
  dorsal muscle scar 221
  duplicature 221, 226, 231, 232, 236
  endopodite 219, 236
  entomodont hinge 224, 236
  epidermis 221
  epipodites 228
  eye spots 224, 229, 232, 236
  eye tubercles 224, 231, 232, 236
  flanges 231, 232, 234
  freshwater ostracods 224, 226, 233, 239
  furca 225, 226, 236
  furrow 238, 239
  head 219, 221
  hinge elements 231, 232, 233, 238
  hypostome 219
  infold 219
  inner lamella 221, 232, 236, 238
  keels 226
  labrum 219
  lamellae 221
  lateral 228, 236
  lateral eyes 221
  left valve 233, 236, 238
  ligament 224
  line of concrescence 223
  lobes 224, 232, 236, 238
  mandibula 219, 224, 226, 233
  mandibular scars 224
  marginal frill 236
  marginal pore canals 224, 226, 231, 232, 233, 236
  marginal zone 223, 224, 233, 236
  maxillula 219, 232
  median groove 224
  median sulcus 221, 236
  merodont hinge 224, 234
  nektonic ostracods 228
  normal pore canals 224, 229
  nuchal furrow 239
  ornament 229, 233, 239, 241
  pelagic ostracods 228, 230, 238
  right valve 228, 233
  rods 224
  rostral incisure 229, 239
  rostrum 239
  selvages 221, 231
  sensilla 224
  setae 225, 226, 238
  sieve pores 224, 229
  sockets 224
  spines 216, 226, 230, 232, 236, 238, 239
  subcentral tubercle 221, 232
  tecnomorphs 225, 238
  teeth 224
  thoracic appendages 226, 233
  valves 219, 221, 225, 226, 229, 231, 232, 234, 236, 238, 242
  ventral lobe 225, 236
  vestibulum 221, 234, 236
  vestment 224
  Zenker's organs 221
outgroups 13
ovule 109
ovum 109
oxygen isotopes 10, 25–8, 139, 182, 207
oxygen minimum zone 62, 155, 179

Pacific Ocean 10, 134, 182, 203, 230, 240
palaeobathymetry 155, 159
Palaeocene 28, 32, 122, 180, 197, 207, 217, 237
palaeoclimate 91, 120, 182, 197, 245, 270
palaeoecology 77, 80, 87, 96, 99, 182, 183, 242, 245
Palaeogene 22, 111, 120, 142, 181, 182
palaeo-oxygen levels 229
palaeotemperature 25, 27–30, 91, 182, 190, 197
Palaeozoic 16, 54, 71, 76, 89, 96, 99, 101, 119, 122, 137, 142, 147, 165, 180, 191, 192, 196, 216, 217, 224, 226, 230, 234, 236, 239, 241, 245, 249, 258, 269
palynofacies analysis 21, 120
palynology 3–4, 46, 120, 123
panspermia hypothesis 39
paraconodonts 266
parapatric speciation 10
paraphyletic groups 13
parasitic 66, 87, 89, 144, 145, 148, 153
parataxa 113
parsimony 12
parthenogenesis 225, 230
peats 123
Pee Dee Belemnite 46
pelagic limestone 215

Pennsylvanian 166
peridinin 80
period 16
peripheral isolates 10
perispore 108
permanent teeth 254
Permian 76, 99, 101, 117, 119, 120, 157, 167, 168, 180, 196, 217, 239, 264, 265, 270
peroxisomes 11
pH 57, 60, 62, 135, 145, 158, 159, 207, 229
Phanerozoic 16, 57, 61, 66
phenetics 12. See also numerical taxonomy
phosphate 153, 203
phosphoric acid 59
phosphorites 65
photic zone 30, 84, 87, 132, 137, 152, 157, 160, 188, 189, 190, 200, 203, 211, 216
photoautotrophy 88
photosymbionts 144, 148, 152, 157
photosynthesis 4–5, 30, 39, 39–45, 59–60, 62, 66, 132, 155, 159, 211
phycocyanin 61, 62
phycoma 71, 77
phylogenetic systematics 12. See also cladistics
phylum 5
picoplankton 62
Pilbara Supergroup 45
placolith 137, 138
planispiral coiling 164, 176
planktonic foraminifera 8–10, 12, 27, 30, 139, 142, 144, 148, 157, 158, 180, 181, 182, 196, 217
planozygote 85
plants 5, 7, 10, 39, 47–8, 88, 61, 104–24
Pleistocene 10, 23, 28, 138, 139, 182, 198, 205, 217, 218, 239, 243
pleomorphism 131
Pliocene 10, 28, 91, 92, 137, 182, 208, 217, 245
pollen 3, 10, 16, 21, 35, 104–23, 274, 276
pollen analysis 120–3
pollen diagram 121–2

pollen rain 110
pollen spectra 121
pollen tube 109
polymorphism 229
polyphyletic groups 13
potassium ferricyanide 278
prasinophytes 71, 75, 77
Precambrian 4, 51, 53, 61, 66, 71, 75–7, 89, 99, 214
precession 28
pre-pollen 116, 119
prismatomorphs 75, 76
progradational 22
prokaryotes 39–43, 47, 61, 82, 88
proloculus 144, 145, 168
Proterozoic 46, 71, 76, 89, 203
protists 5, 188, 215, 226
protoconodonts 266
protoplasm 4
provinces 76, 86, 89, 134, 158, 190, 269
provincialism 89, 99, 205, 245, 269, 270
pseudacellids 215
pseudochitin 35, 96, 99
pseudopodia 4, 11, 92, 142, 144, 148, 149, 153, 171, 183, 188, 200, 211
psychrospheric ostracods 229, 240
pteridophytes 104, 119
pteromorph acritarchs 77

Quaternary 25–8, 31, 88, 93, 120–2, 138, 142, 157, 158, 181, 182, 207, 213, 229, 234, 236, 242, 245
quinqueloculine 169, 170

radiolarian oozes 135, 191
radiolarians 40, 75, 92, 135, 159, 188–99, 207, 212, 217, 274, 275, 276, 277, 278
 aperture 192
 apical 192
 apophyses 196
 bar 188
 basal shell mouth 192
 calymma 188, 189
 central capsule 188, 189, 192, 198
 cephalis 192
 chamber 192
 chamberlets 192

 discoidal lattice 192
 double wall 196
 ectoplasm 188, 198
 endoplasm 188, 189, 198
 extracapsulum 188
 filopodia 188
 intracapsulum 188
 lattice shells 188, 192, 196, 198
 marginal keel 196
 oral teeth 196
 plates 198
 post-abdominal segments 192
 pseudopodia 188
 radial beams 192
 radial elements 192
 radial spines 192, 196
 rays 192
 sagittal ring 192
 spicules 188, 196, 198
 spines 192, 196, 202
 styles 192
 tangential elements 188
 terminal pole 192
radiolarites 191. See also cherts
Recent 16, 33, 75, 77, 84, 86, 101, 121–2, 160, 169, 182, 183, 188, 197, 216, 236, 239, 241
red clays 135
red tides 80, 87
regression 18, 22, 75, 138, 245
regulatory genes 9
relative sea level 21, 75, 86. See also eustacy
rhabdoliths 137
rhodophytes 155
RNA 39, 48, 55, 60, 75, 89
18sRNA 136, 211
r-strategists 155

saccate pollen 119. See also spores and pollen, monosaccate; bisaccate pollen
safranin stain 123
salinity 3, 30, 60–88, 145, 153–4, 157, 180, 182, 203, 207, 217, 218, 219, 224, 228, 229, 233, 242, 243, 270
scaphopods 145
schizont 85, 145
schwagerinid wall 167

scolecodonts 7, 96, 101–2
  carriers 101
  chitinous 101
  dorsal 101
  mandibles 101
  maxillae 101–2
  MI elements 102
  plates 101
sea surface temperature 28, 86, 92, 181, 207
sea-grass communities 182
sedimentary sequences 5–11, 21, 22, 89, 182
seismic stratigraphy 21
semitectate 111. See also tectum
sequence boundary 21–2
serial endosymbiotic theory 47
sexual dimorphism 224–5, 231, 236
sexual reproduction 4, 5, 48–50, 61, 82, 85, 144, 145, 199
shales 22, 35, 46, 65, 75, 77, 96, 123, 140, 145, 191, 198, 208, 214, 269, 274
sibling species 12, 13, 180
Siegenian 116
silica 45, 92, 142, 146, 170, 188, 191, 192, 197, 200, 201, 204, 207, 211, 212
silica cycle 198
siliceous microfossils 274
silicified microbiotas 46
silicoflagellates 5, 92, 189, 191, 196, 197, 211, 214, 276
  apical 211
  apical bar 211, 212
  apical bridge 212
  apical domes 212
  apical plate 212
  apical portion 211
  basal ring 211, 212
  cross bar 212
  flagellum 211
  hemispherical lattice 211, 212
  lateral bars 211, 212
  portals 212
  rods 211
  sculpture 214
  spines 211, 212
  stomatocycsts 214
  struts 212
  windows 212
Silurian 75, 77, 90, 99, 109, 116, 192, 196, 219, 239, 245, 249, 261, 269, 270
sister groups 12
slates 96
smaller benthic foraminifera 27, 142, 144, 148, 152, 160
SMOW (standard mean oceanic water) 46, 56
sodium hypochlorite 275
species 5, 8, 10–12, 18–19, 21, 22–3
spirochetes 43
sponge spicules 7, 145
sporae dispersae 113
spore mother cells 104
spores 10, 16, 35, 73, 75, 104–9, 113–23, 276
spores and pollen
  acavate 108
  alete 73, 106, 109
  amb 106
  apical 104
  archegonia 104
  atectate 108
  cavum 108
  cingulizonate 108
  cingulum 108
  colpate 111
  columellae 111
  contact areas 106, 111
  coronae 108
  distal polar face 106
  ektexine 106, 111, 119
  endexine 106, 111
  equatorial 106, 108, 111
  equatorial flange 108
  exine 106, 111, 119
  exospore 106
  flanges 108
  furrow 111
  germinal apertures 106
  heteropolar 106
  hilum 106
  intectate 111
  inter-radial areas 108
  intine 106
  kyrtomes 108
  laesurae 106
  monocolpate 111
  monosaccate 111, 119
  occulate 111
  polyplicate grains 111, 113
  protonema 104
  proximal face 106
  proximal pole 106
  sculpture 106
  segments 106
  stephanoporate 111
  tectate 111, 120
  tricolpate 111, 120
  triploid 109
  triprojectate 111
  trisaccate 111
sporoderm 106, 116
sporophyte 104, 106
sporopollenin 35, 71, 75, 108
stable isotopes 25–7, 30, 53, 139, 182
standard time units – stu 19
stenohaline 228
stratigraphical column 16
stromatolites 45–6, 54–6, 61, 65–6
  marginal zone 65
  non-skeletal 66
  skeletal 66
strontium sulphate 188
successive last appearance zone 18
sulphate reduction 44, 55, 64
sulphate-reducing bacteria 44, 64
sulphate–reducing zone 64
sulphur bacteria 64, 88
sulphur isotopes 44, 64
sulphur–oxidizing bacteria 64
Swaziland Supergroup 45
symbiosis 189
sympatric speciation 11, 245
symplesiomorphy 12
synapomorphy 12
syngamy 47

tectin 142, 145, 148
tectum 111, 166. See also semitectate
teeth 268
temperature 8, 11, 26–30, 35, 76, 86, 123, 134, 137–8, 142, 153, 155, 180, 181, 182, 190, 203, 207, 216, 229, 230, 242, 243, 269, 274, 276

terrestrial environments   9, 12, 16, 21, 22–3, 45, 88, 142, 157, 177, 207, 208, 213, 219, 228, 233
Tertiary   10, 46, 28, 32, 76, 91, 120, 129, 135, 153, 157, 182, 207, 217, 236, 239, 243, 245
test   142, 153, 155–7, 158, 175, 179, 182
testate amoebae   145
Tethys Ocean   175, 180, 217
tetrabromoethane   277
tetrads   49, 104, 106, 109, 111, 113
tetragonal   106
tetrahedral tetrads   106
thermal alteration index   120
thrombolites   64
tintinnids   99, 215–18, 276
  aboral region   216
  alveoli   216
  aperture   215, 216, 217
  cell mouth   215
  chamber   216
  cilia   215, 216
  collar   217
  crown   215
  lorica   215, 217
  membranelles   215
  peduncle   215
  pellicle   215
  spines   216
Tithonian   218
Toarcian   226
Tommotian Stage   53
tommotiids   53
tonsteins   65
trace fossils   51
transgression   18, 22, 53, 91, 137, 239
transgressive surface   21, 22
transgressive systems tract   21
travertine   66
Tremadoc   76, 261, 269
Triassic   74, 76, 90, 92, 119, 129, 137, 164, 169, 182, 196, 233, 237, 239, 249, 263, 264, 270, 269
trilete mark   106, 116. *See also* Y-mark
trilete spores   106

unicellular condition   4, 5, 61, 92, 129, 138, 199, 211, 213
uninucleate cells   144

vacuoles   4, 5, 62, 144, 149, 189, 215
Valanginian   217
Varangian glaciation   54
vascular tissues   104
vegetative reproduction   50
Vendian glaciation   76
vertebrates   12, 249, 254, 267, 268
vital effects   139, 207

Warrawoona Group   46
water depth   3, 8, 10, 64, 66, 86, 88, 135, 139, 143, 149, 152, 153, 157, 158–60, 179, 181, 182, 188, 189, 191, 197, 200, 229, 230, 236, 242, 269

Y-mark   106. *See also* trilete mark

zone or index fossils   18, 22, 138
zooxanthellae   189

Printed and bound by CPI Group (UK) Ltd, Croydon, CR0 4YY
09/06/2025

14685998-0003